Communications
in Computer and Information Science   1228

Commenced Publication in 2007
Founding and Former Series Editors:
Simone Diniz Junqueira Barbosa, Phoebe Chen, Alfredo Cuzzocrea,
Xiaoyong Du, Orhun Kara, Ting Liu, Krishna M. Sivalingam,
Dominik Ślęzak, Takashi Washio, Xiaokang Yang, and Junsong Yuan

Editorial Board Members

More information about this series at http://www.springer.com/series/7899

Yichun Xie · Yong Li · Ji Yang ·
Jianhui Xu · Yingbin Deng (Eds.)

Geoinformatics in Sustainable Ecosystem and Society

7th International Conference, GSES 2019
and First International Conference, GeoAI 2019
Guangzhou, China, November 21–25, 2019
Revised Selected Papers

 Springer

Editors
Yichun Xie ⓘ
Eastern Michigan University
Ypsilanti, MI, USA

Yong Li
Guangzhou Institute of Geography
Guangzhou, China

Ji Yang ⓘ
Guangzhou Institute of Geography
Guangzhou, China

Jianhui Xu ⓘ
Guangzhou Institute of Geography
Guangzhou, China

Yingbin Deng ⓘ
Guangzhou Institute of Geography
Guangzhou, China

ISSN 1865-0929   ISSN 1865-0937 (electronic)
Communications in Computer and Information Science
ISBN 978-981-15-6105-4   ISBN 978-981-15-6106-1   (eBook)
https://doi.org/10.1007/978-981-15-6106-1

This Springer imprint is published by the registered company Springer Nature Singapore Pte Ltd.
The registered company address is: 152 Beach Road, #21-01/04 Gateway East, Singapore 189721, Singapore

Preface

The 2019 International Geoinformatics Week was hosted in Guangzhou, China, on November 21–25, 2019. Annual meetings of Geoinformatics in Sustainable Ecosystem and Society (GSES) and Geospatial Artificial Intelligence for Urban Computing (GeoAI) are organized by the Guangzhou Institute of Geography in collaboration with multiple universities and research institutes, and in 2019 were held within this Geoinformatics Week. More than 300 scholars from around the world attended this international event and participated in academic cooperation.

GSES 2019 was the 7th annual academic conference and is sponsored by an agreement among universities in China and the USA. This annual conference focuses on Earth observation, eco-environmental evolution, geospatial analysis, and technological advances and their applications in natural resource management and sustainable social development.

GeoAI 2019 was the first international conference organized by the Guangzhou Institute of Geography in collaboration with many research institutions and universities. This will become an annual conference focusing on the development of GeoAI for urban computing research frameworks, theories, and methods.

The coordination with the steering chairs, Chenghu Zhou, Michael Batty, Shih-Lung Shaw, and Xinyue Ye, was essential for the success of the conference. We sincerely appreciate their constant support and guidance. It was a great pleasure to work with such an excellent team. We would like to express our appreciation to all the members of the Organizing Committee and Program Committee for their efforts and support. We are grateful to all the authors who submitted their papers to the 2019 International Geoinformatics Week.

Given the rapidity with which science and technology is advancing in all the fields covered by GSES and GeoAI, we expect that the future GSES and GeoAI conferences will be as stimulating as this one was, as indicated by the contributions presented in these proceedings.

May 2020

Yong Li
Yichun Xie

Organization

International Steering Committee

Chair

Chenghu Zhou — Guangzhou Institute of Geography, China

Honrary Chairs

Michael Batty — University College London, UK
Renzhong Guo — Shenzhen University, China

Co-chairs

Yichun Xie — Eastern Michigan University, USA
Shih-Lung Shaw — University of Tennessee, USA
Xinyue Ye — New Jersey Institute of Technology, USA
Yong Li — Guangzhou Institute of Geography, China

Members

Shuming Bao — China Data Institute, USA
Qian Du — University of Mississippi, USA
Yingling Fan — University of Minnesota, USA
Bin Li — Central Michigan University, USA
Lin Liu — University of Cincinnati, USA
Zhongren Peng — University of Florida, USA
Meipo Kwan — University of Illinois at Urbana Champaign, USA
Bo Huang — Chinese University of Hong Kong, Hong Kong
Jay Lee — Kent State University, USA
Xia Li — East China Normal University, China
Hui Lin — Chinese University of Hong Kong, Hong Kong
Victor Mesev — Florida State University, USA
Zhenjiang Shen — Kanazawa University, Japan
Daniel Sui — University of Arkansas, USA
Zhengzhen Tan — MIT-China Future City Lab, USA
Paul Torrens — New York University, USA
Fahui Wang — Louisiana State University, USA
Shaowen Wang — University of Illinois at Urbana-Champaign, USA
Yehua Dennis We — The University of Utah, USA
Xiaobai Yao — University of Georgia, USA
Alexander Zipf — Heidelberg University, Germany

Program Committee

Chair

Chenghu Zhou   Guangzhou Institute of Geography, China

Co-chairs

Yichun Xie	Eastern Michigan University, USA
Shih-Lung Shaw	University of Tennessee, USA
Xinyue Ye	New Jersey Institute of Technology, USA

Members

Yanwei Chai	Beijing University, China
Chuqun Chen	South China Sea Institute of Oceanology, Chinese Academy of Sciences, China
Min Chen	Nanjing Normal University, China
Andrew Crooks	George Mason University, USA
Anrong Dang	Tsinghua University, China
Chengbin Deng	State University of New York at Binghamton, USA
Qingyun Du	Wuhan University, China
Zhixiang Fang	Wuhan University, China
Song Gao	University of Wisconsin-Madison, USA
Qingfeng Guan	China University of Geosciences, China
Lingqian Hu	University of Wisconsin-Milwaukee, USA
Yingjie Hu	University at Buffalo, USA
Yujie Hu	University of South Florida, USA
Haosheng Huang	University of Zurich, Switzerland
Jie Huang	Chinese Academy of Sciences, China
Qungying Huang	University of Wisconsin-Madison, USA
Bandana Kar	Oak Ridge National Laboratory, USA
Ruren Li	Shenyang Jianzhu University, China
Wenwen Li	Arizona State University, USA
Xiang Li	East China Normal University, China
Zhenglong Li	University of South Carolina, USA
Xingjian Liu	The University of Hong Kong, Hong Kong
Yu Liu	Peking University, China
Ying Long	Tsinghua University, China
Xiaogang Ma	University of Idaho, USA
Kun Qin	Wuhan University, China
Miaoyi Li	Fuzhou University, China
Xiaopin Liu	Sun Yat-sen University, China
Zongyao Sha	Wuhan University, China
Tiyan Shen	Beijing University, China
Lijun Sun	McGill University, Canada
David Wang	New Jersey Institute of Technology, USA
De Wang	Tongji University, China

Yunpeng Wang	Guangzhou Institute of Geochemistry, Chinese Academy of Sciences, China
Hua Wang	Guangdong University of Technology, China
Jiaoe Wang	Chinese Academy of Sciences, China
Zhifeng Wu	Guangzhou University, China
Shengwu Xiong	Wuhan University, China
Yang Xu	The Hong Kong Polytechnic University, Hong Kong
Haowen Yan	Lanzhou Jiaotong University, China
Junyan Yang	Southeast University, China
Ling Yin	Shenzhen Institutes of Advanced Technology, Chinese Academy of Sciences, China
Xiaoru Yuan	Beijing University, China
Wenze Yue	Zhejiang University, China
Bailang Yu	East China Normal University, China
Qingming Zhan	Wuhan University, China
Anbing Zhang	Hebei University of Engineering, China
Xueliang Zhang	Shanghai University of Finance and Economics, China
Yaolong Zhao	South China Normal University, China
Feng Zhen	Nanjing University, China
Jiangping Zhou	The University of Hong Kong, Hong Kong
Suhong Zhou	Sun Yat-sen University, China
Xuesong Zhou	Arizona State University, USA
Qing Zhu	Southwest Jiaotong University, China

Organizing Committee

Chair

| Chenghu Zhou | Guangzhou Institute of Geography, China |

Co-chairs

Yichun Xie	Eastern Michigan University, USA
Shih-Lung Shaw	University of Tennessee, USA
Xinyue Ye	New Jersey Institute of Technology, USA
Yong Li	Guangzhou Institute of Geography, China
Min Ji	Shandong University of Science and Technology, China
Zhigen Zhao	Anhui University of Science and Technology, China

Secretary

| Ji Yang | Guangzhou Institute of Geography, China |

Members

| Meijing Zeng | Guangzhou Institute of Geography, China |
| Wenlong Jing | Guangzhou Institute of Geography, China |

Xulong Liu	Guangzhou Institute of Geography, China
Hao Jiang	Guangzhou Institute of Geography, China
Jianhui Xu	Guangzhou Institute of Geography, China
Liusheng Han	Guangzhou Institute of Geography, China
Chongyang Wang	Guangzhou Institute of Geography, China
Yingbin Deng	Guangzhou Institute of Geography, China
Yingwei Yan	Guangzhou Institute of Geography, China
Xiaoling Li	Guangzhou Institute of Geography, China

Publication Committee

Co-chairs

Xinyue Ye	New Jersey Institute of Technology, USA
Yichun Xie	Eastern Michigan University, USA
Shih-Lung Shaw	University of Tennessee, USA
Bandana Kar	Oak Ridge National Laboratory, USA
Andrew Crooks	George Mason University, USA

Student Paper Competition Committee

Co-chairs

Xinyue Ye	New Jersey Institute of Technology, USA
Yang Xu	The Hong Kong Polytechnic University, Hong Kong
Jie Huang	Chinese Academy of Sciences, China

Sponsor

Guangdong Academy of Sciences, China

Organizers

Guangzhou Institute of Geography, China
Eastern Michigan University, USA
University of Tennessee, USA
New Jersey Institute of Technology, USA

Co-organizers

Cartography and Geographic Information System Committee of Chinese Geographical Society, China
The Geographical Society of Guangdong, China
Guangdong Society of Remote Sensing and Geographic Information Systems, China
State Key Laboratory of Resource and Environmental Information System, Institute of Geosciences and resources, Chinese Academy of Sciences, China
The College of Urban and Environmental Sciences, Peking University, China
State Key Laboratory of Surveying, Mapping and Remote Sensing Information Engineering, Wuhan University, China

School of Geography and Planning, Sun Yat-sen University, China
School of Geography, South China Normal University, China
School of Civil and Transportation Engineering, Guangdong University of Technology, China
School of Architecture and Urban Planning, Shenzhen University, China
School of Land Science and Technology, China University of Geosciences (Beijing), China
School of Geography and Information Engineering, China University of Geosciences (Wuhan), China
Department of Land Surveying and Geo-informatics, The Hong Kong Polytechnic University, Hong Kong
School of Earth Sciences and Engineering, Hohai University, China
The College of Environment and Planning, Henan University, China
School of Mining and Surveying Engineering, Hebei University of Engineering, China

Support Agencies

Guangzhou iMapCloud Intelligent Technology Co. Ltd., China
Esri China Information Technology Co. Ltd., China
National Earth System Science Data Center, China
Guangzhou Huanghuagang Sci-Tech Park, China

Contents

**Remote Sensing Monitoring of Resources and Environment
and Intelligent Analysis**

Intelligent Perceptions and Services of Spatial Information

Ecology, Environment and Social Sustainable Development

The Applications of Geospatial Data in the Sustainable Development of Social Economy

Research on 3D Analysis Method of Sight Line of Mountain Scenic Area Based on GIS—Taking Guangzhou Baiyun Mountain Scenic Area as an Example

Shaoping Guan$^{(\boxtimes)}$ and Hong Yu

School of Design, South China University of Technology, Higher Education
Mega Centre, Panyu District, Guangzhou 510006, China
shpguan@scut.edu.cn, 1019231373@qq.com

Abstract. China's scenic spots are facing the dual pressure of increasing the number of tourists and the decline in the quality of viewing. How to effectively coordinate the contradiction between the development of scenic spots and the quality of viewing, and realize the maximization of resources within the scenic spots and the improvement of visual senses have become common problems faced by the academic and tourism development departments. In the past few decades, the planning methods of scenic spots and the selection methods of scenic areas have been more traditional. It is impossible to accurately simulate and judge when dealing with complex topographical conditions such as mountains, which has a subjective impact on the exploration and development of scenic landscape resources. This method has caused problems such as narrow viewing horizons, incomplete coverage of the tour route, and occlusion of the line of sight. Practice has proved that the traditional scenic area planning theory based on qualitative and the analysis method based on field research are one of the main sources of the above problems, especially in mountainous scenic areas. With the development of social economy, the selection of traditional scenic spots and the planning of scenic spots urgently need a change to adapt to the new development and new requirements of the current viewing experience. This study used the GIS viewshed analysis function and the analytic hierarchy process to identify blind spots in the scenic area by simulating the visibility of different viewing distance ranges of the current scenic spots and tour lines, and then combined with the questionnaire survey results of tourists, planners, scenic spot managers and college landscape teachers, the analytic hierarchy process is used to determine the evaluation system of new viewpoints. Finally, according to the evaluation factor, the weighted superposition in the GIS is used to determine the newly added viewpoints, so as to improve the quality and effective utilization of the scenic spots in the mountain-type scenic areas. The key to the GIS landscape viewshed analysis is that the scenic spots and tour lines planning of the scenic area must be based on objective geographical factors and current resources, and priority is given to the analysis of the current situation quality. Then, according to the distribution of the current visual blind zone, new viewpoints and tour lines will be carried out. Compared with developed countries like the WEST, which have relatively perfect geographic information platforms, China's geographic information data acquisition requires authorization from multiple departments, and the

© Springer Nature Singapore Pte Ltd. 2020
Y. Xie et al. (Eds.): GSES 2019/GeoAI 2019, CCIS 1228, pp. 3–18, 2020.
https://doi.org/10.1007/978-981-15-6106-1_1

data quality is uneven. In response to this problem, this paper used python combined with Amap to obtain the vector boundary of the study area, and imported it into the rivermap downloader X3 to download DEM data, reducing the difficulty of data acquisition. This method can provide data security for the study area simulation and provide an important scientific basis for landscape viewshed analysis. Taking Baiyun mountain scenic area as an example, this paper comprehensively utilizes GIS technology to identify the viewing space of the above process through simulation and analysis of elevation, slope, slope direction and view area analysis, and constructs the best viewing pattern. Moreover, the whole analysis process is taken as the framework to enrich the evaluation system of mountain scenic spots. The results shown that the current viewing area of the 26 scenic spots in the near-view, mid-view and distant-view analysis results are 6.09%, 11.89% and 26.62%, respectively, and the overall view is poor, and the open field is mainly concentrated in the south-central area of the study area, the northeast and southwest sides have narrower views, mainly due to the lesser distribution of viewpoints. The near, middle and distant view along the four main tour lines may be better, but the central area and the northwest side are not transparent due to the topography and geomorphology. On the other hand, due to the lack of travel coverage on the north side of the study area, there is a large area of visual blind spots in the north. In order to solve the above problems, this paper will re-classify the evaluation system of the site selection and the elevation, slope and aspect of the study area. Under the principle of low cost and ecological protection, the site selection of newly-built scenic spots was obtained, and then fine-tuned according to the site investigation, and the location of newly-built scenic spots was finally determined. This method can effectively enhance the quality of the selected viewpoints in the planning process and reduce subjective errors in the case of high data quality. Moreover, this method also provides an essential reference for the analysis of geographical elements in the early stage of mountain-type scenic areas. We should transform from traditional subjective planning ideas into objective planning thinking based on data analysis, respect objective reality and maximize the exploitation of available resources to provide reference for the sustainable development of scenic spots.

Keywords: GIS · Viewshed analysis · Baiyun Mountain · Viewpoints selection

1   Introduction

The "Scenery Scenic Area" (hereinafter referred to as the Scenic Area) is a region where the central and local governments of China manage and protect natural and cultural heritage protected areas [1], and its nature and function are similar to those of the WEST National Park. With the improvement of the social and economic level, people are meeting the increasing material conditions, and the demand for returning to nature to get close to nature is getting higher and higher [2]. In addition, the "State Council's Opinions on Accelerating the Development of Tourism Industry" and the policy standards for the approval of the new "General Planning Standards for Scenic Spots" (GB/T50298-2018) on March 1, 2019 have been promulgated successively.

The construction of China's scenic area and the development of tourism resources have developed rapidly.

The mountain-type scenic area is a scenic area with mountainous landform as the main feature, and it is a large category in the classification of landscape features [3]. In the process of planning and construction, due to the complex terrain of mountainous terrain and the traditional means of planning of some scenic area, the construction and operation conditions have not reached the planned expected results, and even destroy the ecological environment and tourism resources of the scenic area. From December 12 to 13, 2013, the document of the Central Urbanization Work Conference held in Beijing mentioned that "Relying on the unique scenery of the existing landscape, the city should be integrated into the nature, so that residents can see the mountains and rivers and remember their hometown" [4], it is also emphasized that the visibility of landscape visual resources needs to be strengthened in urban and rural planning work.

Correct evaluation of landscape is the basis for the correct design of landscape [5]. The research on landscape visual resource analysis and protection originated in the WEST in the 1960s. In recent years, many foreign scholars have tried to obtain data on land use, vegetation, topography, and infrastructure distribution in the study area by integrating field research, maps and aerial imagery, and then combined with GIS technology for landscape viewshed analysis [6–9]. China's research on GIS landscape visual analysis started late. Yu et al. [10, 11] discussed the application of GIS in landscape visual security, cultural security and ecological security. Zhou et al. [12] used GIS spatial analysis technology. The sensitivity of the forest park attractions was evaluated. Su [13] used the landscape sensitivity S grading to explore the landscape and visual impact of the scenic spots.

The construction of geographic information data platforms in the WEST is relatively complete, and researchers can obtain high-precision basic terrain data by visiting relevant websites. However, China started late in this respect. The data such as topographic maps need to be obtained by the government's multi-sectoral authorization, and the data is generally incomplete, old and difficult to acquire. This paper takes Baiyun Mountain, a national 5A-level scenic area, as the research object, and acquires the appropriate boundary of the research area through Python combined with the Amap open platform API [14], and importing the appropriate amount of boundary into the rivermap downloader X3 to obtain higher precision Baiyun Mountain Scenic Area DEM terrain data, Using GIS technology, combined with the terrain topographic data of the scenic area, the 3D simulation of the visual field of view is carried out, and multi-factor analysis is carried out for the narrow area of the view. The optimization strategy is proposed by the ArcGIS10.5 assignment superposition function, which provides a reference for the 3D viewshed analysis of the mountain-type scenic area.

2 Research Object and Scope

The study area is located inside the Baiyun Mountain Scenic Area in Guangzhou, Guangdong Province (113°17'12.04"East, 23°10'56.92"North), with a total area of about 1846 hm^2 and 17 km from downtown Guangzhou. The terrain of the scenic area is rectangular, about 7 km long from north to south, 4 km wide from east to west, and

382 m above sea level and the highest peak of Mount Moxing is 382 m above sea level. It belongs to the subtropical monsoon maritime climate, warm and humid, the same period of rain and heat, the annual precipitation is 1689.3–1876.5 mm, and the precipitation in the rainy season from April to September accounts for more than 85% of the whole year.

The main reason for choosing Baiyun Mountain Scenic Area as a typical mountain scenic area is that Baiyun Mountain has been one of the famous mountains in South Guangdong since ancient times and has a strong cultural heritage. Baiyun Mountain is a branch of the Jiulian Mountain Range extending to the southwest. It belongs to the hilly terrain of Lingnan. The terrain is in the middle and is inclined from the northeast to the southwest. The mountain has the representativeness of the mountains in the Lingnan area. Due to its constant rain and heat and high precipitation, Baiyun Mountain has a green coverage rate of over 95% and it has abundant natural resources and 876 species of plants, which is convincing to the Lingnan area. Therefore, Baiyun Mountain Scenic Area was selected as a case study for the purpose of reference to other similar Lingnan mountain-type scenic area.

3   Research Methods and Data Sources

3.1   Research Methods

The study mainly analyzes the geographic data information of Baiyun Mountain Scenic Area on the GIS platform. Firstly, the viewing area is analyzed by the viewshed, and secondly, we analyzed the location index of the scenic spots.

Viewshed Analysis. Firstly, the study imported vector data of scenic spots and main tour lines into the simulated ground surface through GIS platform. Secondly, elevation offset and densification are carried out on the data of scenic spots and scenic route. Finally, buffer radius is set to generate visual areas with different visual distances.

Questionnaire. We obtained the bias data of the factors affecting the location of the scenic spots through the questionnaire survey of different residents such as local residents, tourists, scenic spot managers, planners and landscape planning experts in Baiyun Mountain Scenic Area.

Analytic Hierarchy Process. The study imported the data from the questionnaire into the Yaahp10.0 software for statistical analysis, and obtained the best factors affecting the location of the new viewpoint.

Grid Overlay Analysis. The study assigned each impact factor to the raster image, then performed reclassification and spatial superposition.

3.2   Data Sources

The research needs to use DEM resolution digital elevation, road network data and building data. The basic data includes the distribution data of the current situation and the main tour lines data of Baiyun Mountain Scenic Area.

DEM Elevation Data Acquisition and Processing. The study used the Amap open platform API search interface to get the Baiyun Mountain area search list, and obtained the correct Baiyun Mountain area poiid after analysis. And we found the boundary information by poiid to get the json format data, then used the python code to get the coordinate data of the boundary points [14], imported the generated vector boundary data into the water through the rivermap downloader X3 to download the DEM resolution digital elevation. Finally, the DEM digital elevation model, slope analysis map, aspect analysis map and elevation analysis map of Baiyun Mountain Scenic Area were constructed by ArcGIS 10.5.

Building and Road Network Data Acquisition and Processing. In the study, the rivermap downloader X3 were used to obtain the building and road vector data of the study area, and the data was imported into ArcGIS 10.5. We used editing tools to crop and adjust buildings and roads against vector boundaries in study areas.

Acquisition and Processing of Viewpoints and Tour Lines. Through on-the-spot investigation, we selected 26 viewpoints with high traffic volume and high popularity within the research area at the stage of the study, and used the Amap open platform API to obtain POIs of all attractions in Baiyun Mountain Scenic Area. Then, the viewpoint data was imported into ArcGIS 10.5 to be attached to the simulated ground surface, and different height differences were set according to the observation point attribute. Baiyun Mountain viewpoints feature attribute is shown in Table 1.

Table 1. Baiyun Mountain viewpoints feature attribute

Number	Elevation (m)	Height difference (m)	View points
1	63	1.5	Longzang Pavilion
2	103	1.5	Longxing Pavilion
3	121	1.5	Longxiang Pavilion
4	67	2.5	Lianquan Gate
5	89	3.0	Deep Fog of Baiyun
6	119	1.5	Muxia Pavilion
7	192	6.0	Baiyun Cableway
8	304	1.5	Zhaidou Pavilion
9	245	7.5	Shuangxi Villa
10	76	10.5	Mingzhu Tower
11	43	1.5	Zhuyun Pavilion
12	58	1.5	Banshan Pavilion
13	82	1.5	Mucui Pavilion
14	205	1.5	Diecui Pavilion
15	33	4.0	Yuexi College
16	150	1.0	Baiyun Mountain Bungee
17	167	6.0	Baiyun Mountain Trail
18	152	8.5	Nengren Temple

(*continued*)

Table 1. (*continued*)

Number	Elevation (m)	Height difference (m)	View points
19	175	4.5	Tianwang Temple
20	189	4.5	Ciyun Temple
21	210	0.5	Sky Falls
22	210	1.5	Panya Pavilion
23	224	0.5	Mingchun Valley
24	262	0.5	Longhugang
25	333	0.5	Moxing Mount
26	21	0.5	Sun Gallery

According to the relevant information [15], we had obtained four planned main tour lines. Line A was from the Luhu tourist area through the Mingzhu tower tourist area to the west gate of Baiyun Mountain. The route runs through the attractions of Nengren Temple, Mingchun Valley, and Mingzhu Tower. It is about 23.4 km long and is the longest tour route in the study area. Line B is from the Lianquan Gate to the Lawn Resort. The route is shorter and the length is about 1.16 km. The C line is from Kezi Mount Gate to the Peak Square. The route is mainly a climbing plank road with a length of about 3.45 km. The D line is from the Yunxi Ecological Park through the Heyi Mount to the Moxing Mount. The total length of the tour route is 3.19 km. Finally, we used the editing tool in ArcGIS 10.5 to extract the road network vector data, filtered out four tour lines and saved them separately into shp files, then encrypted each tour route and set the visual point spacing to 5 m.

4   Results and Analysis

4.1   Viewshed Analysis of Scenic Spots

According to related research [16], the farthest viewing distance of the human eye is about 1200 m, and the clear viewing distance is 200–300 m. Therefore, in this study, 0–300 m is selected as the near-field line of sight range, 300–500 m is the medium-view line-of-sight range, and 500–1000 m is the distant-range vision range, and we analyze the field of view of three different line-of-sight ranges. The results of the analysis are shown in Table 2.

Table 2. Baiyun Mountain viewpoints visibility

Visual range	Visible area (ha)	Proportion of visible area (%)	Invisible area (ha)	Proportion of invisible area (%)
Close	111.91	6.09	1726.89	93.91
Medium	218.56	11.89	1620.24	88.11
Remote	489.16	26.62	1349.64	73.38

The Result of Near View. As shown in the near-field view analysis chart of Fig. 1, according to the data analysis results, the total visible area within the visible range of 300 m is 111.91 ha, accounting for only 6.09% of the total research area. The overall situation is poor, and the central and southeastern regions are better, which is related to the distribution density of the scenic spots in the region. The southwestern and northern viewpoints of the study area are less distributed, and it is easy to form a blind spot of close-up vision, which affects the viewing experience.

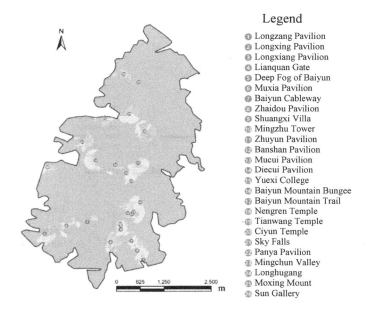

Legend

① Longzang Pavilion
② Longxing Pavilion
③ Longxiang Pavilion
④ Lianquan Gate
⑤ Deep Fog of Baiyun
⑥ Muxia Pavilion
⑦ Baiyun Cableway
⑧ Zhaidou Pavilion
⑨ Shuangxi Villa
⑩ Mingzhu Tower
⑪ Zhuyun Pavilion
⑫ Banshan Pavilion
⑬ Mucui Pavilion
⑭ Diecui Pavilion
⑮ Yuexi College
⑯ Baiyun Mountain Bungee
⑰ Baiyun Mountain Trail
⑱ Nengren Temple
⑲ Tianwang Temple
⑳ Ciyun Temple
㉑ Sky Falls
㉒ Panya Pavilion
㉓ Mingchun Valley
㉔ Longhugang
㉕ Moxing Mount
㉖ Sun Gallery

Fig. 1. Baiyun Mountain near viewshed

The Result of Medium View. As shown in the field of view analysis chart in Fig. 2, according to the data analysis results, the total visible area within the range of 300–500 m is 218.56 ha, accounting for 11.89% of the total research area. From the view point of the Lianquan Gate to the north of Mingzhu Tower, the scenic spots are better. The Diecui Pavilion and Muqing pavilion in the north are in poor condition. The viewable area along the southwestern part of the scenic spot is inconsistently affected by the terrain.

The Result of Distant View. As shown in the visual field analysis chart of Fig. 3, according to the data analysis results, the total visible area within the visual range of 1000 m is 489.16 ha, accounting for 26.62% of the total research area. Most of the viewpoints of the scenic spots can be seen as good, and can form a line of sight corridors with the adjacent scenic spots along the line. However, the two scenic spots in the north of the Diecui Pavilion and Muqing Pavilion are still in poor condition. Due to the small number of viewpoints in the central and western regions, there is a visual blind spot for large-area scenic spots in the visible range of 1000 m.

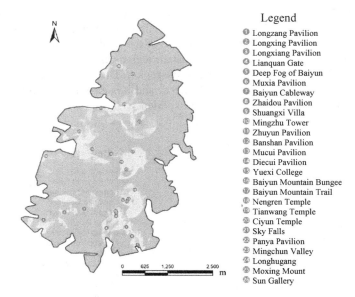

Fig. 2. Baiyun Mountain medium viewshed

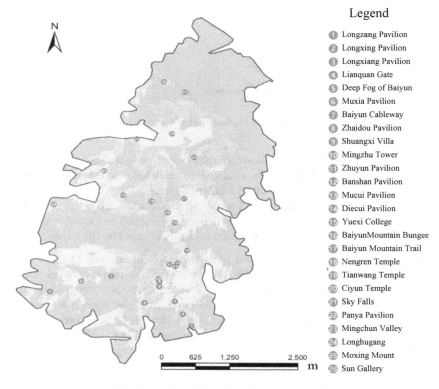

Fig. 3. Baiyun Mountain distant viewshed

4.2   View Analysis of Tour Lines

The four main tour lines after the densification treatment were extracted and analyzed in the range of near, medium and distant range visual range to check whether the visibility in the line of sight along the main tour line was good and whether the scenic spots could be seen in the scenic area, so as to evaluate the main tour line to some extent (see Fig. 4).

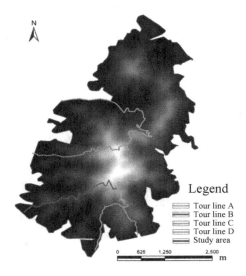

Fig. 4. Baiyun Mountain main tour line distribution map

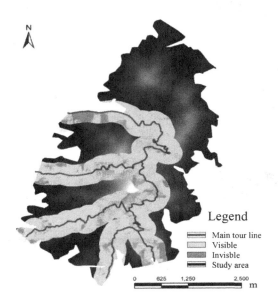

Fig. 5. Baiyun Mountain main tour line near viewshed

The Result of Near View. The field of view analysis of the existing main line is shown in Fig. 5. The results show that 34.86% of the scenes can be seen on the existing main line of the viewing, and most of the close-ups along the line can be observed. The northwest corner of the A line and the southeast side of the C line are of poor quality, mainly due to the topography.

The Result of View at Medium Distance. According to the analysis of the field of view of Fig. 6, in the range of the medium viewing distance, 50% of the scenes can be seen along the main line, and the visual blind area is greatly reduced from the near field of view. There are still poor viewing quality in the west and southeast along the line, and there is a large-area visual blind zone in the core intersection area of the central A, B and D lines.

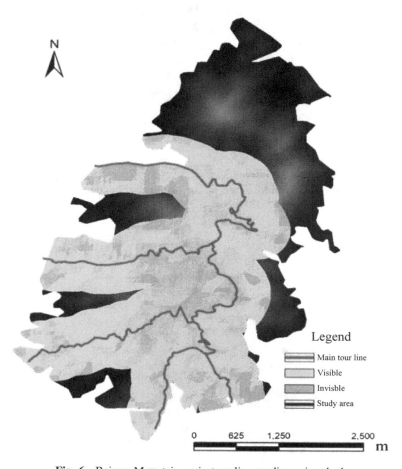

Fig. 6. Baiyun Mountain main tour line medium viewshed

The Result of Distant View. According to the viewshed of Fig. 7, the visible area along the main line is 1184.97 ha, which accounts for 64.44% of the total research area (Table 3). From the data point of view, the distant visible area accounts for the highest proportion, and the visibility along the line is better, but the invisible area of the A, B, and D lines intersecting the area gradually increases, and the terrain is seriously affected, and the far-distance line-of-sight corridor cannot be formed.

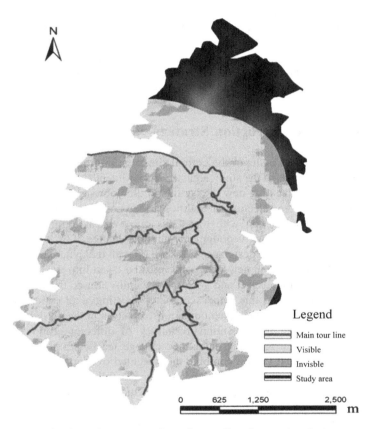

Fig. 7. Baiyun Mountain main tour line distant viewshed

Table 3. Baiyun Mountain main tour line visibility

Visual range	Visible area (ha)	Proportion of visible area (%)	Invisible area (ha)	Proportion of invisible area (%)
Near view	641.07	34.86	1197.73	65.14
Medium view	919.48	50.00	919.32	50.00
Distant view	1184.97	64.44	653.83	35.56

4.3 Summary of Current Viewshed Results

According to the superposition of the current view point and the main tour line of view can provide an overview of most of the scenery of the entire scenic area. The poor quality of the view is mainly concentrated in the following two parts. The first category is located at the boundary of the study area. The main reason is that the boundary part is mainly at the foot of the mountain. There is no distribution of foothills and viewpoints. The other point is that the landscape resources and the urban landscape at the foot of the mountain are less different, so the visibility is poor. The second category is located in the northern part of the Baiyun Mountain Scenic Area. The main reason is that there are only two view points in the current situation, and it is greatly affected by the topography, and the visibility of the near, middle and distant is not ideal, and there is no main line coverage in the north. Therefore, the overall visibility of the northern part of the study area is poor.

5 Planning and Construction Strategy

According to the analysis results of the current situation, in order to fully consider the global visibility experience of the study area, this paper intends to solve the problem of insufficient visibility by creating new viewpoints in the poor sight area in the northern part of the study area. The selection of traditional viewpoints is generally determined by the planners in combination with the status quo vertical terrain, site historical relationship and personal subjective judgment. With the change of time and the internal structure of the scenic area, the design can not keep up with the dynamic changes of the scenic area. The location of the new viewpoint must be taken into account in terms of its openness [17], elevation, aspect and geological stability. Based on the protection of the ecological environment, this paper combines existing exploration routes and infrastructure for secondary development and transformation in the northern part of the study area. Then through the questionnaire survey of different groups of tourists, landscape managers, local residents, landscape college teachers and planners in the study area, using AHP and GIS to combine the weighted overlay analysis of the new site selection sites. The specific analysis methods are as follows.

First, we select three decision-making factors, including height, slope and distance from the road within the scenic spot, through expert interviews and questionnaires. The aspect is the alternative factor and builds the evaluation index system. The data is standardized by the yaahp10.1 technical software platform, and a uniform value range is adopted (Fig. 8). The evaluation is set to 0–9. The larger the value, the more suitable for the new site. Finally, the weights of the factors affecting the newly-built viewpoints are: elevation factor 0.0904, slope factor 0.5559, relative distance factor 0.3537, east/south direction factor 0.8402, and west/north direction factor 0.1598.

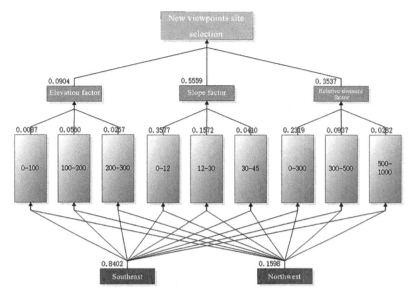

Fig. 8. New viewpoints AHP model

Secondly, we reclassify the elevation, slope and aspect of the study area in the GIS platform and assign values (Fig. 9, Fig. 10 and Fig. 11). For the distance factor inside the scenic area, the whole road network of the study area can be imported into the GIS

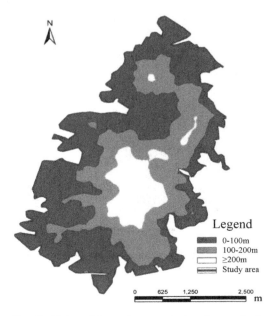

Fig. 9. Baiyun Mountain current elevation analysis

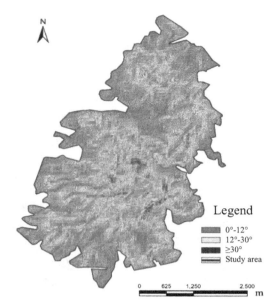

Fig. 10. Baiyun Mountain current slope analysis

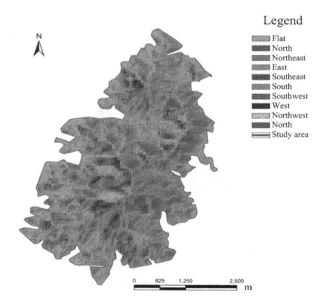

Fig. 11. Baiyun Mountain current slope direction analysis

database, and the Euclidean distance tool can be used for calculation, and then the result is re-assigned using the reclassification function.

Finally, this paper calculates the weighted sum of all the influence factor layers and obtains the optimal site selection range of new scenic spots based on the principle of low cost and ecological protection (Fig. 12). At the same time, according to the specific conditions of the site, fine-tuning, and finally determine the location of the new viewpoint.

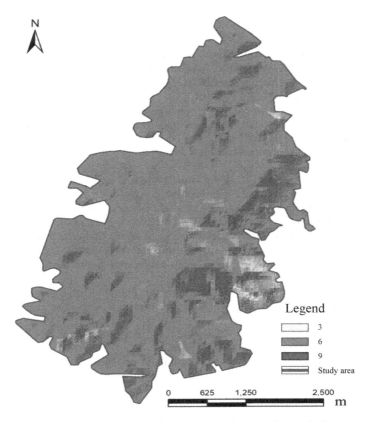

Fig. 12. Baiyun Mountain new viewpoints overlay analysis

6   Conclusion and Discussion

This paper takes the 3D viewshed of Baiyun Mountain Scenic Area as an example, and uses the GIS platform to carry out the "evaluation-simulation" mechanism to comprehensively evaluate and optimize the viewing horizon experience. Its data acquisition, evaluation methods and systems have a strong reference role for mountain-type scenic areas in Lingnan. Effectively help the scenic spot selection and the planning of the park road from the traditional qualitative to the comprehensive analysis method combining qualitative and quantitative. Although this method reduces the threshold for data acquisition, since the research data is not real-time mapping data, the overall accuracy and timeliness will be inaccurate with the current situation. On the other hand, in this case, the Baiyun cableway is a special linear viewing path. Due to the lack of information such as the path of the ropeway and the curvature, it is impossible to determine the field of view simulation along the elevation data of the line, which can be further improved in future research.

References

1. Bulletin on the development of China's scenic spots. http://www.mohurd.gov.cn/zxydt/w020121204199374149771793750.doc. Accessed 31 Oct 2019
2. Faming, L.: Study on the planning method of lilac scenic area based on gis analysis. Chinese Landscape Archit. **34**(07), 123–128 (2018)
3. CJJ/T 121-2008, classification criteria for scenic spots [S]
4. Chinese government network Homepage. http://www.gov.cn/ldhd/2013-12/14/content_2547880.htm. Accessed 31 Oct 2019
5. Feng, Y.: Study on the evaluation method of urban landscape. Urban Plan. J. (06), 46–49 + 57–80 (1999)
6. Garré, S.: The dual role of roads in the visual landscape: a case-study in the area around Mechelen (Belgium). Landscape Urban Plan. **92**(2), 125–135 (2009)
7. Hernández, J.: Integration methodologies for visual impact assessment of rural buildings by geographic information systems. Biosyst. Eng. **88**(2), 255–263 (2004)
8. Hernández, J.: Assessment of the visual impact made on the landscape by new buildings: a methodology for site selection. Landscape Urban Plan. **68**(1), 15–28 (2004)
9. Rogge, E.: Reducing the visual impact of 'greenhouse parks' in rural landscapes. Landscape Urban Plan. **87**(1), 76–83 (2008)
10. Kongjian, Y.: Design of landscape safety pattern and application of geographic information system in sensitive areas—taking Xiangshan ski resort in Beijing as an example. Chinese Landscape Archit. **01**, 11–16 (2001)
11. Kongjian, Y.: Landscape sensitivity basis and case study of landscape protection planning (3rd prize). Urban Plan. (02), 46–49 + 64 (1991)
12. Zhou, R., Liu, M.: Evaluation of scenic spots in forest parks based on landscape sensitivity. Chinese J. Appl. Ecol. (11), 2460–2466 (2008)
13. Wencun, S.: Preliminary discussion on landscape and visual impact assessment of scenic spots in Sanya City. Environ. Sci. Manag. **38**(04), 181–184 (2013)
14. CSDN Homepage. https://blog.csdn.net/qq_912917507/article/details/82869249. Accessed 31 Oct 2019
15. Baiyun Mountain Scenic Area Administration Homepage. http://www.baiyunshancom.cn/publicfiles/business/htmlfiles/byszz/yltj/201006/5655.html. Accessed 31 Oct 2019
16. Song Xiaodong, F., Ye Jiaan, S., Niu Xinyi, T.: Geographic Information System and Its Application in Urban Planning and Management, 2nd edn. Science Press, Beijing (2010)
17. Liu, B.: Quantitative analysis of visual attraction elements in landscape space. J. Nanjing Forestry Univ.: Nat. Sci. Edn. **4**, 149–152 (2014)

A 3D Visualization Method of Global Ocean Surface Based on Discrete Global Grids

Shuxiang Wang[1,2], Kaixiang Wen[1,3(✉)], Li Liu[4], Chen Zhang[1], and You Li[5]

[1] Guangdong Key Laboratory of Geospatial Information Technology and Application, Guangzhou Institute of Geography, Guangzhou 510070, China
690612731@qq.com
[2] Shandong University of Technology, Zibo 255000, China
[3] Guangdong University of Technology, Guangzhou 510000, China
[4] Ludong University, Yantai 264001, China
[5] State Grid Hunan Electric Power Limited Company Maintenance Company, Changsha 410004, China

Abstract. Ocean surface 3D visualization is an important aspect in marine information research and is of great significance to the construction of digital ocean. Due to the limitation of rendering efficiency and loading speed, the existing 3D visualization of ocean surface cannot meet the construction of ocean surface environment in a large area. Therefore, how to realize the global 3D visualization of the ocean surface is one of the key issues to be solved. For this, the paper proposes a 3D visualization method based on discrete global grids. Firstly, a multi-scale grid model for GPU based on the existing global discrete grid model is designed. Secondly, a dynamic wave simulation method based on GPU shader technology is proposed in order to realize the dynamic wave simulation. Finally, the effectiveness of the method is verified by comparative experiments. The experimental results show that the method of the paper achieves the effect of pixel-level ocean-land boundary and supports the dynamic simulation and real-time updating of the wave driven by the wind field. Compared with the existing 3D visualization method of ocean surface, the loading speed is faster and the rendering efficiency is higher. Therefore, the method is more suitable for the application needs of environment construction of the global ocean surface.

Keywords: 3D visualization · Ocean surface · Global discrete grid · Wave rendering

1 Introduction

The ocean covers 70% of the global surface and is an important part of the geographical environment. Ocean surface visualization uses computer graphics to reproduce the real ocean surface environment. It is of great significance to Digital Ocean. With the development of computer graphics technology, the traditional ocean surface visualization based on 2D graphics system has been unable to meet need of visualization and application due to problems such as abstract expression and lack of sense of

© Springer Nature Singapore Pte Ltd. 2020
Y. Xie et al. (Eds.): GSES 2019/GeoAI 2019, CCIS 1228, pp. 19–32, 2020.
https://doi.org/10.1007/978-981-15-6106-1_2

reality. Ocean surface visualization based on 3D graphics technology has become the mainstream of research of ocean surface visualization. However, concerning efficiency, the existing 3D visualization method of ocean surface is mainly constructed for a small-scale regional ocean environment. The limitations of spatial scale and scope limit the development of 3D visualization systems of ocean surface. Therefore, how to build an ocean surface environment of global scale and realize global 3D visualization of the ocean surface has become a key issue to be solved.

Aiming at the above problems, this paper proposes a method for 3D visualization of ocean surface based on global discrete grids. The method includes two parts: First, in order to effectively organize and manage the global ocean surface grid, and make full use the ability of GPU parallel computing to improve the efficiency of ocean surface rendering, a multi-scale grid model of ocean surface for GPU rendering is designed, based on the existing global discrete grid model. Secondly, based on the tessendorf model, a method of wave dynamic simulation with GPU shader technology as the core is proposed in order to realize the function of dynamic simulation and real-time update of ocean wave driven by wind field. At the end of the paper, the effectiveness of the method is verified by comparative experiments.

2 Related Works

The existing research on 3D visualization of ocean surface mainly focuses on three aspects: ocean surface grid organization, wave dynamic simulation and the optimization of ocean surface rendering efficiency.

2.1 Ocean Surface Grid Organization

In order to realize the visualization of the ocean surface by computer, it is necessary to abstract the real ocean surface into an ocean surface grid. For large-scale ocean surface visualization, it is essential to efficiently organize and manage the ocean surface grid. Now, the grid organization method of ocean surface based on the projection grid model is the most widely used. This method realizes the infinite extension effect of the ocean surface by constructing an ocean surface grid in the projection space so that the ocean surface grid can automatically move and zoom with the angle of view. However, since all ocean surfaces are represented by a grid, it is difficult to reflect the characteristics of different ocean areas (such as boundaries of land and ocean, etc.) [1, 2]. In addition, there is a method of ocean surface grid organization based on the LOD grid model. This method does not load all ocean surface into the memory at one time but dynamically performs the loading and unloading of the ocean surface grid according to the viewpoint [3, 4]. However, since the ocean surface in different regions is represented by different grids, the actual application requires grid splicing in real time and the ocean surface rendering efficiency is inevitably affected during the splicing process [4–6].

2.2 Wave Dynamic Simulation

The ocean wave simulation simulates the ocean surface spray by modifying the vertex position of the ocean surface grids. The sense of reality, instantaneity and interactivity directly affect the realistic degree of the ocean surface scene. The wave dynamic simulation is relatively mature. The research in this aspect can be traced back to the 1980s. Reeves, Bailey et al. proposes a method of wave simulation based on the Gerstner model [7, 8]. According to the parameters of the orbit, the method can help to produce realistic wave shapes and subdivide waves of different characteristics through ray tracing and adaptive methods. Later, He and Gary improves it on the basis of this method [9, 10]. Through the A buffer technique, different waveforms are generated according to the steepness and depth of the waves, and the simulation and rendering of the broken waves on the inclined beach are realized. In addition, Paul, Chen et al. proposes a real-time method of wave simulation based on texture mapping [11, 12]. Through random analog sine, wave shape is simulated and mapped, the velocity of wave simulation is faster. Tessendorf and Schlitzer proposes a method of wave simulation based on ocean wave spectrum [13, 14]. The method simulates height of the wave by 2D fast Fourier inverse transform, and simulates the undulating feature of the wave in vertical direction by the form of a height field sequence. And the dynamic simulation of ocean surface spray is simulated. Due to operability and efficiency, the Tessendorf method is the most widely used.

2.3 The Optimization of Ocean Surface Rendering Efficiency

Low ocean surface rendering efficiency can seriously affect the user's visual experience. Especially when performing large-scale ocean surface visualization the ocean surface rendering efficiency directly determines the availability of ocean surface visualization methods. In recent years, scholars have begun a lot of research on how to improve the efficiency of ocean surface rendering. At present, there are mainly two ways to improve the efficiency of ocean surface rendering. One way is to update in parallel the position of the ocean grid point based on the GPU shader which loads the offset value of the wave motion into the GPU for the parallel sampling by the shader in a displacement texture so as to reduce CPU load and improve ocean surface rendering efficiency [15]. Another way is to build an ocean surface visualization system based on the CUDA (Compute Unified Device Architecture). For example, Su et al. achieves parallel reading of ocean surface data based on the CUDA architecture [16]. Zhang et al. realizes the parallel calculation of wave shape based on the CUDA architecture [17]. To some extent, the above research has solved the problem of ocean surface rendering efficiency. If the ocean surface grid is complicated to construct or the memory occupancy is large, it will still affect the overall ocean surface rendering efficiency [18].

In summary, in order to realize the 3D visualization of the ocean surface facing the world, it is necessary to solve the problems of ocean surface grid organization, wave dynamic simulation and ocean surface rendering efficiency optimization. For this, this paper proposes a 3D visualization method of ocean surface based on global discrete grid which organizes the global ocean surface through a global discrete grid model, and uses the GPU shader in combination with the method of wave simulation named Tessendorf to achieve efficient rendering of ocean surface waves.

3 Methods

3.1 Global Multi-scale Ocean Surface Grid Model for GPU

In order to effectively organize and manage the global ocean surface grid and make full use of the GPU parallel computing ability to improve the ocean surface rendering efficiency, this paper explores on the basis of the global longitude and latitude discrete grid model and designs a global multi-scale ocean surface grid model of GPU-oriented rendering. The multi-scale ocean surface grid model is shown in Fig. 1:

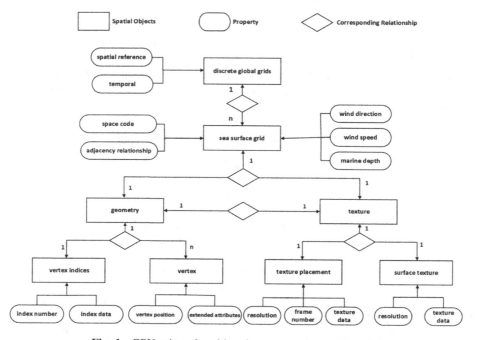

Fig. 1. GPU-oriented multi-scale ocean surface grid model

The ocean surface grid is the basic unit of ocean surface data organization and rendering. It is organized and constructed according to the rules of global latitude and longitude grid splitting. Thus, unique spatial code of each ocean surface grid corresponds to a global discrete grid, which records the number of stages the row number and the column number of the current grid. Spatial coding can be used to retrieve information such as wind speed, wind direction and ocean depth, ocean area distribution, and the adjacency relationship between the grids. The geometric object of the ocean surface grid describes the geometric information of the ocean surface, which consists of vertex and vertex indices. The vertex is the grid point, which is considered as the minimum unit of ocean level fluctuation. In addition to the basic position information, in order to realize the ocean surface dynamic simulation based on GPU, the vertex attribute is extended. Other attributes including latitude and longitude and vertex type are added. The vertex index records the organization relationship among discrete vertices for the transformation of point-to-face primitives.

In order to realize the dynamic simulation of the ocean wave, the texture object of the ocean surface grid is divided into displacement texture and surface texture. The displacement texture records the dynamic fluctuation process of the ocean surface wave. The current wind direction and wind speed are used as parameters and then generated based on the ocean wave modeling algorithm. Finally, it is loaded into the GPU in the form of 3D texture. The surface texture records the color and transparency of the current ocean surface grid, which can be generated by seawater depth or mask data of the recorded ocean area. The depth of the color can reflect the water body shading effect caused by the depth of the ocean. In addition to reflecting the texture of the water, the degree of transparency can also be used to hide the epicontinental water to achieve the effect of ocean and land boundary.

The advantage of this model structure is that the geometry of the ocean surface grid and its specific morphological features (recorded in the displacement texture) are separated from each other and only correlated by the GPU when rendering. When the ocean surface grid is first loaded, only the most basic geometry structure needs to be constructed, and the more complex morphological features can be reused by multiple ocean surface grids by means of resource sharing. Thus the construction speed of the gird is effectively improved.

Figure 2 shows the multi-scale ocean surface rendering process based on the multi-scale ocean surface grid model. The division of latitude and longitude of the ocean area in the virtual earth is carried out. With the transformation of the viewpoint, the spatial data of the corresponding scale in the visible ocean area is dynamically scheduled, and the geometry and texture of the ocean surface grid are constructed by spatial data. The ocean surface grids of different scales in the same ocean area are organized in a top-down quadtree pyramid. As the viewpoint goes forward and the resolution of the grid increases exponentially, the undulating effect of ocean waves is gradually obvious.

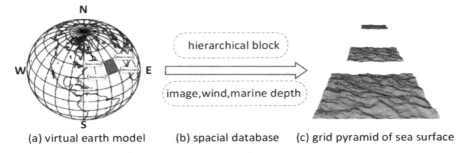

(a) virtual earth model   (b) spacial database   (c) grid pyramid of sea surface

Fig. 2. Flow chart of multi-scale mapping of global ocean surface. a, latitude longitude division of the virtual earth model. b, dynamic scheduling of spatial data of corresponding scales in the visible ocean area. c, organization of the ocean surface grid by top-down quadtree pyramid.

3.2   Wave Dynamic Simulation Method Driven by Wind Field

To a large extent, the sense of reality in ocean surface visualization depends on the dynamic fluctuation of ocean surface spray, especially the dynamic simulation of ocean surface spray driven by wind field, which not only meets the visual needs of users, but also matches the real geographical environment and more in line with the requirements of geographic information simulation applications. In this paper, based on the multi-scale ocean surface grid model, a dynamic simulation method of ocean wave driven by wind field is proposed.

In order to realize the dynamic simulation of ocean surface waves, we need to model the motion pattern of ocean surface waves. This paper uses a wave modeling method based on ocean wave spectrum proposed by Tessendorf [13]. The method considers that the height of the ocean surface wave is superimposed by a series of sine and cosine waves, so it can be solved by inverse transform of 2D fast Fourier. The formula is as follows:

$$h(m, t) = \sum_k \tilde{h}(k, t) \cdot \exp(ik \cdot m) \tag{1}$$

Where, $m = (x, y)$, which represents the horizontal position of the ocean level point. $t = \frac{2\pi * \text{frame}}{\text{totalframe}}$, frame represents the current number of frames, totalframe represents the total number of frames of an ocean wave cycle and $k = (k_x, k_y)$, which is a two-dimensional vector. In order to make the generated ocean surface height field have the characteristics of cyclic repeating of up, down, left and right, $m = \left(\frac{aS}{R}, \frac{bS}{R}\right)$, $k = (\frac{2\pi c}{S}, \frac{2\pi d}{S})$, $-\frac{R}{2} < a, b, c, d < \frac{R}{2}$, S is a numerical constant representing the height field resolution, and R is the number of Fourier samples in one dimension. The influence of wind field conditions on Fourier amplitude changes with time, which is the key to calculate the wave height. However, the literature [13], which describes the law of wave motion in the form of height field sequence, can only simulate the wave fluctuation characteristics in the vertical direction, as shown in Fig. 3(a). In a real geographical environment, the waves tend to tilt under the effect of the wind field, as shown in Fig. 3(b).

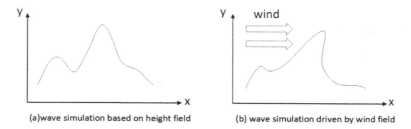

(a)wave simulation based on height field   (b) wave simulation driven by wind field

Fig. 3. Horizontal wave profile. (a), wave simulation based on height field. (b), wave simulation driven by wind field. y represents the vertical direction of the ocean surface, x represents the horizontal direction and the blue arrow represents the wind field. (Color figure online)

In this paper, the Tessendorf method is improved. According to the current fluctuation state of the ocean wave and the current ocean surface wind field condition, the offset value of each wave point in the horizontal direction can be calculated to achieve more realistic wave simulation.

The main steps are as follows: First, find the gradient of formula (1):

$$G_x(m, t) = \sum_k \tilde{h}(k, t) \cdot ik_x \cdot \exp(ik \cdot m) \tag{2}$$

$$G_y(m, t) = \sum_k \tilde{h}(k, t) \cdot ik_y \cdot \exp(ik \cdot m) \tag{3}$$

$G_x(m, t), G_y(m, t)$ represent the gradient values of the height field in the horizontal and vertical directions, which is solved based on two dimensional inverse fast Fourier transform. For the ocean surface height field, since the gradient points to the direction in which the height field grows fastest, the unit vector of the gradient vector can be regarded as the direction in which the ocean wave is most inclined. Then the offset value of the wave in the horizontal direction is:

$$d = (d_x \cdot d_y) = f \cdot (g_x \cdot g_y) \tag{4}$$

f is the dot product of the gradient direction (g_x, g_y) and the current wind field vector (w_x, w_y), reflecting the degree of influence of the wind field on the waves. Therefore, an ocean wave model of describing the current wave motion in the vertical, horizontal and vertical directions is established. The dynamic simulation of the ocean surface can be achieved by applying three directions of offset values to the initial position of the ocean level grid point. In the dynamic simulation mode, the conventional method of modifying the position of the grid point based on the CPU in real time has a large calculation amount and affects the rendering efficiency. The method of pre-building a multi-temporal grid will increase the memory footprint and affect the loading efficiency of the grid which is not suitable for dynamic visualization of the large-scale ocean surface. What's worse, the grid is difficult to modify after construction, and cannot match the real-time wind field. Based on the multi-scale ocean surface grid model, this paper proposes a dynamic simulation method of ocean surface on grounds of GPU which can update the grid point position of ocean surface in parallel through GPU, reduce CPU calculation, and load the offset value of wave motion into GPU for shader sampling in the way of displacement texture. When the wind field conditions change, only the displacement texture needs to be reloaded instead of any modification to the constructed ocean grid. The detailed algorithm flow is shown in Fig. 4:

1. Traverse the virtual earth tile quadtree to determine the current visible ocean area and construct the ocean surface grid of the current ocean area.
2. According to the spatial coding of the grids, obtain the wind direction and wind speed of the nearest ocean surface grids. The average value is taken as the parameter. The offset value of the ocean wave in the horizontal, vertical and vertical directions is calculated according to the ocean wave modeling algorithm. The

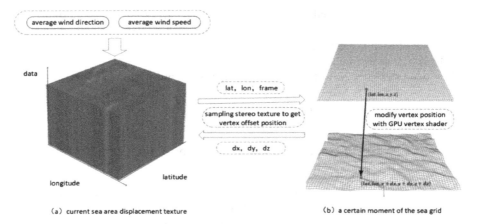

Fig. 4. Ocean surface dynamic simulation process

values are stored respectively in the three channels (r, g and b) of displacement texture. The displacement textures at different moments are combined to obtain the stereo texture and then loaded into the vertex shader. If the wind speed and direction of the current scene do not change much from that of the previous frame, the displacement texture does not need to be updated, which is conducive to reducing the amount of calculation.

3. In the vertex shader, the texture coordinates are determined according to the coordinates (lat,lon) of latitude and longitude of the grid vertex and the current frame number.

$$texcoord = vec3\left(\frac{lat}{hres}, \frac{lon}{hres}, frame\%(totalframe - 1)\right) \tag{5}$$

4. hres is the span of latitude and longitude of the wave height field, m is the surface area of the earth (m° × m°) corresponding to the ocean surface height field, totalframe is the total number of frames of a wave motion cycle, and % is the operation of get the remainder.

5. The obtained texture coordinates are used to sample the stereo texture of the current ocean area to get the offset value D of the vertex and overlay the offset value on the original vertex position.

$$position = ModelViewProjectionMatrix * \left(gl_{vertex} + D_x + D_y + D_z\right) \tag{6}$$

$D_x = D.r*N_x$, $D_y = D.g*N_y$ and $D_z = D.b*gl\_normal$. N_x, N_y are the horizontal and vertical vectors of the vertex when it cuts the earth's surface.gl_position is the vertex coordinates of the final output, gl_ModelViewProjectionMatrix is the vertex transformation matrix, gl_vertex is the original vertex coordinates and gl_normal is the vertex normal vector.

4   Experiment and Discussion

In order to verify the validity and feasibility of the proposed method, the experiment is implemented based on the osgEarth. The software environment includes Windows 7, Visual Studio 2010 and OpenGL. The hardware configuration is Intel(R) Core(TM) i3-2100 dual-core 3.1GHZ CPU, NVIDIA Quadro 600 graphics card, 1 GB video memory and 8 GB memory.

The image in the experimental data uses the ESRI_Imagery_World_2D published by ESRI. Terrain uses the Terrain Tile Service published by ReadMap. The land part uses STRM data with resolution of 90 meters, and the ocean bottom part uses GEBCO (General Bathymetric Chart of the Oceans) to provide global deep ocean measurement data ($30'' \times 30''$). The wind field data is simulated and interpolated according to the wind speed and wind direction information acquired by the ship sensor in real time. The ocean area distribution data (images of the ocean and the land are respectively represented by different colors) are generated and edited by the shoreline data in the electronic chart. According to the multi-scale ocean surface grid model before the experiment, the data is organized in multiple scales to construct a tile pyramid of a quadtree structure. Figure 5, Fig. 6 and Fig. 7 show the comparison of the ocean surface effects of the proposed method at different scales.

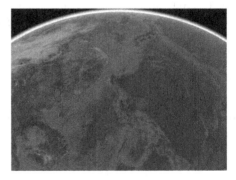

Fig. 5. Ocean surface global view of the method

Fig. 6. Ocean surface partial view of this method

Comparing the three pictures, it can be found that with the progressive view, the ocean surface is gradually clear, the ocean water tone transition is smooth, the effect of macroscopical ocean-land boundary and the microscopic ocean surface wave can be effectively reflected, basically satisfying the application need of global multi-scale ocean surface seamless browsing. Figure 7 and Fig. 8 show the difference between this method and the projected grid method for the ocean close-up view.

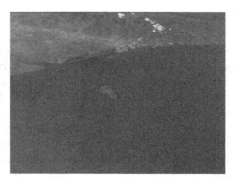

Fig. 7. Ocean surface close-up view of the method

Fig. 8. Ocean surface close-up view of projection grid method

It can be seen from the figure that the method of this paper can quickly call the ocean area distribution data of the current area with the support of the discrete grid framework to achieve the fine boundary between the land and ocean. The projection grid method can only achieve the rough ocean-land boundary effect (the ocean-land boundary scheme provided by Trit) through the surface elevation acquired by RTT technology in real time, which tends to cause the problem of seawater covering the real surface.

Fig. 9. Ocean wave effect with wind speed of 5 m/s

Fig. 10. Ocean wave effect with wind speed of 24 m/s

Figure 9 and Fig. 10 show the comparison of ocean surface wave effects that achieved in different wind field conditions (In Fig. 9, wind speed is breeze of 5 m/s and in Fig. 10 wind speed is gentle breeze of 24 m/s). When the wind speed is high, the undulating effect of the ocean surface is more obvious.

In order to verify the rendering and loading efficiency of the proposed method, the method is compared with the ocean surface visualization method based on projection grid and the existing ocean surface visualization method based on discrete grid under the same experimental conditions. The experimental results are shown in Fig. 11 and Fig. 12:

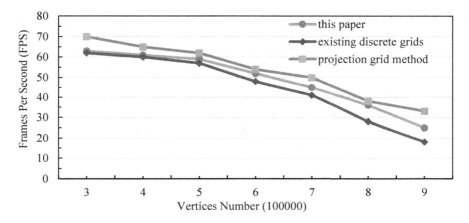

Fig. 11. Comparison of frame rate display of three methods

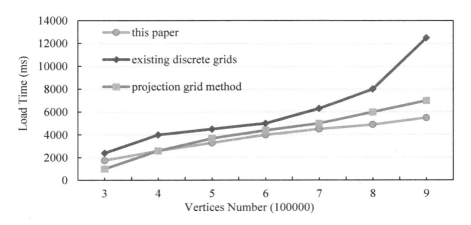

Fig. 12. Comparison of three methods in terms of loading time

It can be seen from Fig. 11 that the ocean surface visualization method based on projection grid implemented by commercial software Triton has obvious advantages in rendering efficiency. The main reason is that the ocean surface visualization method based on the discrete grid needs to simulate the ocean wave when rendering every frame so it has a certain influence on the frame rate. Because of the GPU technology, the improved discrete grid method implemented in this paper is much more efficient than the existing discrete grid method in terms of ocean wave simulation efficiency. Compared with the projected grid method, the method can maintain the frame rate difference within 6 frames/second.

Figure 12 shows that this method is faster than the other two methods. With the increase of the number of grids, the existing discrete grid method has an exponential increase in loading time. The projection grid method does not need to reconstruct a new grid, but its ocean-land boundary method has efficiency problems, which ultimately affects the overall loading speed. In contrast, the method in this paper uses tile quadtree to organize and manage the grids. There is no constraint among the peer grids so it can make full use of multi-threading for fast parallel loading. When 2 threads are enabled, the load of 900,000 grid vertices can be completed in about 5 s.

In order to verify the ability that the ocean surface visualization method can quickly respond to the change of real-time wind field, the update time of the wave effect is compared among the three methods when the wind field conditions change. As shown in Fig. 13:

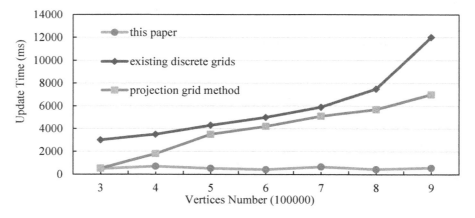

Fig. 13. Comparison of the three methods in terms of update time

As can be seen from the figure above, as the method in this paper can update the whole wave effect only by modifying the displacement texture when the wind field conditions change, the update time can be stable at about 0.5 s and the wave currently being simulated will not be affected because this part of calculation can be performed in the background thread. In the other two methods, the update of the ocean wave is equivalent to reloading the ocean surface grid, which cannot meet the needs of real-time dynamic wind field update in terms of update time and visual experience.

Compared with the existing 3D visualization method of ocean surface, the loading speed is faster and the rendering efficiency is higher. Therefore, the method is more suitable for the application needs of environment construction of the global ocean surface.

5 Conclusion and Future Work

In order to solve the problems of poor rendering efficiency and slow loading speed in the existing ocean surface 3D visualization method based on global discrete grid, this paper proposes an optimization method, which mainly includes two parts: Firstly, this paper expands on the traditional latitude and longitude discrete grid, and designs a global multi-scale ocean surface grid model based on GPU in order to effectively organize and manage the global ocean surface grids, while making full use of GPU parallel computing capabilities. Secondly, this paper proposes a real-time updating method of wave dynamic simulation based on multi-scale ocean grid model, considering that the effect of wave fluctuation driven by wind field is more in line with the needs of geographic information simulation application. Finally, the method was verified by comparison experiments. The experiment results show that the proposed method has stable rendering efficiency, fast loading speed, and has the functions of reflecting the characteristics of different ocean areas, realizing the fine land-ocean boundary and supporting the dynamic update of wind field conditions, which is more in line with the application needs of global ocean surface environment simulation. The next step will be to discuss how to achieve real-time interactive dynamic simulation between global ocean surface and ship navigation, laying the foundation for specific applications of the global ocean.

Acknowledgments. This study was jointly supported by China Southern Power Grid Guangzhou Power Supply Bureau Co., Ltd. Key Technology Project (0877002018030101SRJS0 0002); Guangdong Provincial Science and Technology Program (2017B010117008); Guangzhou Science and Technology Program (201806010106, 201902010033); the National Natural Science Foundation of China (41976189,41976190); the Guangdong Innovative and Entrepreneurial Research Team Program (2016ZT06D336); the Southern Marine Science and Engineering Guangdong Laboratory (Guangzhou) (GML2019ZD0301); the GDAS's Project of Science and Technology Development (2016GDASRC-0211,2018GDASCX-0403,2019GDASYL-0301001, 2017GDASCX-0101,2018GDAS CX-0101,2019GDASYL-0103003).

References

1. Yoo, J.S., Min, K.J., Ahn, J.W.: Concept and framework of 3D geo-spatial grid system. In: Kawai, Y., Storandt, S., Sumiya, K. (eds.) W2GIS 2019. LNCS, vol. 11474, pp. 136–149. Springer, Cham (2019). https://doi.org/10.1007/978-3-030-17246-6_11

2. San, O., Staples, A.E.: An efficient coarse grid projection method for quasigeostrophic models of large-scale ocean circulation. Int. J. Multiscale Comput. Eng. **11**(5), 463–495 (2013)
3. Lv, C., Yu, F.: The application of a complex composite fractal interpolation algorithm in the seabed terrain simulation. **2018**, 1–6 (2018)
4. Li, C., Li, T., Huang, Q.: Research status and prospect for maritime object monitoring technology. J. Phys: Conf. Ser. **1288**, 012064 (2019)
5. Wang, Q., et al.: Grid evolution method for DOA estimation. IEEE Trans. Sig. Process. **66**(9), 2374–2383 (2018)
6. Hong, X., et al.: Simulating and understanding the gap outflow and oceanic response over the Gulf of Tehuantepec during GOTEX. Dyn. Atmos. Oceans **82**, 1–19 (2018)
7. Reeves, W.T.: A Simple Model of Ocean Waves (1986)
8. Bailey, R.J., Jones, I.S.F., Toba, Y.: The steepness and shape of wind waves. J. Oceanogr. Soc. Japan **47**(6), 249–264 (1991)
9. Huaiqing, H., et al.: A way to real-time ocean wave simulation. In: International Conference on Computer Graphics, Imaging and Visualization (CGIV 2005) (2005)
10. Mastin, G.A., Watterberg, P.A., Mareda, J.F.: Fourier synthesis of ocean scenes. IEEE Comput. Graph. Appl. **7**(3), 16–23 (1987)
11. Chapman, P., et al.: Seabed visualization. In: Proceedings Visualization 1998 (Cat. No. 98CB36276) (1998)
12. Chen, H., Li, Q., Wang, G., Zhou, F., Tang, X., Yang, K.: An efficient method for real-time ocean simulation. In: Hui, K.-C., et al. (eds.) Edutainment 2007. LNCS, vol. 4469, pp. 3–11. Springer, Heidelberg (2007). https://doi.org/10.1007/978-3-540-73011-8_3
13. Tessendorf, J.: Simulating Ocean Water. SIGGRAPH2001 Course notes, pp. 47–58. Addison Wesley, Boston (2001)
14. Schlitzer, R.: Interactive analysis and visualization of geoscience data with Ocean Data View. Comput. Geosci. **28**(10), 1211–1218 (2002)
15. Li, H., Quan, W., Xu, C., Wu, Y.: A GPU-based mipmapping method for water surface visualization. In: Proceedings of SPIE 10610, MIPPR 2017: Parallel Processing of Images and Optimization Techniques; and Medical Imaging, 1061003, 6 March 2018
16. Su, T., et al.: Multi-dimensional visualization of large-scale marine hydrological environmental data. Adv. Eng. Softw. **95**, 7–15 (2016)
17. Zhang, F., et al.: Spatial and temporal processes visualization for marine environmental data using particle system. Comput. Geosci. **6**(1), 53–54 (2019)
18. Liu, S., et al.: A framework for interactive visual analysis of heterogeneous marine data in an integrated problem solving environment. Comput. Geosci. **7**(1), 20–28 (2017)

Investigating Evolutions of Metro Station Functions in Shenzhen with Long-Term Smart Card Data

Fan Zhang[1,2], Kang Liu[1,3], Ling Yin[1(✉)], and Fan Zhang[1,4]

[1] Shenzhen Institutes of Advanced Technology, Chinese Academy of Sciences, Shenzhen 518055, China
yinling@siat.ac.cn
[2] University of Chinese Academy of Sciences, Beijing 100049, China
[3] State Key Laboratory of Resources and Environmental Information System, Beijing 100101, China
[4] Shenzhen Institute of Beidou Applied Technology, Shenzhen 518055, China

Abstract. The significances of urban metro stations are far more transport nodes in a city. Their surroundings are usually well developed and attract plentiful human activities, which support various functions in cities such as living and working. Investigating evolutions of metro station functions can help understand the developments of the stations as well as the whole city in a quick, low-cost, continuable and effective way. Using long-term smart card data collected from 2014 to 2018 of Shenzhen, China, this study identifies the functions of metro stations in different years, and reveals the evolution patterns of the functions over years, which is the first attempt as far as we know. The analytical results indicate that the function differentiations among stations have enlarged during those years; the cores of workplaces has shifted from Luohu to Futian and Nanshan district; the cores of residences have shifted to relatively peripheral districts such as Bao'an District, Longhua District and Longgang District; in general, the functions have evolved toward a more home-work-separation pattern. Those findings can help understand the changes of metro station functionalities, which are important clues for the governments to make better and sustainable public-transport and land-use policies.

Keywords: Metro station · Smart card data · Function · Evolution · Long-term

1 Introduction

The development of a metro system promotes land development and land price along the metro lines, thus affecting the spatial structure of a whole city. The governments would comprehensively consider various factors when planning and constructing metro system in order to meet traffic demand, adjust urban structure and coordinate planning and development. The influences of a metro system on the spatial structure of a city are mainly reflected in the layouts and functions of the stations. In recent years, the rise of transit-oriented development (TOD) shows that the governments hope to construct a new urban pattern and guide the direction of urban development through metro

© Springer Nature Singapore Pte Ltd. 2020
Y. Xie et al. (Eds.): GSES 2019/GeoAI 2019, CCIS 1228, pp. 33–53, 2020.
https://doi.org/10.1007/978-981-15-6106-1_3

construction. Because the impact of accessibility along metro lines can change the economic structure around the metro stations, governments will consider how to stimulate the economic development along metro lines and improve the economic benefits when planning metro lines and stations. However, in real construction and operation processes, there are plenty of problems exposed. The land-use development lagging behind metro construction can hardly bring abundant passenger flow, which will cause waste of resources; the mismatch of station layouts and travel demands generate a burst of passenger flow in workday peak hours. Identifying and investigating the evolutions of metro station functions can help understand the developments of stations and reveal the problems mentioned above.

In recent years, urban big data such as mobile phone data, floating car data, geo-tagged social-media data and smart card data (SCD) have become potential and promising data sources for revealing and solving urban problems [25–29]. Compared to traditional data (e.g., travel survey, statistical bulletin, etc.), they have advantages of low acquisition cost, large volume, wide coverage and high spatiotemporal resolution, and can capture more details of dynamic changes. Based on the SCD, researches have mainly focused on travel purpose inference [1–4], travel pattern mining [5–12], home-work separation analysis [13–15], spatial structure recognition [16–18], short-term passenger flow prediction [19] and TOD evaluation [20]. As for the topic of metro station functions, most of current studies have focused on the classifications of metro station functions [21–24]; however, no literature has investigated the evolution of metro station functions in a long-term time period as far as we know.

Therefore, taking using of long-term and large-scale SCD, this study investigates the evolutions of metro station functions in a metropolitan city, Shenzhen, China. Specifically, we define the activities of passengers arriving in the areas around metro stations as station functions, such as living, working, eating and leisure. Due to the mixed land use around stations, functions are also generally mixed with different mixing properties. The changes of mixed proportion of metro station functions, such as enhanced working function and weakened living function, are called the evolution of metro station functions. Through this way, we identify the functions of Shenzhen metro stations in each time period, investigate their evolutions and analyze their spatial characteristics, which can be beneficial for commercial site-selection, real estate investment, and decision-making for future metro planning and construction.

The highlights of this study can be summarized as follows:

(1) With the rapid development of Shenzhen in recent years, especially after the opening of new metro lines in 2016, the functions of many metro stations have changed. In order to depict these dynamic changes, this study provides a framework for identifying metro station functions and their evolutions over time, which can provide policy-decision basis for the governments and help enterprises to select business locations.

(2) The large-scale spatiotemporal data - SCD are used to identify the functions of metro stations and investigate their evolutions, which are of greater data volumes and higher spatiotemporal resolutions than traditional data such as surveys.

(3) Most existing studies have only used short-term SCD (e.g., less than one week) to identify metro station functions at one certain time period, while this study investigates the functions of Shenzhen metro stations using SCD of five years, which can reflect long-term evolutions of the stations.

2   Materials and Methods

2.1   Data

This study uses SCD of 118 Shenzhen metro stations from Line 1 to Line 5. Table 1 shows one record of the SCD statistics at a metro station, which includes the sum of swiping-in (entry) and swiping-out (exit) records from 06:00:00 to 06:59:59 on September 1, 2018, respectively.

Table 1. An example of the basic SCD user statistics at a metro station.

Station	Date-Time	Entry-volume	Exit-volume
ShangTang	2018-09-01 06	329	14

In order to reflect the regular functions of metro stations, we eliminate the data of legal festivals and major events, as well as the abnormal data caused by incomplete data collection, and complement the missing data by the average values of their adjacent months. Table 2 describes characteristics of the processed data.

Table 2. Data description.

Month	Date	Time	Data category
2014.1–2018.9 (57 months)	1st to 7th day (each month)	6:00–23:00 (interval of 1 h)	Entry-volume; Exit-volume

Figure 1 shows the entry and exit volumes of Station GaoXinYuan and BaoAnZhongXin in September 2018. We can see that the time-series in Monday to Thursday, Friday, Saturday and Sunday show four different patterns. Noting that people always engage in some leisure activities after working at Friday night, we choose the average values of entry and exit volumes from Monday to Thursday as representative of weekday, respectively. Moreover, the entry and exit volumes on Sunday are used to represent weekend owing to the overtime phenomenon on Saturday.

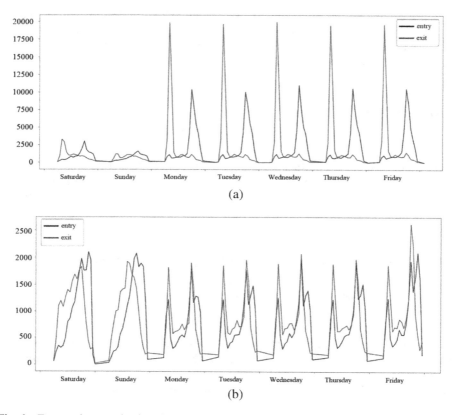

Fig. 1. Entry-volume and exit-volume time-series of (a) GaoXinYuan and (b) BaoAnZhongXin in September 2018.

2.2   Methodology

Identifying Metro Station Functions. The SCD during August 1 to August 7 in each year of 2014 to 2018 are extracted for use as representatives of the corresponding years.

Taking the week of August 1 to August 7, 2018 as an example, firstly we calculate the average entry/exit volumes in each station from Monday to Thursday and use min-max normalization to standardize data between 0 and 1. Secondly, we combine the average entry and exit volume data to a 36-dimensional feature vector, as the time range of data is 6:00 to 23:00 each day. Similarly, we implement min-max normalization of entry and exit volume on Sunday and connect them to a 36-dimensional feature vector. Figure 2 shows the volumes of GaoXinYuan on (a) Monday to Thursday and (b) Sunday in August 2018.

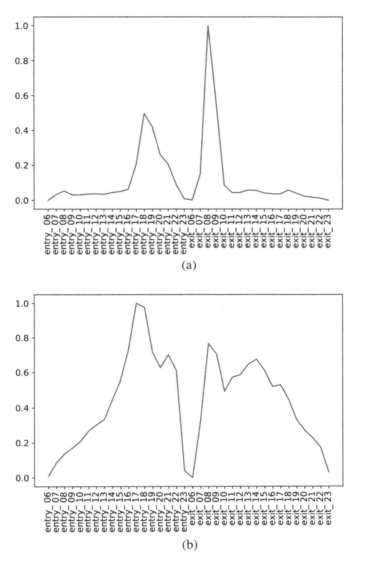

Fig. 2. Entry-volume and exit-volume time-series of GaoXinYuan on (a) Monday to Thursday and (b) Sunday in August 2018.

After that, K-means clustering algorithm is used to cluster the feature vectors of Monday-to-Thursday and Sunday respectively.

Measuring Evolutions of Metro Station Functions. Five features are defined to represent different metro station functions, as shown in Table 3.

Table 3. Feature definition.

No.	Feature	Function
1	Exit-volume of daily average	Station attraction
2	Entry-volume on morning-peak-hours (7:00–9:00)	Living
3	Exit-volume on morning-peak-hours (7:00–9:00)	Working
4	Exit-volume on weekday noon hours (11:00–13:00)	Eating
5	Exit-volume on weekend	Leisure

In order to identify the evolutions of station functions, we first extract the trends of these features (time series) through a naive time series decomposition algorithm based on moving average method, as the derived sequence formed by moving average eliminates the fluctuation caused by the influence of short-term accidental factors on the original time series, and present the basic development trend of the research object. Figure 3 shows an example of decomposition for the time series of one feature, from which we can see that the original time series is decomposed into three components, i.e., trend, seasonality and residuals. On this basis, we calculate the linear-fitting slopes of the extracted trend lines through ordinary least squares (OLS), as shown in Fig. 4, which can describe the evolutions (change degrees) of the functions represented by these features in a quantitative way.

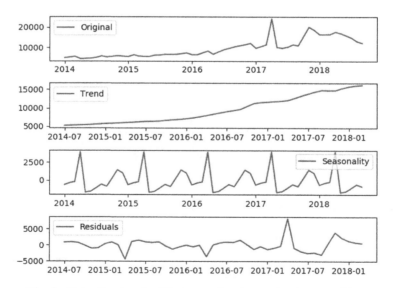

Fig. 3. Extracting trends of features using time-series decomposition.

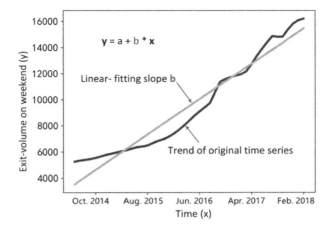

Fig. 4. Extracting change degree (evolutions) of metro station functions by calculating linear-fitting slopes of the defined features.

3   Results

3.1   Metro Station Functions in Each Year and Their Function Type Transition

By clustering the feature vectors which are composed of entry-volume and exit-volume time series between Monday and Thursday, we can obtain the annual station commuting type from 2014 to 2018. We cluster all metro stations into three types, named as living-oriented station, working-oriented station and mixed station, respectively according to the characteristics of time series.

Figure 5 shows the spatial distributions of the metro station commuting types annually from 2014 to 2018. As summarized in Table 4, during the five years, numbers of mixed stations have converted to living-oriented or working-oriented stations, while the rest stations remain the same in their functions. In general, the functions of Shenzhen metro stations have evolved toward a more home-work-separation pattern.

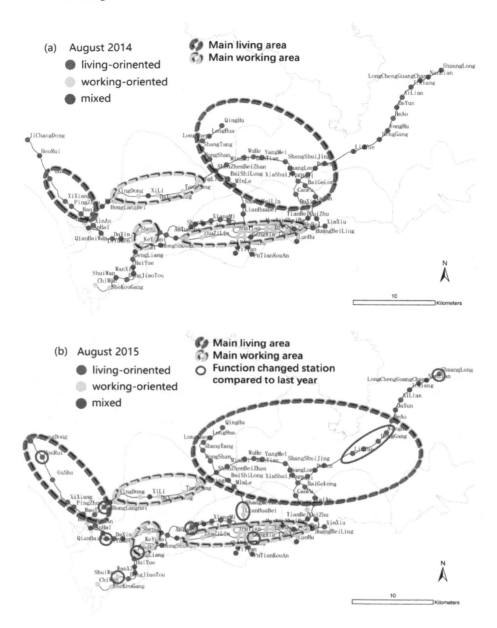

Fig. 5. Spatial distribution of metro station functions. (a) August 2014, (b) August 2015, (c) August 2016, (d) August 2017 and (e) August 2018.

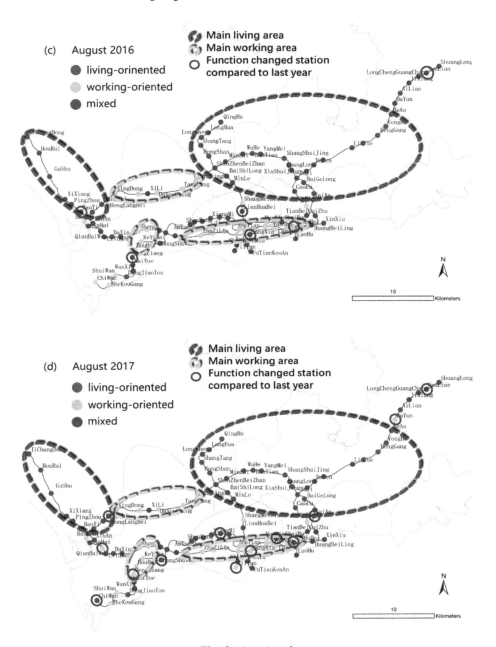

Fig. 5. (*continued*)

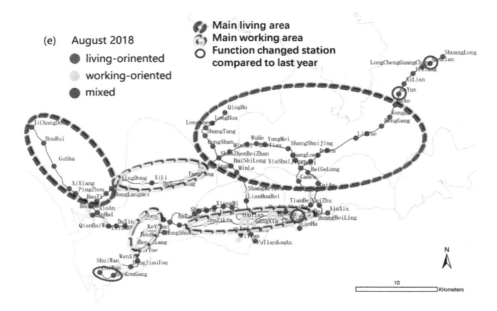

Fig. 5. (*continued*)

From the function changed stations compared to last year in Fig. 5, we can see that DengLiang station had a process of changing from a living-oriented station to a mixed station and then a working-oriented station. This is a typical case of function changed stations. Through the method of this study, we can find the detailed process and time point of each function changed station.

As exit-volume on weekend may be caused by the leisure function or living function around station, we cluster stations into two categories in order to extract

Table 4. Transitions between different function types of metro stations in Shenzhen from 2014 to 2019.

From	To		
	Living-oriented	Working-oriented	Mixed
Living-oriented	–	(count: 1) DengLiang	(count: 2) XiangMeiBei, BaoTi
Working-oriented	(count: 1) LiYuMen	–	(count: 3) ChiWan, SheKouGang, TongXinLing
Mixed	(count: 11) LingZhi, LiuYue, YongHu, HouRui, ShenKang, HengGang, ShuangLong, TangKeng, HongShuWan, HongLangBei, NanLian	(count: 7) ShuiWan, ShangMeiLin, HouHai, GangXia, QianHaiWan, YiTian, LaoJie	–

exit-volume caused by leisure function. The results serve as a supplement to the clustering results of weekday.

Combining with the consequences of weekday and weekend, the detailed evolution of station types among five years are demonstrated in Appendix, and the detailed transitions between different function types are summarized in Table 4.

3.2 Evolutions of Metro Station Functions in Past Five Years

Figure 6, 7, 8, 9 and 10 show the change degrees (i.e. linear-fitting slopes, as more than 90% of the linear regression results are statistically significant) of the five features, which represent the evolutions of station functions. In these figures, green color represents negative value, while red color shows the opposite. The larger the size of circle, the greater the absolute value of linear-fitting slopes (negative or positive change degrees).

Figure 6 maps the change degrees of daily average exit-volume, which represent the evolutions of station attractions. Ten stations with largest increases are GaoXinYuan, CheGongMiao, HouHai, GuShu, WuHe, FuTian, ShenDa, ShangMeiLin, XiLi and HuaQiangBei. These station areas have been developed a lot during the past five years. Half of these metro stations locate at the major Hi-tech and finance workplaces of Shenzhen. Their fast developments reflect the continuous growth of Hi-tech and finance industry in Shenzhen. Contrarily, the change degrees of JiChangDong, LuoHu, ShiJieZhiChuang, HuaQiangLu, TaoYuan, XiangMiHu, HouRui and LaoJie are negative. That is to say, the attractions around these stations are declining. The decline of JiChangDong and HouRui station is due to the relocation of the Shenzhen airport. Interestingly, the rest declining metro stations all locate at the traditionally major attractions of Shenzhen such as old CBDs, traditional eating places, and tourist spots,

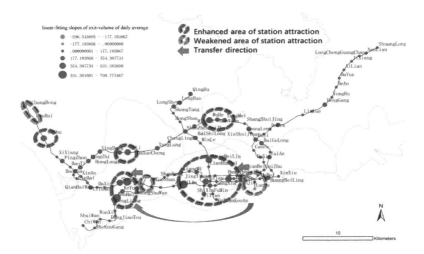

Fig. 6. Change degree (evolution) of exit-volume of daily average (station attraction). (Color figure online)

which implicates a shift pattern of attractions in Shenzhen with the rapidly urban development, as we see in Fig. 6, the circles and arrows.

Figure 7 maps the change degrees of entry-volume on morning-peak-hours, which represent the evolutions of living functions of stations. According to this figure, the change degrees of living functions at JiChangDong, TaoYuan, CheGongMiao, ShiJieZhiChuang, XiangMeiBei, YiTian, XiangMiHu, KeXueGuan, ZhuZiLin, ShaoNianGong, GouWuGongYuan, LianHuaXi, HuiZhanZhongXin, CaoPu, LuoHu and HuaQiangLu are negative, indicating that the living functions around these stations have weakened in general. These stations mainly locate at the traditional centers of Shenzhen, i.e., Luohu District and Futian District. The top 10 stations with positive change degrees are GuShu, WuHe, MinZhi, BuJi, LongSheng, QingHu, XiaShuiJing, PingZhou, DanZhuTou and XiLi, which indicates that the living functions around these stations are greatly enhanced. These stations mainly locate at Bao'an District, Longhua District and Longgang District, which are relatively peripheral districts of Shenzhen city and have relatively lower rental and housing prices. This can be explained by the sharp rises of rental and housing prices (especially after the year of 2015) and quick developments of those districts in recent years. We can see that the living function center shows a trend of transfer from inside to outside Shenzhen customs.

Figure 8 maps the change degrees of exit-volume on morning-peak-hours. We can see that the change degrees of JiChangDong, XiangMiHu, TaoYuan, ShiJieZhiChuang and HuaQiaoCheng are negative, revealing the decline of working functions around these stations. On the contrary, the top 10 stations with positive change degrees are

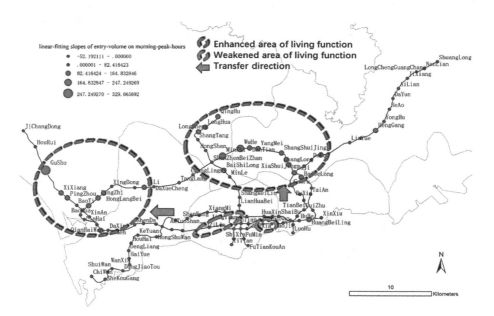

Fig. 7. Change degree (evolution) of entry-volume on morning-peak-hours (living function). (Color figure online)

GaoXinYuan, CheGongMiao, HouHai, ShenDa, FuTian, KeYuan, XingDong, ShangMeiLin, WuHe and TangLang. Similarly, with the Fig. 6 shows, most of these metro stations locate at the major Hi-tech and finance workplaces of Shenzhen, while the rest are newly developed Hi-tech workplaces. From this, we can see that the high-tech industry and the financial industry continue to grow, and the working function center is concentrated to the Science and Technology Park and the Futian Central District.

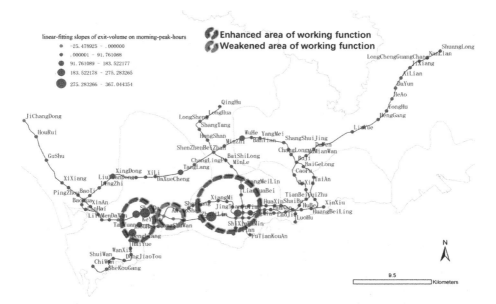

Fig. 8. Change degree (evolution) of exit-volume on morning-peak-hours (working function). (Color figure online)

Figure 9 maps the change degrees of exit-volume on weekday noon hours. The change degrees of JiChangDong, HouRui, HuaQiangLu, LaoJie, LuoHu, XiangMiHu, ShiJieZhiChuang, ShuangLong and LinHai are negative, revealing the decline of eating function around these stations. Most of these station areas are traditional eating centers of Shenzhen. Meanwhile, the change degrees are positive in all other stations, reflecting the enhancement of eating functions in the whole city, especially in station areas of HuaQiangBei, CheGongMiao, ShenZhenBeiZhan, HouHai, ShangMeiLin, FuTian, GaoXinYuan, HuaXin, HuiZhanZhongXin and ShiXia, which mainly locate at the major and newly developed working districts of the city. This pattern indicates that people are shifting their lunch places from traditional eating places to working places on weekdays, which are possibly caused by the fast development of eating industry around working places as well as the fast increase of online order in Shenzhen.

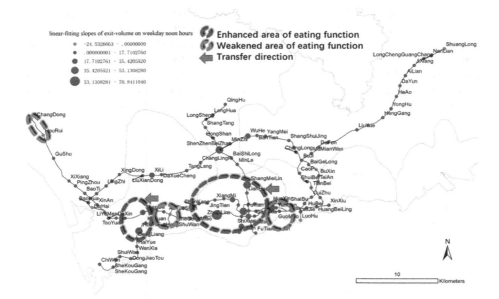

Fig. 9. Change degree (evolution) of exit-volume on weekday noon hours (eating function). (Color figure online)

Figure 10 maps the change degrees of exit-volume on weekend, which reflects the weekend activities around stations. The analysis of this part excludes the stations with living functions on weekdays (ref. Figure 5). The stations of LaoJie, LuoHu, Shi-JieZhiChuang, HuaQiangLu, TaoYuan, QiaoChengDong, XiangMiHu, HongShuWan and SheKouGang have negative change degrees, indicating that the weekend activity flow around these stations is decreasing. Similarly, most of these declining station areas are traditional leisure places. On the contrary, the top 10 stations with large positive change degrees of leisure functions are HouHai, BaoAnZhongXin, ShangMeiLin, CheGongMiao, FuTian, GaoXinYuan, HuaQiangBei, ShenZhenBeiZhan, HuiZhanZhongXin and ShenDa, indicating that recreation facilities have established or developed during those years, making them popular leisure places for the citizens.

3.3   Comparison Analysis on Evolutions of Working and Living Functions

Figure 11 compares the evolutions of working and living functions. The areas with large increase in working functions are mainly concentrated in Futian District and Nanshan District, while the areas with large increase in living functions are mainly located at Bao'an District, Longhua District and Longgang District. Such distribution characteristics reflect the enhancement of home-work separations in Shenzhen in recent years.

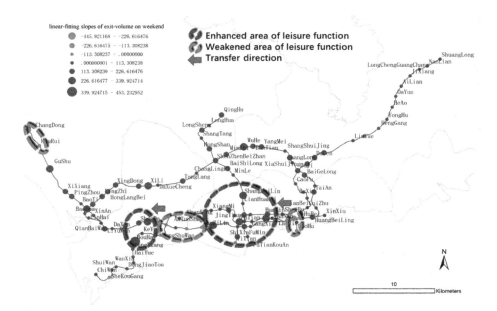

Fig. 10. Change degree (evolution) of exit-volume on weekend (leisure function). (Color figure online)

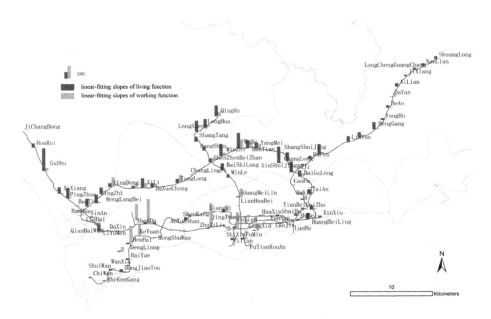

Fig. 11. Comparison of evolutions of working and living functions.

4   Conclusions

The rapid developments of urban metro systems make them as skeletons of cities. Except for acting as traffic nodes, metro stations have become more and more important places in the residents' daily life. Based on long-term and large-scale SCD, this study analyzes the function evolutions of Shenzhen metro stations from 2014 to 2018.

Firstly, the average hourly entry/exit volumes from Monday to Thursday are used to classify the stations into working-oriented, living-oriented and mixed function types. Stations with leisure functions are extracted according to entry/exit volumes on Sunday. Through this way, the functions of metro stations in Shenzhen in each year are obtained.

Then, five features extracted from five kinds of passenger-volume time series are defined for describing metro station functions. We also propose to calculate the linear-fitting slopes of the five features to measure the evolutions (change degrees) of the functions from 2014 to 2018.

The main findings of this study can be summarized as follows:

(1) The difference of metro station functions especially home-work separations in Shenzhen have enlarged in recent years.
(2) The functions of most metro stations have enhanced, while few of them have weakened.
(3) The cores of metro station functions have shifted in space. For instance, the cores of workplaces have shifted from Luohu district (with developed traditional industries) to Futian district (with developed financial industries) and Nanshan district (with developed high-tech industries). The cores of residences have shifted to relatively peripheral districts such as Bao'an District, Longhua District, and Longgang District. People also shift their lunch places from traditional eating places to working places on weekdays. Those conclusions are of great values for deeply understanding the metro stations, which would be beneficial for sustainable development planning of public transport.

This study demonstrated how we can take advantage of the long-term metro passenger data (i.e., the smart card data) to detect the functions and attractions of metro stations as well as the whole city in a quick, low-cost, continuable and effective way.

This study detected the changes of metro station functions using a simple linear-slope-based method, while in the future, we would like to try some more advanced methods such as matrix factorization. In addition, the point-of-interest (POI) data can also be integrated to refine and verify our results.

Acknowledgements. This research is supported by the National Natural Science Foundation of China (No. 41771441, 41901391), China Postdoctoral Science Foundation (No. 2019M653114), a grant from State Key Laboratory of Resources and Environmental Information System, the Basic Research Project of Shenzhen City (No. JCYJ20170307164104491), and the Joint Engineering Research Center for Health Big Data Intelligent Analysis Technology, and Shenzhen Discipline Construction Project for Urban Computing and Data Intelligence.

Appendix

Station	Aug. 2014	Aug. 2015	Aug. 2016	Aug. 2017	Aug. 2018
AiLian	Living-oriented	Living-oriented	Living-oriented	Living-oriented	Living-oriented
AnTuoShan	Mixed	Mixed_leisure	Mixed	Mixed	Mixed
BaiGeLong	Living-oriented	Living-oriented	Living-oriented	Living-oriented	Living-oriented_leisure
BaiShiLong	Living-oriented	Living-oriented	Living-oriented	Living-oriented	Living-oriented
BaiShiZhou	Living-oriented	Living-oriented	Living-oriented	Living-oriented	Living-oriented
BanTian	Living-oriented	Living-oriented	Living-oriented	Living-oriented	Living-oriented
BaoAnZhongXin	Mixed_leisure	Mixed_leisure	Mixed_leisure	Mixed_leisure	Mixed_leisure
BaoHua	Working-oriented_leisure	Working-oriented_leisure	Working-oriented_leisure	Working-oriented_leisure	Working-oriented_leisure
BaoTi	Living-oriented_leisure	Living-oriented_leisure	Mixed_leisure	Mixed_leisure	Mixed_leisure
BuJi	Living-oriented	Living-oriented	Living-oriented	Living-oriented	Living-oriented
BuXin	Living-oriented	Living-oriented	Living-oriented	Living-oriented	Living-oriented
CaoPu	Living-oriented	Living-oriented	Living-oriented	Living-oriented	Living-oriented
ChangLingPi	Working-oriented_leisure	Working-oriented_leisure	Working-oriented_leisure	Working-oriented_leisure	Working-oriented
ChangLong	Living-oriented	Living-oriented	Living-oriented	Living-oriented	Living-oriented
CheGongMiao	Working-oriented_leisure	Working-oriented_leisure	Working-oriented_leisure	Working-oriented_leisure	Working-oriented_leisure
ChiWan	Working-oriented	Working-oriented	Working-oriented	Living-oriented	Mixed
CuiZhu	Mixed	Mixed	Mixed	Mixed	Mixed
DaFen	Living-oriented	Living-oriented	Living-oriented	Living-oriented	Living-oriented
DaJuYuan	Working-oriented_leisure	Working-oriented_leisure	Working-oriented_leisure	Working-oriented_leisure	Working-oriented_leisure
DanZhuTou	Living-oriented	Living-oriented	Living-oriented	Living-oriented	Living-oriented
DaXin	Living-oriented	Living-oriented	Living-oriented	Living-oriented	Living-oriented
DaXueCheng	Working-oriented	Working-oriented	Working-oriented	Working-oriented	Working-oriented
DaYun	Mixed_leisure	Mixed_leisure	Mixed	Working-oriented	Mixed
DengLiang	Living-oriented	Living-oriented	Mixed	Working-oriented	Working-oriented
DongJiaoTou	Living-oriented	Living-oriented	Living-oriented	Living-oriented	Living-oriented
FanShen	Living-oriented	Living-oriented	Living-oriented	Living-oriented	Living-oriented
FuMin	Mixed	Mixed	Mixed	Mixed	Mixed
FuTian	Working-oriented_leisure	Working-oriented_leisure	Working-oriented_leisure	Working-oriented_leisure	Working-oriented_leisure
FuTianKouAn	Mixed_leisure	Mixed_leisure	Mixed_leisure	Mixed_leisure	Mixed_leisure
GangXia	Mixed	Working-oriented	Mixed	Working-oriented	Working-oriented
GangXiaBei	Working-oriented	Working-oriented	Working-oriented	Working-oriented	Working-oriented_leisure
GaoXinYuan	Working-oriented_leisure	Working-oriented_leisure	Working-oriented_leisure	Working-oriented_leisure	Working-oriented_leisure
GouWuGongYuan	Working-oriented_leisure	Working-oriented_leisure	Working-oriented_leisure	Working-oriented_leisure	Working-oriented_leisure
GuoMao	Working-oriented_leisure	Working-oriented_leisure	Working-oriented_leisure	Working-oriented_leisure	Working-oriented_leisure
GuShu	Living-oriented	Living-oriented	Living-oriented	Living-oriented	Living-oriented
HaiShangShiJie	Working-oriented_leisure	Working-oriented_leisure	Working-oriented_leisure	Working-oriented_leisure	Working-oriented_leisure
HaiYue	Mixed_leisure	Mixed_leisure	Mixed_leisure	Mixed_leisure	Mixed_leisure
HeAo	Living-oriented	Living-oriented	Living-oriented	Living-oriented	Living-oriented
HengGang	Mixed	Living-oriented	Living-oriented	Living-oriented	Living-oriented

(continued)

(continued)

Station	Aug. 2014	Aug. 2015	Aug. 2016	Aug. 2017	Aug. 2018
HongLangBei	Mixed	Mixed	Mixed	Living-oriented	Living-oriented
HongLing	Mixed	Mixed	Mixed	Mixed_leisure	Mixed
HongShan	Living-oriented	Living-oriented	Living-oriented_leisure	Living-oriented	Living-oriented
HongShuWan	Mixed_leisure	Mixed_leisure	Mixed_leisure	Living-oriented_leisure	Living-oriented_leisure
HouHai	Mixed_leisure	Working-oriented_leisure	Working-oriented_leisure	Working-oriented_leisure	Working-oriented_leisure
HouRui	Mixed	Living-oriented	Living-oriented	Living-oriented	Living-oriented
HuangBeiLing	Mixed	Mixed_leisure	Mixed	Mixed	Mixed
HuaQiangBei	Working-oriented_leisure	Working-oriented_leisure	Working-oriented_leisure	Working-oriented_leisure	Working-oriented_leisure
HuaQiangLu	Working-oriented_leisure	Working-oriented_leisure	Working-oriented_leisure	Working-oriented_leisure	Working-oriented_leisure
HuaQiaoCheng	Mixed_leisure	Mixed_leisure	Mixed_leisure	Mixed_leisure	Mixed_leisure
HuaXin	Working-oriented_leisure	Working-oriented_leisure	Working-oriented_leisure	Working-oriented_leisure	Working-oriented_leisure
HuBei	Working-oriented	Working-oriented	Working-oriented	Working-oriented	Working-oriented
HuiZhanZhongXin	Working-oriented_leisure	Working-oriented_leisure	Working-oriented_leisure	Working-oriented_leisure	Working-oriented_leisure
JiChangDong	Living-oriented	Living-oriented	Living-oriented	Living-oriented	Living-oriented
JingTian	Living-oriented	Living-oriented	Living-oriented	Living-oriented	Living-oriented
JiXiang	Mixed	Mixed	Mixed	Mixed	Mixed
KeXueGuan	Working-oriented_leisure	Working-oriented_leisure	Working-oriented_leisure	Working-oriented_leisure	Working-oriented_leisure
KeYuan	Working-oriented_leisure	Working-oriented_leisure	Working-oriented_leisure	Working-oriented_leisure	Working-oriented_leisure
LaoJie	Mixed_leisure	Mixed_leisure	Working-oriented_leisure	Mixed_leisure	Working-oriented_leisure
LianHuaBei	Mixed	Working-oriented	Mixed	Mixed	Mixed
LianHuaCun	Mixed	Mixed_leisure	Mixed_leisure	Mixed	Mixed
LianHuaXi	Working-oriented	Working-oriented_leisure	Working-oriented_leisure	Working-oriented_leisure	Working-oriented_leisure
LingZhi	Mixed	Living-oriented	Living-oriented	Living-oriented	Living-oriented
LinHai	Mixed_leisure	Mixed_leisure	Mixed_leisure	Mixed_leisure	Mixed
LiuXianDong	Working-oriented_leisure	Working-oriented	Working-oriented	Working-oriented	Working-oriented
LiuYue	Mixed	Living-oriented	Living-oriented	Living-oriented	Living-oriented
LiYuMen	Working-oriented_leisure	Living-oriented	Living-oriented	Living-oriented	Living-oriented
LongChengGuangChang	Mixed	Mixed	Mixed	Mixed	Mixed
LongHua	Living-oriented	Living-oriented	Living-oriented	Living-oriented	Living-oriented
LongSheng	Living-oriented	Living-oriented	Living-oriented	Living-oriented	Living-oriented
LuoHu	Mixed_leisure	Mixed_leisure	Mixed_leisure	Mixed_leisure	Mixed_leisure
MinLe	Living-oriented	Living-oriented	Living-oriented	Living-oriented	Living-oriented
MinZhi	Living-oriented	Living-oriented	Living-oriented	Living-oriented	Living-oriented
MuMianWan	Living-oriented	Living-oriented	Living-oriented	Living-oriented	Living-oriented
NanLian	Mixed	Mixed	Living-oriented	Mixed	Living-oriented
PingZhou	Living-oriented	Living-oriented	Living-oriented	Living-oriented	Living-oriented
QianHaiWan	Mixed_leisure	Mixed_leisure	Mixed_leisure	Working-oriented_leisure	Working-oriented_leisure
QiaoChengBei	Working-oriented	Working-oriented	Working-oriented	Working-oriented_leisure	Working-oriented_leisure
QiaoChengDong	Working-oriented_leisure	Working-oriented_leisure	Working-oriented_leisure	Working-oriented_leisure	Working-oriented_leisure
QiaoXiang	Mixed_leisure	Mixed_leisure	Mixed_leisure	Mixed_leisure	Mixed_leisure

(continued)

(continued)

Station	Aug. 2014	Aug. 2015	Aug. 2016	Aug. 2017	Aug. 2018
QingHu	Living-oriented	Living-oriented	Living-oriented	Living-oriented	Living-oriented
ShaiBu	Mixed_leisure	Mixed_leisure	Mixed_leisure	Mixed_leisure	Mixed_leisure
ShangMeiLin	Mixed	Working-oriented	Working-oriented	Working-oriented_leisure	Working-oriented_leisure
ShangShuiJing	Living-oriented	Living-oriented	Living-oriented	Living-oriented	Living-oriented
ShangTang	Living-oriented	Living-oriented	Living-oriented	Living-oriented	Living-oriented
ShaoNianGong	Working-oriented_leisure	Working-oriented_leisure	Working-oriented_leisure	Working-oriented_leisure	Working-oriented_leisure
SheKouGang	Working-oriented_leisure	Working-oriented_leisure	Working-oriented_leisure	Working-oriented_leisure	Mixed_leisure
ShenDa	Working-oriented_leisure	Working-oriented_leisure	Working-oriented_leisure	Working-oriented_leisure	Working-oriented_leisure
ShenKang	Mixed_leisure	Living-oriented	Living-oriented	Living-oriented	Living-oriented
ShenZhenBeiZhan	Mixed_leisure	Mixed_leisure	Mixed_leisure	Mixed_leisure	Mixed_leisure
ShiJieZhiChuang	Living-oriented_leisure	Living-oriented_leisure	Living-oriented_leisure	Living-oriented_leisure	Living-oriented_leisure
ShiMinZhongXin	Working-oriented_leisure	Working-oriented_leisure	Working-oriented_leisure	Working-oriented_leisure	Working-oriented_leisure
ShiXia	Mixed	Mixed	Mixed	Mixed	Mixed
ShuangLong	Mixed	Living-oriented	Living-oriented	Living-oriented	Living-oriented
ShuiBei	Working-oriented	Working-oriented	Mixed_leisure	Working-oriented	Working-oriented
ShuiWan	Mixed	Working-oriented	Working-oriented	Working-oriented	Working-oriented
TaiAn	Living-oriented	Mixed	Mixed	Living-oriented	Living-oriented
TangKeng	Mixed	Living-oriented	Living-oriented	Living-oriented	Living-oriented
TangLang	Working-oriented_leisure	Working-oriented_leisure	Working-oriented_leisure	Working-oriented	Working-oriented
TaoYuan	Mixed	Mixed	Mixed_leisure	Mixed_leisure	Mixed_leisure
TianBei	Working-oriented_leisure	Working-oriented_leisure	Working-oriented_leisure	Working-oriented_leisure	Working-oriented_leisure
TongXinLing	Working-oriented_leisure	Working-oriented_leisure	Working-oriented_leisure	Mixed_leisure	Mixed_leisure
WanXia	Mixed	Mixed_leisure	Mixed	Mixed	Mixed
WuHe	Living-oriented	Living-oriented	Living-oriented	Living-oriented	Living-oriented
XiangMeiBei	Living-oriented	Living-oriented	Living-oriented	Mixed_leisure	Mixed_leisure
XiangMi	Living-oriented	Living-oriented	Living-oriented	Living-oriented	Living-oriented
XiangMiHu	Working-oriented_leisure	Working-oriented_leisure	Working-oriented_leisure	Working-oriented_leisure	Working-oriented_leisure
XiaShuiJing	Living-oriented	Living-oriented	Living-oriented	Living-oriented	Living-oriented
XiLi	Mixed	Mixed	Mixed	Mixed	Mixed
XinAn	Living-oriented	Living-oriented	Living-oriented	Living-oriented	Living-oriented
XingDong	Working-oriented	Working-oriented	Working-oriented	Working-oriented	Working-oriented
XinXiu	Living-oriented	Living-oriented	Living-oriented	Living-oriented	Living-oriented
XiXiang	Living-oriented	Living-oriented	Living-oriented	Living-oriented	Living-oriented
YangMei	Living-oriented	Living-oriented	Living-oriented	Living-oriented	Living-oriented
YanNan	Working-oriented_leisure	Working-oriented_leisure	Working-oriented_leisure	Working-oriented_leisure	Working-oriented_leisure
YiJing	Mixed_leisure	Mixed_leisure	Mixed	Mixed_leisure	Mixed
YiTian	Mixed	Mixed	Mixed	Working-oriented	Working-oriented
YongHu	Mixed	Living-oriented	Living-oriented	Living-oriented	Living-oriented
ZhuZiLin	Working-oriented	Working-oriented	Working-oriented_leisure	Working-oriented	Working-oriented

References

1. Lu, K., Khani, A., Han, B.: A trip purpose-based data-driven alighting station choice model using transit smart card data. Complexity **2018**, 1 (2018)
2. Alsger, A., Tavassoli, A., Mesbah, M., Ferreira, L., Hickman, M.: Public transport trip purpose inference using smart card fare data. Transp. Res. Part C: Emerg. Technol. **87**, 123 (2018)
3. Jung, J., Sohn, K.: Deep-learning architecture to forecast destinations of bus passengers from entry-only smart-card data. IET Intell. Transp. Syst. **11**, 334 (2017)
4. Tamblay, S., Galilea, P., Iglesias, P., Raveau, S., Muñoz, J.C.: A zonal inference model based on observed smart-card transactions for Santiago de Chile. Transp. Res. Part A: Policy Practice **84**, 44 (2016)
5. Gan, Z., Yang, M., Feng, T., Timmermans, H.: Understanding urban mobility patterns from a spatiotemporal perspective: daily ridership profiles of metro stations. Transportation **47**, 315–336 (2018)
6. Wang, Z., Hu, Y., Zhu, P., Qin, Y., Jia, L.: Ring aggregation pattern of metro passenger trips: a study using smart card data. Physica A: Stat. Mech. Appl. **491**, 471 (2018)
7. Kim, J., Corcoran, J., Papamanolis, M.: Route choice stickiness of public transport passengers: measuring habitual bus ridership behaviour using smart card data. Transp. Res. Part C: Emerg. Technol. **83**, 146 (2017)
8. Ma, X., Liu, C., Wen, H., Wang, Y., Wu, Y.: Understanding commuting patterns using transit smart card data. J. Transp. Geogr. **58**, 135 (2017)
9. Briand, A., Côme, E., El Mahrsi, M.K., Oukhellou, L.: A mixedture model clustering approach for temporal passenger pattern characterization in public transport. Int. J. Data Sci. Anal. **1**, 37 (2016)
10. Goulet-Langlois, G., Koutsopoulos, H.N., Zhao, J.: Inferring patterns in the multi-week activity sequences of public transport users. Transp. Res. Part C: Emerg. Technol. **64**, 1 (2016)
11. Long, Y., Liu, X., Zhou, J., Chai, Y.: Early birds, night owls, and tireless/recurring itinerants: an exploratory analysis of extreme transit behaviors in Beijing, China. Habitat Int. **57**, 223 (2016)
12. Kieu, L.M., Bhaskar, A., Chung, E.: Passenger segmentation using smart card data. IEEE Trans. Intell. Transp. Syst. **16**, 1537 (2015)
13. Zou, Q., Yao, X., Zhao, P., Wei, H., Ren, H.: Detecting home location and trip purposes for cardholders by mining smart card transaction data in Beijing subway. Transportation **45**, 919 (2018)
14. Huang, J., Levinson, D., Wang, J., Zhou, J., Wang, Z.: Tracking working and living dynamics with smartcard data. Proc. Natl. Acad. Sci. **115**, 12710 (2018)
15. Long, Y., Thill, J.: Combining smart card data and household travel survey to analyze workings–living relationships in Beijing. Comput. Environ. Urban Syst. **53**, 19 (2015)
16. Kim, K.: Identifying the structure of cities by clustering using a new similarity measure based on smart card data. IEEE Trans. Intell. Transp. Syst. 1 (2019)
17. Maeda, T.N., Mori, J., Hayashi, I., Sakimoto, T., Sakata, I.: Comparative examination of networking clustering methods for extracting community structures of a city from public transportation smart card data. IEEE Access **7**, 53377 (2019)
18. Gong, Y., Lin, Y., Duan, Z.: Exploring the spatiotemporal structure of dynamic urban space using metro smart card records. Comput. Environ. Urban Syst. **64**, 169 (2017)

19. Li, Y., Wang, X., Sun, S., Ma, X., Lu, G.: Forecasting short-term subway passenger flow under special events scenarios using multiscale radial basis function networkings. Transp. Res. Part C: Emerg. Technol. **77**, 306 (2017)
20. Yang, J., Chen, J., Le, X., Zhang, Q.: Density-oriented versus development-oriented transit investment: decoding metro station location selection in Shenzhen. Transp. Policy. **51**, 93 (2016)
21. Tang, T., et al.: FISS: function identification of subway stations based on semantics mining and functional clustering. IET Intell. Transp. Syst. **12** 558 (2018)
22. Wang, J., Kong, X., Rahim, A., Xia, F., Tolba, A., Al-Makhadmeh, Z.: IS2Fun: identification of subway station functions using massive urban data. IEEE Access **5**, 27103 (2017)
23. El Mahrsi, M.K., Come, E., Oukhellou, L., Verleysen, M.: Clustering smart card data for urban mobility analysis. IEEE Trans. Intell. Transp. Syst. **18**, 712 (2017)
24. Zhou, Y., Fang, Z., Zhan, Q., Huang, Y., Fu, X.: Inferring social functions available in the metro station area from passengers' staying activities in smart card data. ISPRS Int. J. Geo-Inf. **6**, 394 (2017)
25. Liu, K., Qiu, P., Li, M., Liu, X., Lu, F.: Exploring urban travel routes' characteristics from a geometric perspective. Comput. Environ. Urban Syst. **74**, 50–61 (2019)
26. Liu, K., Gao, S., Qiu, P., Liu, X., Yan, B., Lu, F.: Road2Vec: measuring traffic interactions in urban road system from massive travel routes. ISPRS Int. J. Geo-Inf. **6**, 321 (2017)
27. Xie, J., Yin, L., Mao, L.: A modeling framework for individual-based urban mobility based on data fusion. In: 2018 26th International Conference on Geoinformatics, pp. 1–6. IEEE (2018)
28. Liu, K., Gao, S., Lu, F.: Identifying spatial interaction patterns of vehicle movements on urban road networks by topic modelling. Comput. Environ. Urban Syst. **74**, 50 (2019)
29. Zhao, Z., et al.: The effect of temporal sampling intervals on typical human mobility indicators obtained from mobile phone location data. Int. J. Geogr. Inf. Sci. **33**, 1471 (2019)

Analysis of Landscape Pattern Spatial Scale in Middle and Upper Reaches of Meijiang River Basin

Yuchan Chen[1,2,3], Zhengdong Zhang[2(✉)], Chuanxun Yang[1,4],
Yang Yang[2], Chen Zhang[1,3], Liusheng Han[1], Ji Yang[3(✉)],
and Xiangyu Han[5]

[1] Key Lab of Guangdong for Utilization of Remote Sensing and Geographical
Information System, Guangdong Open Laboratory of Geospatial Information
Technology and Application, Guangzhou Institute of Geography,
Guangzhou 510070, China
[2] School of Geography, South China Normal University,
Guangzhou 510631, China
zhangzdedu@163.com
[3] Southern Marine Science and Engineering Guangdong Laboratory
(Guangzhou), Guangzhou 511458, China
yangji@gdas.ac.cn
[4] University of Chinese Academy of Sciences, Beijing 100049, China
[5] State Grid Chengdu Electric Supply Company, Chengdu 610000, China

Abstract. The scale research of landscape pattern is an important basis for the study of spatiotemporal evolution of landscape pattern and the scientific and reasonable allocation of landscape pattern. This paper takes the middle and upper reaches of Meijiang river as the research area. Combined with spatial grain size analysis and extent size analysis, the spatial scale effect was studied to select the optimal research scale in this research area. The results show that most of the appropriate grain size of landscape pattern indexes are in range of 90–150 m, and the optimal grain size scale is 150 m in Middle and Upper Reaches of Meijiang River Basin. At the grain size scale of 150 m, the spatial self-correlation of landscape pattern index over 50% is the highest at the extent size of 300 m, and the optimal extent size scale is 300 m in this basin. This paper determines the optimal scale of landscape pattern in the middle and upper reaches of Meijiang river basin, which is of great significance to the ecological balance and sustainable development of the basin.

Keywords: Landscape pattern spatial scale · Grain scale effect · Extent scale effect

1 Introduction

The study of landscape pattern is one of the research emphases and hotspots in watershed ecology. The landscape pattern of the watershed can directly or indirectly affect the ecological and hydrological processes of the basin through the patches of

© Springer Nature Singapore Pte Ltd. 2020
Y. Xie et al. (Eds.): GSES 2019/GeoAI 2019, CCIS 1228, pp. 54–67, 2020.
https://doi.org/10.1007/978-981-15-6106-1_4

different spatial arrangement, sizes and shapes [1], thus significantly affecting the water environment quality of the basin [2–5]. The 21st century, the scope and intensity of intervention of human activities on the watershed landscape pattern increase [6]. Under the influence of frequent human activities, the landscape pattern of the watershed changes dramatically, the area of forest and grassland decreases sharply, and the area of construction land expands dramatically, which causes the abnormal ecological process of the watershed, causing serious water environment problems such as soil erosion and water eutrophication [7].

Rational allocation of watershed landscape pattern is a key means to restore the normal operation of watershed ecosystem, control soil erosion and alleviate non-point source pollution [8–10]. However, due to the significant spatial scale dependence of landscape pattern [11, 12], too large or too small research scale cannot accurately reflect the characteristics of quantity, area, shape and heterogeneity of landscape pattern. Based on this calculation simulation will inevitably lead to invalid landscape pattern optimization results. Therefore, the scale study of landscape pattern is an important basis for scientific and rational optimal allocation of landscape pattern, and prevention and treatment of ecological environment in the watershed, which is of great significance to the ecological balance and sustainable development of the watershed.

Landscape scale study includes extent and grain effect study. At present, the theory and method of grain research have been relatively mature [13, 14]. It is considered that the appropriate grain of landscape should retain the original characteristic trend of the landscape, and avoid the redundancy of detailed information as much as possible. Based on this principle, researchers analyze the characteristics of landscape pattern under different landscape granularity one by one, identify the segmentation points of granularity with obvious changes in landscape pattern characteristics, and select the optimal research granularity in the range before the segmentation points [15]. In the field of landscape pattern scale, the results of extent study are relatively few compared with grain study [16]. Existing studies are mostly based on the spatial autocorrelation characteristics of landscape patches, combined with gradient loop method, equal-side sliding window method and other extent DIVISION methods, to analyze the extent effect of landscape pattern [17, 18]. Although the research on grain and extent effect of landscape pattern has made some achievements, few researches can analyze grain and extent effect of landscape pattern at the same time. However, grain and extent are two aspects of landscape scale, which exist and restrict each other at the same time. Single study on either of them cannot fully reflect the scale characteristics of landscape pattern, which leads to incomplete scale study of landscape pattern.

After China's reform and opening up, the landscape pattern of the middle and upper reaches of Meijiang river basin has changed dramatically and soil erosion is serious. Therefore, the landscape pattern of the basin urgently needs scientific optimization. Therefore, taking the middle and upper reaches of Meijiang river basin as an example, we study the grain effect of that basin, and on the basis of grain analysis, study the extent effect of the basin for exploring the optimal research scale. This research can provide the basis for the research of landscape pattern characteristics and scientific development of landscape pattern optimization.

2   Research Area and Data

The Meijiang river basin is the main source of Han river, which is located in the east of Guangdong (Fig. 1). The longitude and latitude are 115°13'–116°33' E, 23°15'–25° 18' N. Meijiang river basin is located in the transitional zone of subtropical and subtropical climate zone, belonging to the subtropical monsoon climate zone. In summer, the precipitation of that basin is large and concentrated. The terrain of the river basin is complex and fluctuates violently. The elevation of the upper reaches of the river is over 1000 m, while the elevation of the middle reaches drops to below 200 m. The precipitation and topographic features of that basin form serious soil erosion. At the same time, after the reform and opening up, human activities have become increasingly frequent. The native vegetation has been destroyed, the landscape area of farmland and urban land has expanded, and the landscape pattern tends to be broken in that basin.

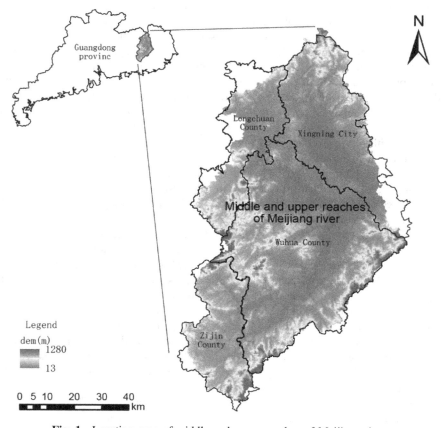

Fig. 1. Location map of middle and upper reaches of Meijiang river

Unreasonable land development by human beings further aggravates the soil erosion in the basin. The middle and upper reaches of meijiang river basin in this research refers to the basin area above Shuikou hydrologic station. Wuhua river, which a river in that range, is one of the rivers with the most serious soil erosion in Guangdong province. Rational allocation of watershed landscape pattern is a key means to effectively control soil erosion. The scale study of landscape pattern can accurately reflect the characteristics of landscape pattern of river basin, which is an important basis for scientific and rational allocation of landscape pattern of river basin and prevention of soil erosion. It is of great practical significance to study and analyze the scale characteristics of landscape pattern in the middle and upper reaches of Meijiang river basin.

This research mainly uses the land use data interpreted by Landsat2015 remote sensing image. The scale of the data is 1:100,000. The resolution is 30 × 30 m. According to the land use and land cover property, 6 primary type landscapes and 14 secondary type landscapes have been divided in the research area.

3   Research Method

3.1   Landscape Pattern Index Analysis Method

Landscape pattern index can quantitatively describe the quantity, structure and heterogeneity characteristics of landscape pattern by endowing the spatial landscape data with certain statistical properties, and can quantitatively analyze the landscape pattern characteristics under different spatial scales based on the spatial data of landscape pattern at different scales. Based on the relevant research results and the actual situation of the research area, the representative landscape pattern index was selected from the aspects of landscape quantity, area shape structure and horizontal structure heterogeneity. Finally, 17 landscape pattern indexes were selected, and the selected landscape pattern indexes were calculated quantitatively with FRAGSTATS. The selected landscape pattern index is shown in Table 1.

Table 1. Landscape pattern index selection

	Landscape pattern index
Number characteristic	Number of Patches (NP), Patch Density (PD), Patch Richness (PR), Shannon's Diversity Index (SHDI)
Area-shape characteristic	Patch Area (AREA_MN), Fractal Dimension Index (FRAC_MN), Largest Patch Index(LPI), Landscape Shape Index (LSI)
Horizontal structural heterogeneity characteristic	Aggregation Index (AI), Patch Cohesion Index (COHESION), Contagion Index (CONTAG), Contiguity Index (CONTIG_MN), Landscape Division Index (DIVISION), Euclidean Nearest-Neighbor Distance (ENN_MN), Proximity Index (PROX_MN), Similarity Index (SIMI_MN), Splitting Index (SPLIT)

3.2   Landscape Grain Effect Analysis Method

This research used the first scale domain analysis method to determine the appropriate grain of the research area. That is, the scale domain is divided according to the scale turning point in the process of landscape index change, and the medium to large granularity is selected in the range before the first turning point of landscape index change [4]. According to relevant research results, this grain size selection can not only guarantee the original characteristic trend of the landscape, but also avoid the calculation redundancy, so it is an appropriate granularity of the research area. With reference to the existing research results, this research took 30, 60, 90, 120, 150, 180, 210, 240, 270, 300, 330, 360, 390, 420 and 450 m as the minimum unit length, and the grain effect of research area was analyzed in these 15 different sizes.

3.3   Landscape Extent Effect Analysis Method

According to landscape ecology, different combinations of land use form different landscape patterns, and regional landscape patterns have high spatial autocorrelation. Reasonable amplitude value should not only ensure less computation, but also ensure that the spatial information is not excessively missing and the spatial autocorrelation is strong. Based on this landscape ecology principle and combined with the previous particle size research, this research used landscape pattern analysis software and geostatistical software to analyze the spatial autocorrelation characteristics of landscape pattern under different extent values, obtain the extent effect of landscape pattern in the research area, and determine the optimal extent scale of landscape pattern.

The research steps of extent analysis are as follows: Firstly, according to the actual situation of the basin, a range of 300–3000 m extent is selected for scale analysis. The interval of scale is 300 m, with a total of 10 extent scales; Secondly, the sliding window module in FRAGSTATS is used to control the magnitude of watershed extent and calculate the values of each landscape pattern index of the landscape level under 10 extent scales; Thirdly, ArcGIS was used to generate 5000 random points in the research area, and corresponding to the coordinates of the random points, the values of each landscape pattern index calculated under each amplitude were assigned to theses random points; Fourthly, the coordinates and the corresponding landscape pattern index values of these random points were input into the GS+ software, and the semi-variable function in the software was used to calculate the Nugget [11] and Still (Co+C) of the landscape pattern index at each extent scale; Finally, the spatial autocorrelation of landscape pattern index under different extent scales was analyzed by the ratio of Nugget to Still, and then the extent effect of landscape on research area was analyzed.

4   Results Analysis

4.1   Landscape Grain Effect Analysis

Through the calculation of landscape pattern index, the grain effect of landscape pattern in quantity, area shape structure and horizontal structure heterogeneity was obtained. The calculation results are shown in Figs. 2, 3 and 4.

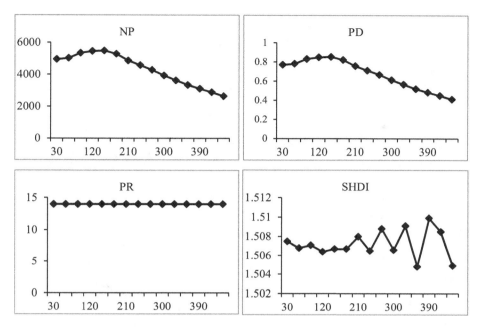

Fig. 2. Grain effect of landscape pattern index of number characteristic

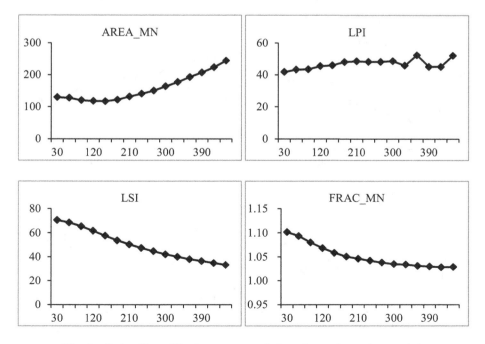

Fig. 3. Grain effect of landscape pattern index of area-shape characteristic

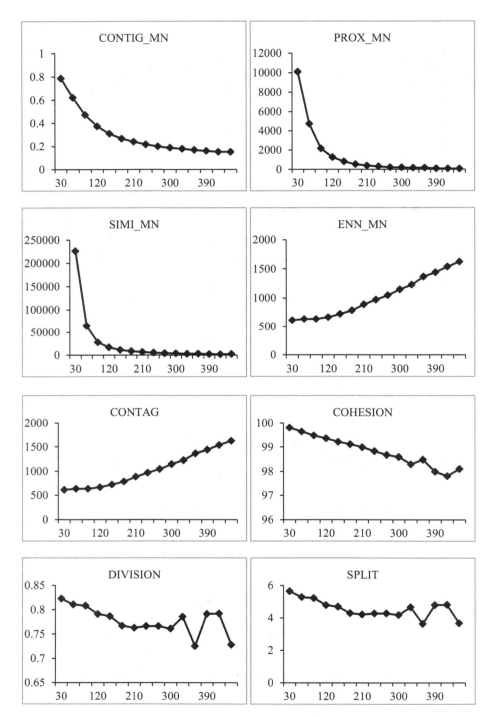

Fig. 4. Grain effect of landscape pattern index of horizontal structural heterogeneity characteristic

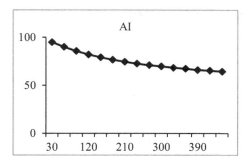

Fig. 4. (*continued*)

As can be seen from Fig. 2, in the four indexes of number characteristic, the grain effect of SHDI is the most significant, NP and PD is weaker, and PR is no. NP and PD increases when the grain size less than 150 m and decrease when the grain size greater than 150 m. SHDI fluctuates sharply with increase of grain size when the grain size greater than 180 m.

As can be seen from Fig. 3, the grain scale effect of the area characteristic index, that is AREA_MN and LPI, is significant. And the shape characteristic index, that is LSI and FRAC_MN is weak particularly. In area characteristic indexes, AREA_MN increases when the grain size less than 150 m and decrease when the grain size greater than 150 m; LPI fluctuates sharply with increase of grain size when the grain size greater than 300 m. In shape characteristic indexes, LSI increases as the grain size increases; there is only a slight turning point on FRAC_MN at the grain size of 420 m.

As can be seen from Fig. 4, the grain effect of heterogeneity characteristic indexes is quite different. CONTIG_MN, PROX_MN, SIMI_MN and AI show a decreasing trend with the increase of grain size, and ENN_MN and CONTAG show an increasing trend with the increase of grain size. The above index granularity doesn't show the grain scale effect. COHESION, DIVISION, SPLIT show significant grain scale effect, and the index value vary violently with the increase of grain size after the point of 300 m grain size.

According to Figs. 1, 2 and 3, the turning points, the first scale domain and the suitable grain size range of landscape pattern indexes in the middle and upper reaches of Meijiang river basin can be obtained (Table 2).

It can be seen from Table 2, except SHDI, all the first scale domain of the indexes is in the range of 30 to 150 m. it is considered that 30 to 150 m is the first scale domain of the grain size effect in this basin. And most of the grain size of landscape pattern indexes are in the appropriate range of 90–150 m.

Table 2. Landscape indexes granularity effect in Middle and upper reaches of Meijiang River basin

Landscape characteristics	Landscape pattern index	Turning point	First scale domain	Suitable grain size range
Number characteristic	NP	150	30–150	90–150
	PD	150	30–150	90–150
	PR	/	/	/
	SHDI	60	30–60	30–60
Area-shape characteristic	AREA_MN	150	30–150	90–150
	LPI	300	30–300	240–300
	LSI	/	/	/
	FRAC_MN	420	30–420	360–420
Horizontal structural heterogeneity characteristic	CONTIG_MN	/	/	/
	PROX_MN	/	/	/
	SIMI_MN	/	/	/
	AI	/	/	/
	ENN_MN	/	/	/
	CONTAG	/	/	/
	COHESION	330	30–330	270–330
	DIVISION	210	30–210	150–210
	SPLIT	300	30–300	240–300

4.2   Landscape Extent Effect Analysis

According to the method of extent analysis in 3.3, the spatial distribution of selected landscape pattern index, the coordinates of random point and the corresponding values of landscape pattern index in different extent was obtained. Then we calculated the Nugget and Still of landscape pattern index used to coordinates and the coordinates values by GS+, and calculated the ratio of Nugget to Still of landscape pattern index of number characteristic, area-shape characteristic and horizontal structural heterogeneity characteristic.

Figure 5 shows that the Co/(Co+C) of landscape pattern index of number characteristic. In this figure, the variation of Co/(Co+C) of the NP, PD, PR and SHDI are not significant in the range of extent size of 300–2400 m. The index of NP and PD are mutated at 2400, SHDI is mutated at 2700 m. There are not mutation points in PR. It is considered that the extent scale effect of number characteristic of landscape pattern index is not obvious in the range of 300–2400 m. the index of NP and PD show strong spatial scale dependence at 2400 m and SHDI at 2700 m. There is basically not spatial scale dependence in PR.

Figure 6 shows that the Co/(Co+C) of landscape pattern index of area-shape characteristic. In this figure, FRAC_MN and LPI show slight variation in the value of Co/(Co+C). The variation of the AREA_MN changes violently, and the mutation point at the extent size of 1200 m. In addition, the big transitions occur at the 2100 and 2400 m. The variation of Co/(Co+C) of the LSI is not significant in the range of extent

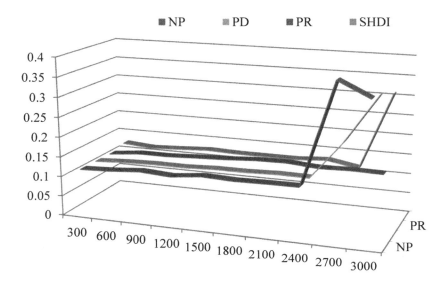

Fig. 5. The Co/(Co+C) of landscape pattern index of number characteristic

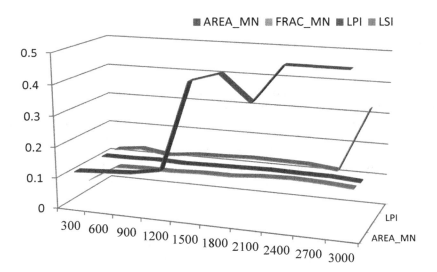

Fig. 6. The Co/(Co+C) of landscape pattern index of area-shape characteristic

size of 300–2700 m, then mutates at 2700 m. Therefore, the index of AREA_MN is considered to have strong spatial scale dependence and effect, LSI have spatial scale dependence at the 2700 m, and spatial scale effect of the FRAC_MN and LPI is not obvious.

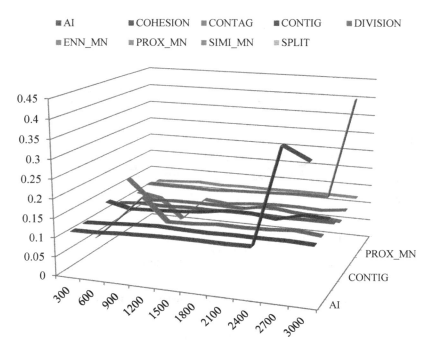

Fig. 7. The Co/(Co+C) of landscape pattern index of horizontal structural heterogeneity characteristic

Figure 7 shows that the Co/(Co+C) of landscape pattern index of horizontal structural heterogeneity characteristic. In this figure, the values of Co/(Co+C) of AI and SIMI_MN show slight variation in smaller extent scale, and mutation occurs after reaching a certain extent size. The index of AI at 2400 m and SIMI_MN at 2700 m. The Co/(Co+C) of the CONTAG, ENN_MN and PROX_MN have opposite trends. The above indexes show slight variation in larger extent scale, and mutation occurs in a certain smaller extent size. The mutation point of the CONTAG at the extent size of 900 m, the ENN_MN at 600 m and the PROX_MN at 900 m. Other indexes (COHESION, CONTIG_MN, DIVISION and SPLIT) show weak trends of the value of Co/(Co+C). It can be seen that the AI and SIMI_MN have stronger spatial extent scale effect in the larger extent size, and the CONTAG, ENN_MN and PROX_MN in the smaller extent size. The spatial scale effect of the COHESION, CONTIG_MN, DIVISION and SPLIT is very weak.

5   Discussion

In the grain scale effect analysis of this paper, the indexes of NP, PD, SHDI, AREA_MN, LPI, FRAC_MN, COHESION, DIVISION and SPLIT presents obvious spatial scale effect, other indexes not obvious. Both the first scale domain and the suitable grain size range of the indexes, which have obvious grain size effect, are very different. This shows that the spatial grain scale effect of each index are very different. For a basin, it is necessary to take into account the spatial scale effect of all the indexes to select the appropriate landscape pattern grain size scale. Considering all the indexes results, 150 m is taken as the research grain size of the research area.

In the extent scale effect analysis of this paper, the ratio of Nugget to Still of 10 landscape pattern indexes reaches the minimum at the extent of 300 m. The 10 indexes are AI, AREA_MN, COHESION, CONTAG, DIVISION, frac_mn, NP, PD, PR, and the SPLIT. The scales of other indexes reaching the minimum ratio are: CONTIG_MN at 600 m, LSI and PROX_MN at 900 m, SHDI at 2700 m, LPI and ENN_MN at 3000 m. Most landscape pattern indexes show significant spatial autocorrelation in a relatively small extent, and the index of SHDI, LPI and ENN_MN show significant spatial autocorrelation in a relatively large extent. This is because with the increase of the extent, the landscape information in each "window" tends to be complete, and the diversity in the number of patch types, the maximum patch area and the distance of Nearest-Neighbor in each "window" becomes less, so that the spatial autocorrelation of SHDI, LPI and ENN_MN is enhanced. Since more than 50% of the spatial autocorrelation of landscape pattern index is the most significant at the extent of 300 m, we think the extent should be selected at 300 m to analyze the landscape characteristics in this river basin.

In this paper, the causes of grain and extent spatial scale effect are not deeply explored. We think this is an important direction for future research, and that's where our next research will go.

6   Conclusion

There have been many studies on the grain scale effect of landscape pattern, but few studies on the extent scale effect, even the combination of grain scale effect and extent scale effect. In this paper, based on the analysis of grain scale effect, the extent scale effect is studied. The following conclusions are obtained through this study:

(1) Most of the appropriate grain size of landscape pattern indexes are in range of 90–150 m, and the optimal grain size scale is 150 m in Middle and Upper Reaches of Meijiang River Basin.

(2) At the grain size scale of 150 m, the spatial self-correlation of landscape pattern index over 50% is the highest at the extent size of 300 m, and the optimal extent size scale is 300 m in Middle and Upper Reaches of Meijiang River Basin.

This paper applies the research method of scale effect of landscape pattern to analyze the grain size and extent size effect of landscape pattern in the middle and upper reaches of meijiang river basin. The results provide the basis for the study of landscape pattern characteristics and spatial and temporal evolution of the river basin and are of great significance to the sustainable development of the river basin ecosystem.

Acknowledgments. The authors would like to thank the Guangdong Innovative and Entrepreneurial Research Team Program (2016ZT06D336), Guangdong Provincial Science and

Technology Program (2017B010117008), Guangzhou Science and Technology Program (201806010106, 201902010033), the National Natural Science Foundation of China (41976189, 41976190), the GDAS's Project of Science and Technology Development (2016GDASRC-0211, 2018GDASCX-0403, 2019GDASYL-0301001, 2017GDASCX-0101, 2018GDASCX-0101), and Southern Marine Science and Engineering Guangdong Laboratory (Guangzhou) (GML2019 ZD0301) for providing financial support.

References

1. Fu, B., Qiu, Y., Chen, L.: The principle and application of landscape ecology (2000)
2. Fu, B.-J., Zhao, W.-W., Chen, L.-D., Liu, Z.-F., Lü, Y.-H.: Eco-hydrological effects of landscape pattern change. Landscape Ecol. Eng. **1**, 25–32 (2005)
3. Ma, L., Bo, J., Li, X., Fang, F., Cheng, W.: Identifying key landscape pattern indices influencing the ecological security of inland river basin: the middle and lower reaches of Shule River Basin as an example. Sci. Total Environ. **674**, 424–438 (2019)
4. Zhou, T., Wu, J., Peng, S.: Assessing the effects of landscape pattern on river water quality at multiple scales: a case study of the Dongjiang River watershed, China. Ecol. Indicators **23**, 166–175 (2012)
5. Chen, Y., et al.: Identifying risk areas and risk paths of non-point source pollution in Wuhua River Basin. Acta Geographica Sinica **73**, 1765–1777 (2018)
6. Board, O.S., National Academies of Sciences, Engineering & Medicine. Environmental Engineering for the 21st Century: Addressing Grand Challenges. National Academies Press (2019)
7. Zhang, X., Zhou, L., Zheng, Q.: Prediction of landscape pattern changes in a coastal river basin in South-Eastern China. Int. J. Environ. Sci. Technol. **16**, 1–10 (2019)
8. Wu, J., Lu, J.: Landscape patterns regulate non-point source nutrient pollution in an agricultural watershed. Sci. Total Environ. **669**, 377–388 (2019)
9. Xiao, R., Wang, G., Zhang, Q., Zhang, Z.: Multi-scale analysis of relationship between landscape pattern and urban river water quality in different seasons. Sci. Rep. **6**, 25250 (2016)
10. Yue, D., Wang, J., Liu, Y., Li, H., Xie, H., Wang, D.: Landscape pattern optimization based on RS and GIS in Northwest of Beijing. Acta Geographica Sinica-Chinese Edn. **62**, 1223 (2007)
11. Thompson, P.L., Isbell, F., Loreau, M., O'Connor, M.I., Gonzalez, A.: The strength of the biodiversity–ecosystem function relationship depends on spatial scale. Proc. Roy. Soc. B: Biol. Sci. **285**, 20180038 (2018)
12. Riitters, K.H., O'neill, R., Jones, K.: Assessing habitat suitability at multiple scales: a landscape-level approach. Biol. Conserv. **81**, 191–202 (1997)
13. Wu, J., Qi, Y.: Dealing with scale in landscape analysis: an overview. Geograph. Inf. Sci. **6**, 1–5 (2000)
14. Wu, J., Shen, W., Sun, W., Tueller, P.T.: Empirical patterns of the effects of changing scale on landscape metrics. Landscape Ecol. **17**, 761–782 (2002)
15. Chen, L., Zhao, W., Fu, B.: The effects of grain change on landscape indices. Quat. Sci. 3 (2003)

16. Thompson, C.M., McGarigal, K.: The influence of research scale on bald eagle habitat selection along the lower Hudson River, New York (USA). Landscape Ecol. **17**, 569–586 (2002)
17. Zhang, R.: Study on dynamics of land use landscape pattern and spatial extent effect in Tianshui (2018)
18. Du, G., Zhang, S., Zhang, Y.: Analyzing spatial auto-correlation of population distribution: a case of Shenyang city. Geograph. Res. **26**, 383–390 (2007)

Design of Distribution Network Pre-planning System Based on GIS

Dawen Huang[1(✉)], Qiugen Pei[2], Yong Xiao[1], and Wensheng Lu[1]

[1] Zhaoqing Power Supply Bureau of Guangdong Power Grid Co., Ltd.,
Zhaoqing 526000, China
hdwgmy@163.com
[2] Guangdong Power Grid Co., Ltd., Guangzhou 510000, China

Abstract. Aiming at the demand of distribution network pre-planning, this paper analyses the current situation and existing problems of distribution network pre-planning, and proposes a distribution grid pre-planning system based on GIS. Based on the design concept and technical route of the grid planning system in the early stage of distribution network, the functional modules and overall architecture of the system are designed. Developing system design based on GIS, maximizing the integration of grid information, providing more accurate and diversified decision-making basis for distribution network planning business personnel, and improving the quality and accuracy of distribution network pre-planning.

Keywords: GIS · Distribution network · Pre-planning · Grid-based · Micro-services · CIM

1 Introduction

The overall process of power distribution network construction project can be divided into three parts: pre-planning design, distribution network construction, and late completion inspection. Among them, the pre-planning and design work is very important in the process of power distribution network construction. In terms of quality, the pre-planning has a direct impact on the stability and optimization of the distribution network power supply system. In terms of investment, it has an important impact on the number of personnel, equipment and materials inputs, including the overall project cost estimation [1]. The pre-planning design mainly carried out preliminary investigations on urban land use planning and power grid development status, and accurately predicted the load distribution. Based on Pre-planning scheme, the designer will select the optimal location of the substation and select the optimal wiring mode, which will make the cost budget and benefit estimation work more scientific and reasonable. It will greatly improve the effectiveness and feasibility of power distribution network work, in order to provide customers with safer power consumption, and promote the overall power grid construction and optimization [2].

The geographic information system (referred to as GIS) can fully integrate the geographical location and its related attribute data, and integrate the relevant information graphics. It provides users with spatial geographic data query, spatial statistical

© Springer Nature Singapore Pte Ltd. 2020
Y. Xie et al. (Eds.): GSES 2019/GeoAI 2019, CCIS 1228, pp. 68–80, 2020.
https://doi.org/10.1007/978-981-15-6106-1_5

analysis and visual support, analysis, management and decision-making, making the planning process and results more intuitive and interactive. The power network line in GIS has linear geographic features. GIS can succinctly display the equipment load of the distribution network and the distribution of users, automatically generate various relevant statistical analysis charts, and effectively provide scientific and effective data for power system operation management and production decision. Therefore, GIS has become one of the most powerful spatial analysis support technologies in power network planning informatization [3–5].

In order to solve the current problems of distribution network planning, such as reference data cannot be displayed simultaneously, a lack of technological application for spatial data query, spatial statistical analysis, data sharing management and so on, this paper proposes a method of distribution network planning based on multi-source power network business data and spatial data, combined with power network CIM model automatic analysis method, and powerful spatial analysis technology of GIS.

The method can automatically update the geographical position and attribute information of the network frame, combine relevant business data with geospatial data, and assist the network frame planner to scientifically and effectively evaluate the current network frame transfer capability, power supply security level, and power supply reliability. It can simulate the wiring operation in the distribution network frame planning, automatically estimate the line load transfer capability after network frame planning, and realize the intelligent distribution network planning.

The method improves the quality and accuracy of the distribution network planning and ensures the effectiveness and feasibility of the distribution network construction work. The system has been piloted and applied in the distribution network planning business of Zhaoqing Power Supply Bureau of Guangdong Power Grid Co., Ltd.

2   Present Situation and Problems in Distribution Network Pre-planning

The pre-planning method of the distribution network, which strengthens the problem-oriented and precise investment concept, will help promote the scientific development of distribution network construction. It is of great significance for building network frame, clarifying the construction standards, and optimizing the construction plan. With urban power network construction expansion, the phenomenon of network frame information and geospatial information detachment in the early planning process of distribution network is also increasing, which leads to the failure of the traditional distribution network pre-planning method to be applied to the current power distribution network project. The pre-planning method of traditional distribution network has limitations in the current era of intelligent informationization, resulting in distribution network planning within a single region and between regions, and the overall distribution network planning and construction and urban space structure lack coordination [4]. The specific problems in the pre-planning methods of traditional distribution networks are as follows.

2.1   Lack of Integrated Design Concept and Depth Analysis

Distribution network planning requires a large amount of reference data, including municipal planning data, power network status map, power grid structure, attribute data, distribution network equipment operation data, marketing data, etc. These data are scattered in different systems, and data information is difficult to obtain and analyze [6].

Therefore, when formulating the scheme, the analysis indicators are relatively simple and cannot effectively describe the complex distribution network operation environment. Lack of comprehensive depth analysis of multiple influencing factors and quantitative analysis of investment benefits led to the failure to select an optimal planning solution.

2.2   Poor Scalability of the Distribution Network

Part of the planning scheme has poor coordination with the urban spatial structure in the location selection of the network frame equipment. The division of the power supply range of the substation is unreasonable. And there is an uncoordinated distribution network planning within a single area and between the various areas. Some new planning loads have no power supply path, which results in limited expansion capacity of the distribution network and difficulty in expanding the power supply range [7].

2.3   Difficult to Optimize the Network Structure of the Distribution Network

The purpose of network frame planning is to balance the power of each partition and plan the power supply range of each substation to determine the optimal power network frame configuration scheme, while meeting economic and reliability requirements. Therefore, distribution network planning is a complex optimization problem with multiple objectives, uncertainty and nonlinearity [5].

2.4   Lack of Necessary Information Support Means

The data collection and finishing is mainly based on offline reporting. The program planning relies on simple calculation, empirical analysis, lack of information-based means of automatic data acquisition and application of technical means such as spatial data query, spatial statistical analysis, and data sharing management. Quality and accuracy are difficult to guarantee [8, 9].

3   Technical Route

Based on the current pre-planning situation of the distribution network, this paper proposes the design concept and technical route of the pre-grid planning system for distribution network.

3.1 Docking with the Business System

By integrating the three major system data of Distribution Network GIS System, Measurement System and Marketing System, the layout data, user quantity data and load data will be organically combined and assisted in visual analysis. The standardized business process and data format are used to realize the main functions including pre-multi-source survey data integration and sharing, line simulation planning, and planning program feasibility analysis, which will assist the distribution network planner to complete the distribution network planning with high efficiency.

3.2 Adopting Micro-services Architecture

The system needs to integrate confidential data such as control detailed planning data and land use planning data, which requires high security performance. At the same time, according to the planning business needs of the distribution network, the planning scheme involves multi-source data fusion, interactive editing of grid devices, real-time rendering of data analysis, etc., which requires high performance and speed of the computer, so the overall architecture of the system adopts micro-services architecture.

3.3 Formulating and Sharing the Planning Scheme

Before the planning scheme is formulated, the planner obtains the network equipment data, the attribute relationship data, the measurement data, and the marketing data from the server according to the feeder line and the load time. The planner only needs to extract the data involved in the solution-related feeder groups. Therefore, the data for scheme formulate is relatively lightweight. The planning scheme will be saved to the local lightweight database SQLite management call. After the planning scheme is completed, the data will be shared with the server for other business personnel to refer to or review on the Local Area Network.

3.4 Unified Integrating Application Environment

Our system uses a unified distributed spatial database system. The server publishes map browsing, data query, spatial analysis, data management and other services in the form of Web Services. The geographic service adopts the WMS and WFS service interfaces conforming to the OGC standard, which realizes seamless integration with the heterogeneous spatial data map services of other GIS platforms, reduces the network transmission burden, and improves the client access speed. Geographic and business data are transmitted in a specific encrypted form to ensure the security of the transmitted data.

4 System Design

4.1 System Function Design

According to the demand analysis of the pre-planning of the distribution network, business application modules include multi-source data visualization integration, planning solution management, grid planning, and planning feasibility analysis are designed. The main function design of the system is shown in Fig. 1.

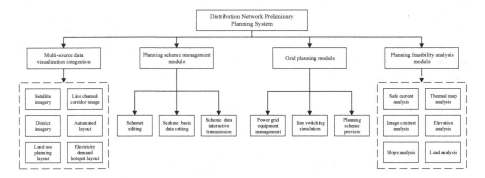

Fig. 1. Main function design of the system

Multi-source Data Visualization Integration. The pre-planning design of the distribution network needs to refer to multi-source data. In addition to the data in the three systems of GIS System, Measurement System, and Marketing System, it is still necessary to consult various forms of data to assist the planning and design, and improve the rationality of the scheme formulation. Multi-source data mainly includes satellite image, line channel corridor image, district image, automated layout map, land use planning layout, power demand hotspot layout and other data. Based on GIS visualization technology, multi-source data integration is displayed on the same map to realize "one map of planning information", which helps to improve the scientific, normative and planning efficiency of grid planning decisions.

Planning Scheme Management Module. The planning scheme management module includes proposal content editing, proposal basic data setting, and proposal data interactive transmission. The storage form of the proposal includes both local and server forms. The planning scheme supports offline editing and also supports sharing of the solution within the LAN. Different operators can provide feedback on the proposed scheme, reducing the situation that the single-person operation leads to insufficient consideration of the program. In addition, according to different privilege levels, users can perform different operations on the server-side proposal data, which is beneficial to proposal data sharing, multi-department collaborative operation, and unified organization and control of scheme data.

Grid Planning Module. Based on the grid the grid composed of feeder groups, the distribution network planning is beneficial to optimize the current complex interlaced network structure and solve the problem of uneven distribution of substation and line load. The traditional distribution network planning method is to collect data from multiple parties, including GIS System, feeder line diagram, metering system data table, Marketing System data table, etc. for manual integration and recalculation. Now users simply select the grid from the business logic or geospatial to display the grid asset device on the map. The system supports users to create new equipment, transform equipment, split lines, merge lines, set segment points, etc. on the map. And it realizes the balance of the substation and each line load, high reliability or minimum network loss, and can intelligently calculate the planning result according to the adjustment of users. The module mainly includes power grid equipment management, line switching simulation, and planning scheme preview.

(1) *Power Grid Equipment Management.* In the pre-planning process of the distribution network, it is necessary to judge whether it can perform the line switching according to the attribute information of the grid equipment. Different grid devices contain multiple attribute data, and some attribute information will affect the planning scheme feasibility, data statistics, etc., such as wire type, wire diameter, station residual interval, transformer type, and so on. The power grid devices that support users viewing and editing mainly include: lines, automation switches, tie switches, towers, cable distribution boxes, switch boxes, distribution transformers.

(2) *Line Switching Simulation.* The system supports users to simulate line switching operation, splitting and merging feeders in different grids, re-planning the grid structure, and re-calculating the statistical data of feeder load, number of users, transformer capacity, line length, safe current, etc., to generate automation layout. Automation layout is based on substation as the starting point and automation switch as the sectional point. In the form of graphic and text integration, it can display the feeder segment situation intuitively for users, simulate the transfer results of fast feedback lines, and assist users to judge the rationality of the scheme. The scheme meets the technical requirements of distribution network planning, such as no more than 6 loop nodes of medium voltage cable lines, no more than 6 main segments of medium voltage overhead lines, and appropriate allocation of sectional switches in longer branch lines. If the planning scheme cannot meet the transfer demand, the transmission line can be simulated repeatedly until the network frame planning is reasonable.

(3) *Preview Planning Scheme.* After the completion of the scheme, the scheme report can be automatically generated, including the technical parameters of the scheme, the estimated number of planned construction, the automated layout map, the load forecast, etc., which assist users to further check the rationality of the scheme from a macro perspective. At the same time, the scheme report realizes the comparison of technical economy between different planning schemes, objectively reflects the advantages and disadvantages and feasibility of the planning scheme, and provides users with more reasonable and optimized planning schemes to improve the scientific planning.

Planning Feasibility Analysis Module. In the process of formulating the plan, it is necessary to take into account factors such as the actual geographical environment and load line conditions to avoid rework caused by insufficient analysis of multiple influencing factors after the plan is formulated. The planning feasibility module can combine spatial geographic information and load users and other related data on the map to support real-time rendering analysis such as safe current analysis, thermal map analysis, image contrast analysis, elevation analysis, slope analysis, load analysis, etc.

It supports real-time rendering analysis such as safe current analysis, thermal map analysis, image contrast analysis, elevation analysis, slope analysis, load analysis, etc., to assist users in evaluating the reliability, economy and feasibility of the planning solution.

4.2   System Whole Framework Design

Starting from the main function design of the above system, the pre-planning system of the distribution network is established on the basis of the information standardization system, the security protection system framework and the system support platform. The system adopts the loosely coupled design of data layer, service platform layer, business logic layer, and presentation layer architecture, as shown in Fig. 2.

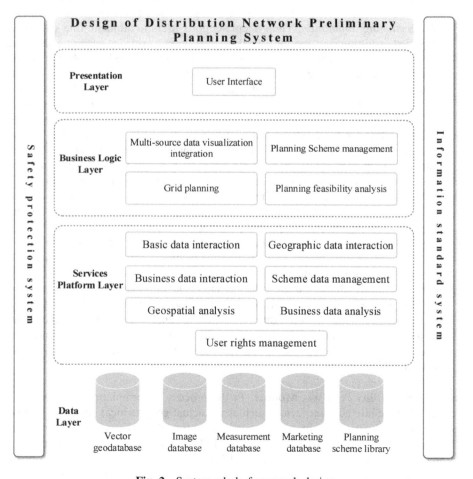

Fig. 2. System whole framework design

Data Layer. Data layer is responsible for integrating GIS System, Measurement System, Marketing System and other pre-planning data to realize automatic update, data storage, data organization, data management and data sharing of multi-source data. It established a unified multi-source data interaction standard specification, realized internal data sharing in the local area network, and provided effective and complete data support guarantee for the formulation of the pre-planning of the distribution network.

Services Platform Layer. The service platform layer is responsible for providing planning service interface services for clients, and implementing interactive operations with geographic data, measurement data, marketing data, and solution data in the data layer through a unified and standardized interface. It plays the role of data organization, scheduling and management. The service platform layer mainly includes services such as data acquisition, data analysis, and data feedback.

Business Logic Layer. The business logic layer is responsible for invoking the service interface provided by the service platform layer, and obtaining the basic data, geographic data, and service data required by the solution, and then performing the simulation operation of the plan. The application layer provides convenient and efficient analysis functions to assist planners in judging the rationality of the solution. The completed planning scheme content and its basic data are fed back to the server through the business service interface to realize data sharing.

Presentation Layer. The presentation layer is responsible for presenting the business data and analysis results in an intuitive visualization. It is the layer at which the user interacts through the client. Users can interact with the functions of the business logic layer in the presentation layer to implement functions such as pre-planning data viewing, element drawing, planning scheme development, and intelligent planning analysis.

5   Major Innovations of the System

5.1   Functional Design Based on Distribution Network Planning "Six Steps"

The main functions of the system are designed based on the six steps of the current distribution network planning business process, including current situation collection, load forecasting, scheme formulating, planning Scheme library, planning project library, and statistical analysis of indicators, as shown in Fig. 3. Through standardizing the business process of distribution network planning in the system, the standardization and unification of the distribution network planning operation process is realized. Users are guided to comprehensively consider various influencing factors to ensure the scientific nature of the planning scheme.

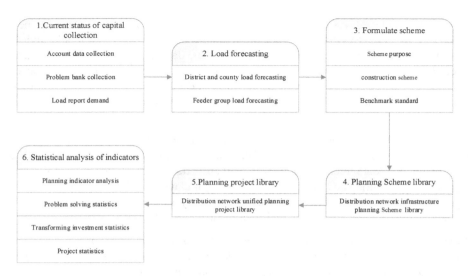

Fig. 3. Six steps of the distribution network planning business process

5.2   CIM Automatically Extracts Network Topology Information

By analyzing the CIM model in the GIS System, the spatial location information and relationship attribute information of the network equipment including the distribution line equipment and the internal equipment of the station can be dynamically obtained, thereby constructing a unique topological relationship data structure. The topology information can effectively describe the context, affiliation, and segmentation information of the device. From the data level, the system reorganizes the decentralized independent equipment reconstruction into a complete and continuous logical structure, which provides a good data foundation for the network pre-planning system of the distribution network. This makes it easy for users to view the information of any segment transformer in the planning process, and quickly count the load and user number data in the segment.

5.3   Automatic Calculation of Segmentation Information and Generation of Automated Layout Maps

With the automatic switch as the segmentation point, the load, number of users and line length of each segment of the feeder are automatically calculated. At the same time, the automatic layout map is automatically generated according to the segmentation information, which assists users in quickly analyzing whether the indicator meets the standard and judges the automatic switch, and the rationality of the installation location, to build a more economical and reasonable network structure, to ensure the implementation of the planning program.

6   Typical Applications of the System

6.1   Intelligent Analysis of Terminal Voltage Quality Problem

The terminal voltage quality problem is one of the important problems in the distribution network, which is mainly caused by long supply radius and heavy load of the line. Through analyzing the grid structure and automatically calculating the power supply radius and line load, the system is able to assist users to judge the case of too long power supply radius, heavy load and overload, as shown in Fig. 4.

Fig. 4.   Intelligent analysis of too long power supply radius and heavy load

6.2   Cutover Load Simulation

In case of heavy load and overload and large branches problems of the lines, where the feeder group cannot meet the requirements of power transfer and supply, it shall be necessary to cutover load. As shown in Fig. 5, part of cutover load from the blue line with heavy load, is transferred to the orange line with light load. After recalculating the segment statistics data, it is able to check whether it fulfills the requirements of the available power transfer and supply.

Fig. 5. Cutover load simulation

6.3 Automatic Section Location Optimization Simulation

In case of centralized load distribution and large branches problems, as shown in Fig. 6, the positions of automatic switch and interconnection switch shall be added or

Fig. 6. Before automatic section location optimization

adjusted. The line is divided into automatic sections with the switch on the automatic column as the dividing point. After analyzing the load of each automatic section through the system, users can set automatic switch in the sections with load concentration problems, in order to realize the average number of users in each section and meet the requirement of no more than 3000 users in each section in the technical guidelines for distribution network planning, as shown in Fig. 7.

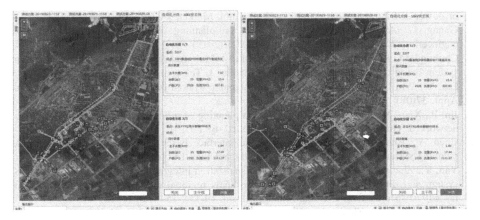

Fig. 7. After automatic section location optimization

7   Conclusion

In summary, GIS can carry out the maximum information integration work. Therefore, GIS is suitable for the design of network pre-planning system for distribution network, providing more accurate and diversified decision basis for distribution network planning business personnel to ensure the safety and reliability of distribution network operation. Through the integration of various professional system information exchange channels related to the distribution network planning, the integration of geographic information, network topology information, equipment operation information and customer marketing information of the distribution network will be realized, so that the distribution network planning reference data can be uniformly displayed and upgraded. It improved the quality and accuracy of the pre-planning of the distribution network. At the same time, through the visual simulation analysis and planning data sharing, the formulation of the distribution network planning and construction plan is more reliable and reasonable. At the same time, the work efficiency of distribution network planning is greatly improved and the planning cost is reduced.

In this paper, based on the network pre-planning system of GIS distribution network, a series of designs are carried out. Based on the design concept and technical route of this system, the main functions and the overall architecture design were carried out. At the same time, it elaborates on the main innovation points and typical applications of the system, and hopes to provide support for the distribution network planning work, thus helping the national power grid construction and sustainable social and economic development.

References

1. Yang, S.: Analysis on the importance and quality improvement measures of the pre-planning of power distribution network project. J. Technol. Market. **24**(10), 268–269 (2017)
2. Guang, L.: Analysis of the importance of pre-planning of power distribution network engineering. J. Shandong Ind. Technol. **02**, 205 (2019)
3. Chuhong, O.: Research on GIS application of distribution geographic information application system based on distribution network automation. J. Telecom. World **04**, 154–155 (2017)
4. Lin, Y., Wu, L.: Design of power system planning system based on GIS. J. Tech. Autom. Appl. **37**(08), 142–145 + 156 (2018)
5. Jin, P., Xiao, X.: Discussion on optimization planning of urban medium voltage distribution network. J. Guangdong Sci. Technol. **22**(18), 67–68 (2013)
6. Chen, Y.: Research and application of data interface based on GIS distribution network planning system. J. New Technol. New Products China **13**, 44–45 (2017)
7. Yan, Y., Zhang, L., Gao, S.: Research on visualization integration application of power GIS system and digital virtual analysis system of new regional distribution network based on IEC61970 specification. J. China Manag. Inform. **19**(15), 154–156 (2016)
8. Feng, Z., Gan, Y., Zhang, X., Wang, A.: Application practice of small GIS in distribution network planning. J. Electric Power Inf. Technol. **10**(07), 98–101 (2012)
9. Luo, Y.: Application of distribution GIS system in distribution network planning. J. Technol. Pioneers **23**, 100 (2013)

New Approaches for Earth Observation Data Acquisition and Processing

Validation and Comparison Among Different VIIRS Cloud Mask Products

Yulei Chi[1(✉)] and Tianlong Zhang[2]

[1] College Geomatics,
Shandong University of Science and Technology, Qingdao 266590, China
cyllilly@126.com
[2] University of the Chinese Academy of Sciences, Beijing 100049, China

Abstract. Universal Dynamic Threshold Cloud Detection Algorithm (UDTCDA) is a recently proposed cloud detection algorithm for remote sensing image with the support of a prior land surface reflectance database. In the visible and near infrared bands, the overall accuracy of cloud detection is still low due to the similar spectral characteristics of some bright surfaces and clouds. Aimed at this problem, this paper proposes an improved VIIRS dynamic threshold cloud detection algorithm (I-DTCDA) on the basis of the characteristics of multi-channels, wide coverage and short revisit period for visible infrared Imaging Radiometer (VIIRS) satellite. The distribution characteristics of thin clouds and other surface features from visible to thermal infrared bands under different spatial and temporal conditions were analyzed, and the detection model of thin clouds and ice/snow was improved. VCM cloud products which used the reflectance and brightness temperature of 16 bands from visible to thermal infrared channels were adopted. The pixel cloud phase logic was established to improve the accuracy of cloud detection. The accuracy of three kinds of cloud detection algorithms was evaluated by visual interpretation. The results illustrated that the I-DTCDA algorithm can identify the clouds over different surface features in remote sensing images with higher precision. What's more, the thin cloud detection results and overall accuracy of I-DTCDA algorithm are higher than those of UDTCDA and VCM algorithms.

Keywords: Cloud detection · UDTCDA · VCM · I-DTCDA · Evaluation

1 Introduction

Cloud detection is a key work for remote sensing image preprocessing and the cloud detection result affects the retrieval of remote sensing parameters and qualitative analysis of images. In recent years, the cloud detection algorithms mainly include threshold methods, statistical methods, object-oriented methods [1] etc. The threshold method is widely used due to the simplicity of operation and the stability of the algorithm, and it achieves considerable accuracy in multi-source satellite cloud detection [2]. Because of the low cloud amount and irregular shape of the thin cloud and broken cloud, the mixed pixels composed of the thin/broken cloud and the underlying surface remained problem that the spectral characteristics of cloud are similar to those of the other features [3, 4]. In fact, it is difficult to determine the

© Springer Nature Singapore Pte Ltd. 2020
Y. Xie et al. (Eds.): GSES 2019/GeoAI 2019, CCIS 1228, pp. 83–92, 2020.
https://doi.org/10.1007/978-981-15-6106-1_6

appropriate threshold by spectral differences, especially in the highlighted surface, the detection of thin cloud is very difficult and the detection accuracy of thin cloud is poor. Therefore, the detection of thin cloud and the broken cloud are facing enormous challenges in optical images.

Visible infrared Imaging Radiometer (VIIRS) is a new earth observation satellite developed from the Advanced Very High Resolution Radiometer (AVHRR) and the MODerate Resolution Imaging Spectroradiometer (MODIS). The spectral channel of the VIIRS covers the range from visible to thermal infrared (0.3–14 μm) with 22 bands. The VIIRS sensor improves the observation accuracy of the edge by sampling synthesis technology, and it adds the Day and Night (DNB) bands for continuous observations. Therefore, the data of VIIRS is of great significance for cloud detection.

Domestic and foreign scholars have done a lot of research on the cloud detection work of VIIRS data. Based on the unique data characteristics of the VIIRS sensor, the VCM cloud mask products are an improvement of the MOD35 cloud mask product. According to surface features and solar illumination, the VCM algorithm tests the reflectance and brightness temperature of spectral channel from visible to thermal infrared by group. The final cloud confidence analysis results were created by aggregating all test results [5]. Although the fixed threshold method has its certain limitations, the VCM cloud mask products are widely used in various remote sensing applications as the typical algorithm of fixed thresholds for its high cloud detection accuracy. Based on the VCM algorithm and the automatic cloud amount assessment model of Landsat_7, Piper and Bahr [6] proposed a fast cloud mask algorithm called VIBCM (VIIRS I-Band Cloud Mask). The test threshold for the reflectance and brightness temperature of the I band is used for cloud detection. This method is simple in operation with high precision relatively. However, the algorithm only uses the brightness temperature of single-band, which has certain defects in cloud detection on ice and snow surfaces. Parmes et al. [7] proposed an automatic cloud detection algorithm for VIIRS data, with an higher overall accuracy (OA > 94%). However, Parmes's algorithm does not consider mixed pixels. Besides, the detection accuracy of thin clouds on surface of ice and snow is relatively low.

Although domestic and foreign researchers have developed the above various cloud detection algorithms, there are still many defects. The problem of mixed pixel including clouds and ice, water, desert and bare land in the image is not taken into account in most methods. To solve this problem, Sun et al. [8] has proposed UDTCDA algorithm supported by a priori surface database with high universality, which the dynamic cloud detection thresholds model was established for visible to near infrared channels after considering atmospheric effects. On the basis of UDTCDA, the method of ice and snow recognition was improved in I-DTCDA and the difference of brightness temperature between thin clouds and typical features in the thermal infrared channel was analyzed simultaneously. A fixed threshold method was introduced to improve the accuracy of thin cloud detection over different surface. Finally, we compared the cloud detection results of the three algorithms in the global typical region. The results obtained by remote sensing visual interpretation method is used to verify and evaluate the accuracy of cloud detection results. The advantages and disadvantages of these algorithms are analyzed.

2 Data Source and Preprocess

2.1 VIIRS Data

The VCM cloud mask used in this paper is a 48-bit medium-resolution cloud mask. It uses twelve M-band data of 750 m resolution and four I-band data of 375 m resolution for spectral test and sea surface spatial test, which determine the overall cloud confidence and overall cloud confidence over water surfaces respectively [9]. In the VCM algorithm, the different test backgrounds are divided by the surface features and solar lighting conditions. The cloud detection tests and the determination of the appropriate threshold were completed independently in each test background. Each cloud detection test includes three thresholds: a high cloud-free confidence, a low cloud-free confidence, and a midpoint threshold. VCM divides its cloud confidence test into five categories, and the final overall cloud confidence is the cube root of the product of test probability used [10]. In addition, VCM uses dual-gain bands to improve cloud detection over desert regions and the cloud phase logic that can identify pixels which contain multiple cloud layers of ice and water clouds was established, simultaneously.

However, both the UDTCDA and I-DTCDA algorithms use the scientific dataset of VIIRS level 1 for cloud detection experiments, and the differences between them are mainly reflected in the choice of bands. Actually, the difference between clouds and ice/snow is mainly concentrated in the short-wave infrared band, the reflectance of the cloud is high and the reflectance of ice/snow is close to zero. In addition to ice/snow, the surface reflectance of clouds and other typical features (including soil, water, vegetation, towns and so on) is quite different in the visible range, and the difference in the near-infrared band is small. Therefore, the UDTCDA algorithm mainly uses the M3, M4, M5, M7, and M10 bands. The I-DTCDA algorithm adds two bands of M15 and M16 based on the obvious difference in brightness temperature between cloud and the surface of highlighted and dark.

2.2 MODIS Data

All of the UDTCDA and I-TDCDA algorithms select a MOD09A1 surface reflectance product that has undergone rigorous atmospheric correction to construct a database of a priori surface reflectance. Assuming that the surface reflectance remains constant for a month, we used the sub-minimum reflectance value synthesis technique based on the selection of blue, green, red and near-infrared bands. The lowest reflectance of each pixel in the four images from a given month was chosen to be the pixel of the one-month series. The method can effectively reduce the influence of cloud/cloud shadows and atmospheric aerosols. Therefore, the data of surface reflectance obtained by this method can accurately express real surface information (Fig. 1).

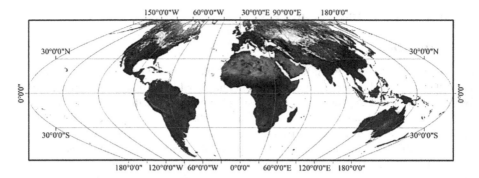

Fig. 1. Surface reflectance data of false color synthesis of global MOD09 in January 2017

3   Methodology

3.1   Correction of Spectral Response

The spectral response of the channels between different sensors is different, and the difference between the blue and red bands is more obvious [11]. Therefore, the error of spectral response caused by using the MOD09A1 product to support VIIRS data requires further correction. In this paper, 50 measured geodetic spectral data such as vegetation, soil, water body and artificial surface in the ENVI surface spectrum database are collected, and the reflection spectrum curve of the surface object and the spectral response function of the sensor are convoluted to obtain the surface reflectance data [12]. There is a strong linear relationship between the surface reflectance of VIIRS and MODIS after the spectral correction. The result is shown in Fig. 2 ($R^2 > 0.99$, $P < 0.01$).

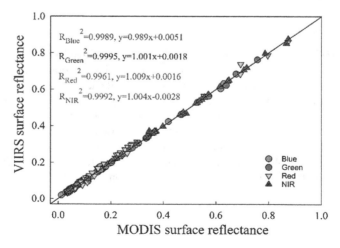

Fig. 2. Comparison of surface reflectance in different channels between VIIRS and MODIS satellites

3.2   Construction of Dynamic Threshold Model

Based on the atmospheric radiation transmission model, the apparent reflectance and the surface reflectance have a strong functional relationship under specific observation conditions and atmospheric conditions [13]. The geometric relationship between apparent reflectance and surface reflectance under specific observation conditions and atmospheric conditions was simulated using the 6S model [14]. The dynamic threshold cloud detection model of visible and near-infrared channel is constructed, and the union of the four single cloud detection results is taken as the final cloud detection result.

$$\rho_i^* = 0.80242 * \rho_i + 0.033842 * \cos \alpha * \cos \beta + 0.161312 \tag{1}$$

$$\rho_i^* = 0.80374 * \rho_i + 0.021677 * \cos \alpha * \cos \beta + 0.139347 \tag{2}$$

$$\rho_i^* = 0.87731 * \rho_i + 0.012633 * \cos \alpha * \cos \beta + 0.123108 \tag{3}$$

$$\rho_i^* = 0.9138 * \rho_i + 0.0061565 * \cos \alpha * \cos \beta + 0.111383 \tag{4}$$

where ρ_i^* is the calculated apparent reflectance, which is the dynamic threshold of cloud detection for different channels; ρ_i is the surface reflectance, which is provided by the constructed monthly synthetic prior surface reflectance database; β is the satellite zenith angle.

3.3   Identification of Thin Clouds and Ice/Snow

In order to improve the situation that thin clouds are recognized incorrectly on the surface with high reflectance of desert, bare land/rock, ice/snow, and the dark surface of water body, the characteristics of reflectance and brightness temperature of thin clouds, thick clouds and other surfaces were randomly extracted based on the UDTCDA algorithm (Fig. 3). It can be found that reflection capabilities of the desert are similar to the reflection capabilities of the bare land. The difference between their reflectance and the reflectance of the thin cloud is small. Meanwhile, the reflectance of the thin cloud fluctuates greatly, which makes the thin clouds over the desert and bare land easily identified by mistakes. In the thermal infrared channel, the occlusion of the thin cloud on the surface reduces the energy of the surface emission and reflection, and the overall brightness temperature is lower than other features. However, the difference of brightness temperature between bare land and desert is significantly lower than other features, and the difference of brightness temperature between water and thin clouds is also very easy to distinguish.

$$0.15 < R_{M3} < 0.4 \cap BT_{M15} \langle 280 \cap BTD \rangle 1.8 \tag{5}$$

where R_{M3} represents the reflectance of the blue band, BT_{M15} represents the brightness temperature of the M15 band, and BTD represents the difference of brightness temperature of the M15 and M16 bands.

In order to reduce the detection error caused by ice and snow, pixels with a normalized snow index (NDSI) greater than 0.4 and a M7 band reflectance greater than

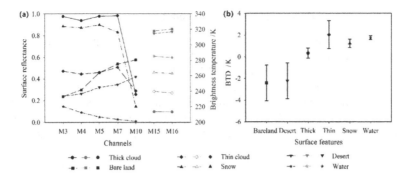

Fig. 3. Distribution of reflectance and brightness temperature for different surface features

0.21 are considered to be ice/snow [15, 16]. At the same time, considering the characteristics of brightness temperature, the pixels with a bright temperature greater than 265 K are considered to be ice and snow.

4   Results and Analysis

In order to highlight the effectiveness at a global scale, we compared the cloud detection results obtained by UDTCDA and I-DTCDA using VIIRS data with VCM products under the same spatial and temporal conditions. Vegetation, water, desert, bare land and snow/ice areas were selected to be the typical land surface features in visual interpretation, and the thick clouds were also evaluated in this section. We chose four images which distributed globally for each selected surface feature.

Cloud detection experiments were performed on the acquired VIIRS data using three algorithms. Figure 4 shows the cloud mask results of UDTCDA, I-DTCDA and VCM algorithm. The pixels of thick cloud ignore the effect of the underlying surface and those pixels has a higher reflectance overall. Therefore, thick clouds can be considered as pure pixels, and the detection of thick clouds is relatively easy. Figure 4 shows the higher thick cloud detection precision of each algorithm, but the VCM cloud mask improves the overall continuity using the corrosion-like expansion method, which may cause some error detection.

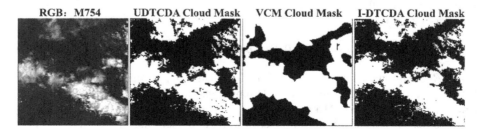

Fig. 4. Comparison results of thick clouds

RGB:M754   UDTCDA Cloud Mask   VCM Cloud Mask   I-DTCDA Cloud Mask

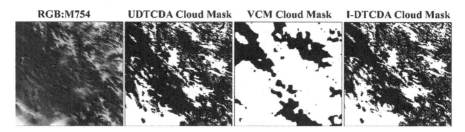

Fig. 5. Comparison results over ice/snow

Ice/snow is a bright surface, UDTCDA algorithm used NDSI to identify clouds over ice/snow surface based on the support of a priori surface reflectance data. Based on the study above, cloud mask of I-DTCDA utilizes the difference of brightness temperature between cloud and snow as the detection results over the snow/ice areas. However, VCM uses the snow/ice database to determine the cloud detection test set and the test threshold. As shown in Fig. 5, I-DTCDA algorithm has a better and effective cloud detection results, and the edge details of the thin cloud are complete. This is because of that the brightness temperature has high sensitivity to the edge of the thin cloud, which can produce more accurate cloud detection results. The VCM product displays the cloud detection result by probability, and the pixels of low probability will have a higher probability of being detected incorrectly.

Deserts and bare land are also bright surfaces, their reflectance is less different from the reflectance of thin clouds. That makes cloud detection more difficult. As shown in Fig. 6, the I-DTCDA algorithm can remove the misunderstanding over bright surfaces and identify thin clouds accurately. This method analyzes the difference of brightness

RGB:M754   UDTCDA Cloud Mask   VCM Cloud Mask   I-DTCDA Cloud Mask

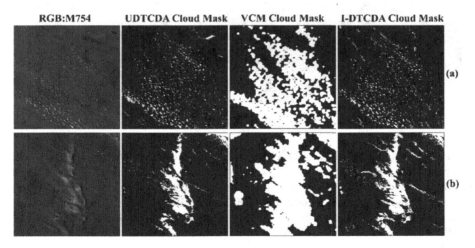

(a)

(b)

Fig. 6. Comparison results over desert and bare land

temperature between thin cloud and other surface features. Meanwhile, the fixed threshold method which is used introduced to help identify thin/broken clouds. Therefore, the cloud detection accuracy of I-DTCDA is higher than that of UDTCDA and VCM algorithms. The position of the VCM cloud mask is accurate, but the cloud amount will be higher than the practical cloud amount, which is related to its corrosion-like expansion operation. UDTCDA will miss some thin clouds due to the lack of brightness temperature.

For dark surfaces such as vegetation and water cover areas, they have a large difference in reflectance from clouds. Due to the existence of mixed pixels, the traditional fixed threshold is difficult to identify thin/broken clouds with complex reflectance, especially on the surface of water with extremely low reflectance, which is easy to occur the missing judgment phenomenon of thin cloud. Figure 7(a) and (b) show the different cloud mask results for surfaces of vegetation and water respectively. I-DTCDA algorithm combines the advantages of UDTCDA and VCM algorithms. A priori surface reflectance database was used to remove atmospheric effects and build dynamic thresholds for visible to near-infrared channels. Meanwhile, the fixed threshold of brightness temperature was increased to aid in the identification of thin clouds. Therefore, I-DTCDA has a better detection effect.

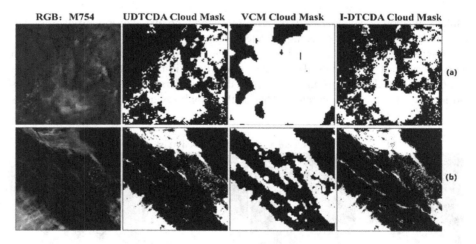

Fig. 7. Comparison results over vegetation and water

The results of UDTCDA, I-DTCDA and VCM cloud masks were analyzed and compared by visual interpretation method whose results are currently closest to the real results of cloud detection. Figure 8 shows the verification results of cloud detection accuracy between different algorithms of 20 scene VIIRS images selected randomly, where the detection index is the correlation coefficient (R^2), root mean square error (RMSE) and average absolute deviation (MAE) between cloud amounts. It can be found that the whole performance of the VCM cloud algorithm is the worst, and the performance of the UDTCDA algorithm is better than the VCM algorithm. In contrast, the I-DTCDA algorithm has the best performance, and its cloud detection results have

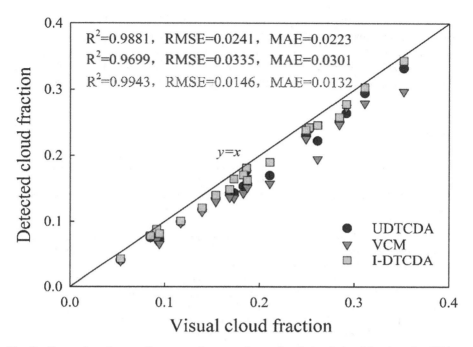

Fig. 8. Comparison in overall accuracy between detected and visual cloud fractions for different cloud algorithms

the highest consistency with the results of visual interpretation. The correlation coefficient (R^2) between cloud amount of VCM and the cloud amount of visual interpretation reached 0.9943, the estimated deviation was the smallest, the average absolute deviation was 0.0132, and the root mean square error was 0.0146. The I-UDTCDA algorithm is more universal and has better overall cloud detection accuracy.

5   Conclusion

In this paper, the I-DTCDA algorithm is briefly introduced and the results of three types of cloud masks are compared and analyzed by visual interpretation results. The following conclusions can be drawn from the comparison results:

1) I-DTCDA cloud mask has better detection results than that of the UDTCDA and VCM on bright surfaces such as ice/snow, desert and bare land, especially for thin cloud coverage areas. The method increases the brightness temperature sensitive to thin clouds and improve the ice and snow recognition model.
2) Due to atmospheric influences, the UDTCDA algorithm will generate excessive thresholds in the dark surface such as water and vegetation, which will cause a large omission error of cloud. For thinner clouds, VCM cloud masks will have higher false positives. In contrast, the I-DTCDA cloud mask is superior to the other two algorithms.

3) By comparison of the results of visual interpretation, the correlation coefficients between cloud amount in VCM, UDTCDA and I-DTCDA results and real cloud amount are 0.9699, 0.9881 and 0.9943, respectively. The root mean square error is 0.0335, 0.0241, 0.0146, and the average absolute deviation is 0.0301, 0.0223, and 0.0132. Therefore, I-DTCDA algorithm has better cloud detection overall accuracy.

References

1. Goodman, A.H., Henderson-Sellers, A.: Cloud detection and analysis: a review of recent progress. Atmos. Res. **21**(3–4), 203–228 (1988)
2. Jedlovec, G.J., Haines, S.L., Lafontaine, F.J.: Spatial and temporal varying thresholds for cloud detection in GOES imagery. IEEE Trans. Geosci. Remote Sens. **46**(6), 1705–1717 (2008)
3. Zhu, Z., Wang, S., Woodcock, C.E.: Improvement and expansion of the Fmask algorithm: cloud, cloud shadow, and snow detection for Landsats 4–7, 8, and Sentinel 2 images. Remote Sens. Environ. **159**, 269–277 (2015)
4. Wu, W., Luo, J., Hu, X.: A thin-cloud mask method for remote sensing images based on sparse dark pixel region detection. Remote Sens. **10**(4), 617 (2018)
5. Revision, J.: Joint Polar Satellite System (JPSS) Ground Project 474–00061 (2012)
6. Piper, M., Bahr, T.: A rapid cloud mask algorithm for Suomi NPP VIIRS imagery EDRS. In: The International Archives of Photogrammetry, Remote Sensing and Spatial Information Sciences, vol. XL-7/W3, pp. 237–242 (2015)
7. Parmes, E., Rauste, Y., Molinier, M.: Automatic cloud and shadow detection in optical satellite imagery without using thermal bands—application to Suomi NPP VIIRS images over Fennoscandia. Remote Sens. **9**(8), 806 (2017)
8. Sun, L., Wei, J., Wang, J.: A universal dynamic threshold cloud detection algorithm (UDTCDA) supported by a prior surface reflectance database. J. Geophys. Res. Atmos. **121** (12), 7172–7196 (2018)
9. Godin, R.: Joint polar satellite system (JPSS) VIIRS cloud mask (VCM) algorithm theoretical basis document (ATBD). Joint Polar Satellite System (JPSS) Ground Project Code, 474, 474-00033 (2014)
10. Kopp, T.J., Thomas, W., Heidinger, A.K.: The VIIRS cloud mask: progress in the first year of S-NPP toward a common cloud detection scheme. J. Geophys. Res.: Atmos. **119**(5), 2441–2456 (2014)
11. Yulei, C., Lin, S., Jing, W.: Improved dynamic threshold cloud detection algorithm for Suomi-NPP visible infrared imaging radiometer. Acta Opt. Sin. **39**(5), 0528005 (2019)
12. Wei, W., Wenbin, Z.Q.: Effects of spectral response differences on NDVI. Remote Sens. Inf. **4**, 91–98 (2015)
13. Johnson, E.C.: A parallel decomposition algorithm for constrained nonlinear optimization. Rensselaer Polytechnic Institute (2001)
14. Vermote, E.F., Tanre, D., Deuze, J.L.: Second simulation of the satellite signal in the solar spectrum, 6 s: an overview. IEEE Trans. Geosci. Remote Sens. **35**(3), 675–686 (2002)
15. Salomonson, V.V., Appel, I.: Estimating fractional snow cover from MODIS using the normalized difference snow index. Remote Sens. Environ. **89**(3), 351–360 (2004)
16. Hutchison, K.D., Iisager, B.D., Mahoney, R.L.: Enhanced snow and ice identification with the VIIRS cloud mask algorithm. Remote Sens. Lett. **4**(9), 929–936 (2013)

A Semi-supervised Regression Method Based on Geography Weighted Regression Model

Ruren Li[1], Hongming Wang[1], Guangchao Li[2(✉)], and Haibo Qi[3]

[1] School of Transportation Engineering, Shenyang Jianzhu University,
Shenyang 100168, China
[2] College of Geoscience and Surveying Engineering,
China University of Mining and Technology, Beijing 100083, China
rurenli@163.com
[3] School of Mining and Geomatics Engineering,
Hebei University of Engineering, Handan 056038, China

Abstract. Aiming at the problem that the geometric weighted regression method has low prediction accuracy when the amount of training data is small, this paper combines the geometric weighted regression model with semi-supervised learning theory to make full use of semi-supervised learning that uses unlabeled samples to participate in training process to enhance the performance of the regression model, a semi-supervised regression model based on geo-weighted regression model is proposed, that is Self-GWR and CO-GWR. Based on the data of elevation, Aerosol Optical Depth (AOD), temperature, wind speed, humidity and pressure of Beijing-Tianjin-wing area, this paper uses geo-weighted regression model, self-trained geography weighted regression model and collaborative training geography weighted regression model respectively to practice. The experimental results show that CO-GWR effectively improves the accuracy of regression model through two regression models, and the accuracy of Self-GWR is slightly lower than that of GWR model, which indicates that the model may accumulate errors in the self-learning process, resulting in poor regression accuracy finally.

Keywords: GWR · Semi-supervised learning · Collaborative training · AOD

1 Introduction

In geospatial spatial analysis, the observed data is obtained in different geographical locations. The global spatial regression model assumes that the regression parameters are independent of the geographical location of the sample data, and in the actual problem research, it is often found that the regression parameters vary with the geographical location. If the global spatial regression model is still used, the regression parameter estimates obtained will be the average of the regression parameters in the whole study area, and cannot reflect the real spatial characteristics of the regression parameters [1, 2]. When spatial data has autocorrelation, the GWR (geographically weighted regression) model provides an estimation method superior to the ordinary linear regression model (OLS). The OLS method only provides estimation of global

© Springer Nature Singapore Pte Ltd. 2020
Y. Xie et al. (Eds.): GSES 2019/GeoAI 2019, CCIS 1228, pp. 93–102, 2020.
https://doi.org/10.1007/978-981-15-6106-1_7

parameters, and the GWR model allows decomposition into local parts. The parameter estimation profoundly explains the relationship between certain types of indicators and spatial impact factors of geospatial data, which is unmatched by the OLS method [3]. The Geographically Weighted Regression (GWR) is a typical local model. The GWR model considers that the regression coefficient changes with the spatial position and can reflect the local relationship between the dependent variable and multiple independent variables [4]. Spatial non-stationary [5]. GWR modeling is performed using labeled samples. The accuracy of the GWR model is related to the number of labeled samples. The less the labeled samples, the lower the accuracy of the model. However, some labeled samples are difficult to obtain in large quantities, and unlabeled samples can be obtained in large quantities. Therefore, how to improve the accuracy of the GWR model with unlabeled samples is of great significance in the case of few labeled samples [6].

Semi-supervised learning samples with labeled and unlabeled samples can make use of a large number of unlabeled samples that are easily accessible, so that the workload of labeled samples is alleviated and a more efficient learning model is obtained [7]. Collaborative training uses two-view to train two classifiers to mark samples to each other to expand the training set. Using unlabeled samples to improve learning performance is an important method for semi-supervised learning [8]. Yang Y et al. [9] demonstrated that collaborative training using unlabeled sample-assisted training improves the learning performance when the labeled samples are small. The main direction of the current research is the clustering problem. The research on the regression problem is relatively rare, mainly because the clustering hypothesis in semi-supervised learning does not hold on the regression problem, and the confidence level is difficult to calculate in the regression analysis. In response to the above problems, the cooperative regression calculation method generates different k-nearest neighbor regression models based on different k-values or distance metrics, and selects unmarked samples with high confidence according to the prediction consistency for marking [10]. Zhao Yangyang [11] proposed a synergistic spatiotemporal geographic weighted regression method. The results show that the performance of spatiotemporal geo-weighted regression models with different kernel functions is not as high as that of geospatial-weighted regression models. Chai Yan [12] proposed a method for extracting boundary vectors based on the characteristics that support vectors are generally located at the boundary of two types of sample sets, and improved the SVM (support vector machines) algorithm. Ma and Wang et al. [13] proposed a regression model based on SVM cooperative training, which is suitable for the model when dealing with a large number of inputs, which alleviates the error accumulation problem caused by using only a single regression model, and improves the pan of the regression model. Ability. Brefeld et al. [14] proposed a semi-supervised least squares regression method to apply collaborative training to the normalized risk minimization problem in Hilbert space.

In summary, there are relatively few studies on semi-supervised regression learning methods, and there are fewer studies in the field of geographic information combined with semi-supervised regression learning. In this paper, for the problem of geographically weighted regression in the case of a small number of labeled samples, the prediction accuracy of the regression model is low. A model combining semi-supervised learning with geographically weighted regression is proposed, namely self-training geographically weighted regression model (Self-GWR) and collaborative training

geography. This method makes full use of the advantages of semi-supervised learning, improves the regression accuracy of GWR under small sample data by unlabeled sample assistance, and can study the non-stable characteristics of space, which is more suitable for the analysis application of space field. The weighted regression model (CO-GWR) was used as the test data for the elevation, aerosol optical thickness, temperature, wind speed, humidity, and pressure in the Beijing-Tianjin-Wing region. By comparing with the conventional geographically weighted regression method, the mean absolute error (MAE), root mean square error (RMSE), Akaike information (AIC), and goodness of fit (R2) are used as evaluation indicators to verify the method. Effectiveness.

2 Methods

2.1 Geographically Weighted Regression

The geographically weighted regression was proposed by Fortheringham et al. [15] of the University of St. Andrews in the United Kingdom based on the regression of spatial coefficient of variation using the idea of local smoothness. Geographically weighted regression is an extension of ordinary linear regression, which introduces the geographic location of the sample points into the regression parameters. The formula is as follows:

$$y_i = \beta_0(u_i, v_i) + \sum_{k=1}^{d} \beta_k(u_i, v_i)x_{ik} + \varepsilon_i \tag{1}$$

In Eq. (1), the n variables of the dependent variable y (y_i, x_{i1}, x_{i2}, $\cdots$ x_{id}) and the independent variable ($x_1, x_2, \cdots x_d$) at the data point (u_i, v_i), ($k = 1, 2, \cdots,$ d)$\beta_k(u_i, v_i)$ are the unknown parameters at the observation point (u_i, v_i). For independent and identically distributed error terms, it is usually assumed to be subject to $N(0, \delta^2)$ distribution.

The regression parameters in the geographic weighted regression model (GWR) are related to the geographical location of the sample data, and the degree of influence (space weight) can be represented by a distance function, which is referred to as the kernel function. Commonly used kernel functions are Gaussian kernel functions, Bisquare kernel functions, and so on. The key to the GWR model is to choose the kernel function and determine its optimal bandwidth. The study found that the bandwidth sensitivity of different kernel functions is different, and the change of bandwidth will have a greater impact on the results [6]. Therefore, the regression model can be distinguished by kernel function and bandwidth.

If the bandwidth of the kernel function is too large, the regression parameter estimation is too large, and too small will cause the regression parameter estimation to be too small [1]. In order to reduce the error caused by bandwidth discomfort, this paper uses Clevel cross-validation method [16] proposed by Cleveland to calculate the optimal bandwidth. The CV method is calculated as:

$$CV = \frac{1}{n} \sum_{i=1}^{n} \left[y_i - \widehat{y_{\neq i}}(b) \right]^2 \tag{2}$$

In formula (2), when the regression parameter $\widehat{y_{\neq i}}(b)$ is estimated, the regression point itself is not included. Only the regression parameter calculation is performed according to the data around the regression point, and the different bandwidth and CV are drawn into the trend line, and the minimum CV(b) value can be found and Its corresponding optimal bandwidth is b.

2.2   Semi-supervised Learning

Self-training Geographically Weighted Regression. In semi-supervised learning, Fralick et al. [17] proposed a self-training learning method. In each round of training, the best sample from the previous round of predictions is added to the current set of labeled samples. The results produced continue to train themselves [18]. Based on the geographically weighted regression model and the semi-supervised self-training theory, this paper proposes a self-training geographic weighted regression model (Self-GWR).

The steps of the self-training geo-weighted regression model algorithm are as follows: 1 determining the labeled samples, unlabeled samples, initializing the GWR model parameters, using the Gaussian kernel function in the GWR model; 2 using the GWR model to perform regression prediction on the unlabeled samples; The highest degree of data is added to the labeled sample set of the regression model and the sample is removed from the unlabeled sample; 4 iteratively trains the GWR model until the unlabeled sample is trained to a certain number.

Collaborative Training Geographically Weighted Regression. Collaborative training is based on a small number of labeled samples and a large number of unlabeled samples. Through continuous iteration, different learners learn from each other [19], which is a semi-supervised learning method. The core idea of collaborative training is: First, use a labeled sample to train a classifier on each view. Then, each classifier selects a number of high-confidence samples from unmarked samples for marking, and puts these new tags. The sample is added to the training set of another classifier so that the other party can update with these new tag samples. This process of "learning from each other and making progress together" is iteratively continued until the two classifiers are not changing or reaching a predetermined number of learning rounds [20].

Based on the theory of geographically weighted regression model and cooperative training regression algorithm [13], combined with semi-supervised learning collaborative training theory, a collaborative training geographic weighted regression model (CO-GWR) is proposed. The CO-GWR model not only integrates the characteristics of the GWR model in geographic applications, but also compensates for the few defects of the sample. The algorithm steps are as follows: 1. Determine the labeled samples, unlabeled samples, initialize the GWR model parameters h1, h2, and h1 is based on Gaussian. The GWR model of the kernel function, h2 is the GWR model based on the Bisquare kernel function; 2 using each regression model to predict the regression of the unlabeled sample set, and selecting the labeled sample set added to the regression model with the highest confidence in the prediction result. And the sample is removed from the unlabeled sample; 3 repeat the second step operation until the unlabeled sample is trained to a certain number; 4 finally take the average of the two model prediction results as the final prediction result.

Confidence. Confidence is the degree to which the unlabeled data in the regression model affects the accuracy of the regression model during training. The higher the confidence, the better the consistency of the prediction and regression models, and the closer to the true value. Therefore, the sample selected by the regression model through high confidence should be a sample that makes the regression model more consistent with the labeled sample [20]. Confidence is based on MAE and its calculation method is:

$$\xi X_{x \int u} = \sum_{x_L} \left[(y_L - \hat{y}_L)^2 - (y_L - \hat{y}_L')^2 \right] \tag{3}$$

In Eq. (3), to mark the true value of the sample, y_L is the predicted value of the labeled sample on the original regression model, y_L is the predicted value of the labeled sample on the new regression model, and the new regression model refers to adding the unlabeled Regression model re-established after the sample. At the $\xi X_{x \in u} > 0$ time, order. $N(x, u, v) = arc\max(\xi X_{x \in u})$. The most unmarked sample $N(x, u, v)$ with the highest confidence. $\xi X_{x \in u} > 0$ shows that the performance of the regression model is improved after the unlabeled samples are added. The maximum confidence indicates that the performance of the model is the largest, that is, the selected sample is the one with the highest confidence among the unlabeled samples participating in the training.

3   Overview of the Study Area

The study area uses the Beijing-Tianjin-Hebei region with a geographical range of 35.5° N −43°N and 113°E −120°E, including 11 prefecture-level cities in Beijing, Tianjin and Hebei Province. It is located in the heart of China's Bohai Sea and is China. The largest and most dynamic region in the north has a land area of more than 200,000 and a total population of more than 100 million. With the rapid development of the economy, air pollution is becoming more and more serious, and Beijing-Tianjin-Hebei is also one of the heavily polluted areas. Therefore, controlling air pollution and optimizing the ecological environment are important tasks in the Beijing-Tianjin-Hebei region. The changes in atmospheric aerosol types and contents are closely related to climate change and atmospheric environmental pollution. The research on AOD is of great significance for the analysis and prevention of atmospheric environmental pollution.

4   Application to Simulated Data

4.1   Data Source

This paper takes geospatial data (elevation), meteorological data (wind speed, temperature, humidity, air pressure) and AOD data in the Beijing-Tianjin-Hebei region as research objects. The geospatial data comes from the Geospatial Data Cloud website (http://www.gscloud.cn/), and the SRTMEDM 90 M resolution raw elevation data is selected; the meteorological data is from the China Meteorological Science Data Sharing Service Network (http://www.escience.gov.cn) A total of 110 meteorological monitoring sites with geographic location information for one day; AOD data from the Terra MODIS C06 secondary aerosol product, the frequency is one day, and the spatial resolution is 3 km. In this paper, MODIS Collection 6 MYD04_3K data set parameter named "Optical_ Depth_ Land_ And_ Ocean", band 2 550 nm Class 2 AOD data, AOD data selected May 2015 data as the research object. The meteorological monitoring stations were randomly divided into marked monitoring stations and unmarked monitoring stations in a 1:1 ratio (Fig. 1).

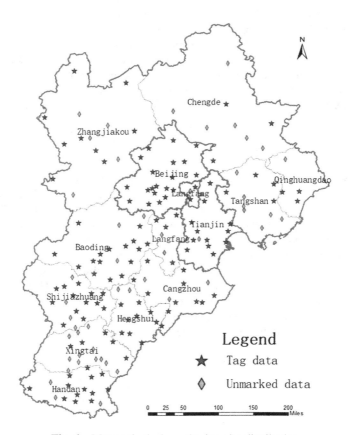

Fig. 1. Meteorological monitoring site distribution

4.2   Data Preprocessing

For the spatial and temporal consistency of the data, the geospatial data, meteorological data, and AOD data are preprocessed. The preprocessed partial data is shown in Table 1, and the preprocessing flowchart is shown in Fig. 2. Firstly, the meteorological data is processed to obtain the daily mean value of 110 meteorological data and its monthly mean value. Secondly, the geospatial data is processed, including the projection coordinate transformation of the image data, and the spatial data value of the meteorological site for the elevation image data. Finally, for AOD data processing, a 5 km × 5 km grid covering the whole area is created for the Beijing-Tianjin-Hebei region, and the position coordinates of the grid center point are extracted, and the grid center point represents the spatial position of the grid. The MODIS image data is processed in batches, the value of the AOD of the grid center point is obtained by resampling, and the monthly mean value of the AOD data of the grid center point is calculated, and the monthly mean value is interpolated by Kriging to extract the value of the AOD at the weather station.

Table 1. Partial data display

ID	Site number	Latitude	Longitude	Elevation	Average wind	Average temperature	Average humidity	Air pressure	AOD
1	53392	41.85	114.6	1418	2.1167	17.1833	26.6667	853.4	0.5282
2	53893	36.7667	114.95	42	1.8885	23.6167	53.25	1003.9296	0.5377
3	54809	36.55	115.2833	42	1.7167	21.9833	56.3333	1004.2833	0.5709
4	53899	36.4833	114.9667	48	2.1417	22.1083	58.3333	1003.7167	0.4822
5	53693	38.0333	114.1333	215	2.8417	23.975	48.3333	978.6333	0.4044
6	54412	40.7333	116.6333	285	1.75	14.2	84	971.95	0.3699
7	54541	39.9	119.2333	28	2.4	23.5083	39.6667	1005.6667	0.5824
8	54301	41.6667	115.6667	1412	2.175	17.45	32.6667	854.7917	0.0865

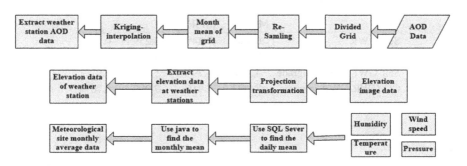

Fig. 2. Data preprocessing flow chart

4.3   Results and Analysis

The correlation coefficient between each index is calculated, and the correlation coefficient matrix is shown in Table 2. It can be seen from Table 2 that the five indicators considered in this paper are related to AOD to a certain extent. Humidity, pressure and temperature are positively correlated with AOD, and elevation and wind speed are negatively correlated with AOD, including elevation, temperature and pressure. The correlation coefficient with AOD is more than 0.62. (The following data retains 4 significant figures)

Table 2. Correlation coefficient matrix

	AOD	Elevation	Air temperature	Humidity	Wind speed	Air pressure
AOD	1	−0.6238	0.6454	0.3951	−0.2053	0.6225
Elevation		1	−0.9398	−0.7152	0.5270	0.9921
Air temperature			1	0.5347	−0.4799	0.9388
Humidity				1	−0.5259	0.7247
Wind speed					1	−0.5373
Air pressure						1

In order to evaluate the prediction effect of the research method, this paper compares it with the GWR model, and calculates the four evaluation indexes of MAE, RMSE, AIC and R2 of each model, and the degree of improvement between the models. MAE reflects the possible error range of the estimated value, RMSE reflects the inversion sensitivity and extremum effect of the interpolation function, and the AIC criterion is based on the concept of entropy, which can weigh the complexity of the estimated model and the superiority of the model fitting data. R2 can measure the pros and cons of the model fit. The calculation results are shown in Table 3.

Table 3. Comparison of forecast results

Mode	MAE	RMSE	AIC	R^2
Self-GWR	0.3604	0.4497	162.2593	0.6513
CO-GWR	0.3137	0.3971	143.3187	0.7386
Self-GWR/GWR upgrade	−4.71%	−4.29%	−5.06%	−5.91%
CO-GWR/GWR upgrade	8.86%	7.91%	7.20%	6.70%

5   Conclusions

From Table 2, we can see that the GWR model is 0.3442, 0.4312, 154.4414 and 0.6922, the Self-GWR model is 0.3604, 0.4497, 162.2593 and 0.6513, and the CO-GWR model is 0.3137, 0.3971, 143.3187 and 0.7386, respectively, among the evaluation indexes MAE, RMSE, AIC and R2 obtained by each model. Compared with

GWR model, CO-GWR model increased percentages by 8.86, 7.91, 7.20 and 6.70, respectively. The experimental results show that the accuracy of the Self-GWR model is the worst among the three models, mainly because the model in the self-training process, the error occurred in the previous stage will accumulate and amplify in the subsequent learning process, so that the model final accuracy is poor. The CO-GWR model has the highest precision and the best effect, which fully demonstrates that the collaborative training method not only plays the role of unlabeled samples in semi-supervised learning, but also alleviates the deficiencies of self-training model error accumulation and improves the generalization performance of the model. The prediction accuracy of the regression model is improved.

This paper combines semi-supervised learning with a geographically weighted regression model, namely self-training geographic weighted regression model (Self-GWR) and collaborative training geographic weighted regression model (CO-GWR). Both models can use unlabeled samples to improve the performance of the regression model, and can also analyze the non-stationary features of the geographical phenomenon, so that the semi-supervised learning method can be better applied in the field of spatial analysis. By comparing with the conventional GWR model, it shows that the CO-GWR model effectively improves the accuracy of the regression model by means of two regression models "mutual learning", while the accuracy of the Self-GWR model is slightly lower than that of the GWR model, indicating that the model is Accumulation of errors may occur during self-learning, resulting in poor final regression accuracy. In the follow-up study, the above research methods will continue to be improved. For example, for collaborative training, two different regression models can be used for analysis. For self-training, outliers in the data can be eliminated to achieve better learning results.

References

1. Yan, W.: Basic Theory and Application of Geographically Weighted Regression. Tongji University, Shanghai (2007)
2. Yang, Y.: Study on the Non-stationary Geographically Weighted Regression Method of Time and Space. Wuhan University, Wuhan (2016)
3. Tang, Q., Xu, W., Ai, W.: Study on spatial differentiation of house price and its influencing factors based on geographically weighted regression. Econ. Geogr. (02), 52–58 (2012)
4. Ren, Q, Wang, L., Li, H.: Analysis of spatial differences of regional economic development in China. Geogr. Geogr. Inf. (01), 110–116 (2017)
5. Wu, B., Barry, M.: Geographically and temporally weighted regression for modeling spatio-temporal variation in house prices. pp. 383–401. Taylor & Francis Inc. (2010)
6. Zhao, Y., Liu, J., Xu, S., et al.: A geographically weighted regression method based on semi-supervised learning. J. Surv. Mapp. (01), 123–129 (2017)
7. Ma, L., Wang, X.: Semi-supervised regression based on cooperative vector machine cooperative training. Comput. Eng. Appl. (03), 177–180 (2011)
8. Guo, X., Wang, W.: An improved collaborative training algorithm: compatible co-training. J. Nanjing Univ. (Nat. Sci.) (04), 662–671 (2016)

9. Yang, Y., Liu, J., Xu, S., et al.: An extended semi-supervised regression approach with co-training and geographical weighted regression: a case study of housing prices in Beijing. ISPRS Int. J. Geo-Inf. **5**(1), 4 (2016)
10. Zhou, Z.H., Li, M.: Semi-supervised regression with co-training. In: International Joint Conference on Artificial Intelligence (2005)
11. Zhao, Y., Liu J., Yang, Y., et al.: A method for estimating PM2.5 concentration in synergistic space-time geographically weighted regression. J. Surv. Mapp. (12), 172–178 (2016)
12. Chai, Y., Wang, Y., Zhang, J.: Support vector machine algorithm under boundary vector. J. Liaoning Techn. Univ. (Nat. Sci. Ed.) (02), 202–205 (2017)
13. Lei, M.A., Wang, X.: Semi-supervised regression based on support vector machine co-training. Comput. Eng. Appl. **25**(2) (2011)
14. Efron, B., Gong, G.: A leisurely look at the bootstrap, the jackknife, and cross-validation. Am. Stat. (1), 36–48 (1983)
15. Fotheringham, A.S., Charlton, M., Brunsdon, C.: Measuring Spatial Variations in Relationships with Geographically Weighted Regression. Springer, Heidelberg (1997)
16. Cleveland, W.S.: Robust locally weighted and smoothing scatterplots. J. Am. Stat. Assoc. **74** (368), 829–836 (1979)
17. Fralick, S.: Learning to recognize patterns without a teacher. IEEE Trans. Inf. Theory **13**(1), 57–64 (2003)
18. Liu, J., Liu, Y., Luo, X.: Semi-supervised Learning Method. Chin. J. Comput. **08**, 1592–1617 (2015)
19. Zhou, Z.H., Li, M.: Semisupervised regression with cotraining-style algorithms. IEEE Trans. Knowl. Data Eng. **19**(11), 908–913 (2007)
20. Zhou, Z.: Semi-supervised learning based on bifurcation. Acta Autom. Sinica **11**, 1871–1878 (2013)

Quality Evaluation of Sea Level Height Data of Jason-3 Satellite Based on Jason-2

Min Ji[1], Huadong Ma[1(✉)], Gang Li[1], Liguo Zhang[2],
and Qingsong Shi[1]

[1] Shandong University of Science and Technology, Qingdao 266590, China
18263829076@163.com
[2] Shandong Institute of Land Surveying and Mapping, Jinan 250102, China

Abstract. This paper used the Jason-3 satellite 001-023 cycle geophysical data GDR and the simultaneous on-orbit Jason-2 satellite 281–303 cycle GDR data for data analysis. The data quality of Jason-3 satellites was evaluated by data screening and analysis of two satellite data, comparative analysis of geophysical parameters, analysis of sea level height discrepancy at crossover points, and analysis of along-trace sea level anomaly. The results show that the four geophysical parameters of Jason-2 and Jason-3 are slightly different. The value distribution curve has the same characteristics as the average daily variation trend. The mean value of sea level height discrepancy at Jason-3's self-crossover is-0.1595 cm and the standard deviation is 4.989 cm, the mean value of sea level height discrepancy of Jason-2 in the same period is -0.079 cm, and the standard deviation is 4.979 cm. The mean value of sea level height discrepancy between Jason-3 and Jason-2 is −2.59 cm, and the mean value of standard deviation is 5.742 cm. The mean value and standard deviation of the along-trace sea level anomaly of Jason-3 and Jason-2 without correction are 2.2304 cm and 0.1171 cm respectively. These data fully show that the accuracy of Jason-3 in the verification and calibration stage meets the accuracy requirements and meets the requirements of marine scientific research and application.

Keywords: Radar altimeter · Jason-3 · Crossover points analysis · Analysis of along-trace sea level anomaly · Jason-3 quality assessment

1 Introduction

Satellite observations of sea level height are of great significance to physical oceanography research, and can provide global ocean circulation information [1], Grid sea level anomaly (SLA) data are widely used in the analysis of Marine physical process, comparison of simulation data [2] and model assimilation process [3]. The satellite radar altimeter can directly measure the global sea level. At present, many series of altimeter satellites have been launched at home and abroad, including ENVISAT (2002–2012), Jason-1 (2001–2013), and Jason-2 (2008), HY-2A (2011), Jason-3 (2016), etc. With the launch of more and more altimeter satellites, the coverage density of global sea altimeter data will increase and the time length will be longer.

© Springer Nature Singapore Pte Ltd. 2020
Y. Xie et al. (Eds.): GSES 2019/GeoAI 2019, CCIS 1228, pp. 103–121, 2020.
https://doi.org/10.1007/978-981-15-6106-1_8

How to achieve the seamless integration of many satellite altimeter data and maintain the long-term altimeter datum has become an important research topic [4].

When two satellite altimeters are in orbit at the same time, the quality of the data from the other satellite can be assessed using more accurate altimeter sea level data as a reference. TOPEX/Poseidon, Jason-1 and Jason-2 are altimeter satellites with high accuracy and cross calibration. Ocean observation in a long time series can provide low-frequency and large-scale ocean signals, which can be used as a reference for other altimeter satellite missions [5]. Faugere et al., Ablain et al., took ENVISAT and Jason-2 [6], Jason-1 and Jason-2 [7] for cross-evaluation, respectively; Guo et al. used TOPEX/Poseidon, Jason-1 and Jason-2 satellite altimetry data from 1993 to 2012 for 20 years to study the temporal and spatial changes of sea level in China sea [8]; Wen et al. took Jason-3 data and Hongze lake as examples to study the screening method of satellite height measurement data [9]; Xie et al. simulated and analyzed the significant wave height in the sea area of Dongsha island based on SWAN model [10]; Peng et al. evaluated the quality of HY-2A data based on Jason-2 data simultaneously in orbit [11]. At present, scholars at home and abroad have made accurate evaluation on the data quality of TOPEX/Poseidon, Jason-1, Jason-2 and other satellites, which have been widely used in various fields of the ocean. However, there are few published articles on the quality evaluation of Jason-3 satellite data, which can not provide quality assurance for the practical application of Jason-3 satellite data, and can not fully reflect the scientific value of jason-3 satellite data.

In this paper, GDR data of Jason-2 satellite in orbit with Jason-3 satellite were used for comparative analysis, firstly, the abnormal values in Jason-3 satellite data were eliminated, and then the data quality of geophysical parameters such as significant wave height, backscattering coefficient, ionospheric delay and wet tropospheric delay were compared with that of Jason-2 satellite data. Finally, the discrepancy of sea level height at the crossover points of Jason-3 and Jason-2 satellite data were compared and the along-trace sea level anomaly were analyzed. The comparative analysis of sea level anomaly has achieved the goal of evaluating the quality of height sea level data from Jason-3 satellite. The results of this study can provide a basis for the application of Jason-3 satellite GDR data.

2   Data Source, Analysis and Preprocessing

2.1   Data Source and Analysis

Jason-3 satellite is a marine terrain satellite jointly developed by CNES, NASA, Eumetsat and other agencies, which is mainly used to monitor sea level height and climate change. The sea level height error of Jason-3 satellite data is expected to be 2.5 cm [12]. The payloads on the satellite are mainly radar altimeter and microwave radiometer. Jason-3 satellite is the continuity of T/P and Jason-1/2 series satellites. It adopts the same orbit design. The orbit height is 1336 km, the period is 10 d, and the inclination angle is 66°. The geoid adopted by Jason-3 is EGM96, and the reference ellipsoid is the same as Jason-2. The main geophysical parameters of Jason-3 satellite are significant wave height, backscatter coefficient, ionospheric delay and radiometer wet tropospheric delay.

Since the first half of Jason-3 satellite's 000th cycle has not yet entered the predetermined orbit, resulting in 45% data loss, the first 23 cycles of Jason-2 and Jason-3 satellites are in the same orbit. After 24 cycles, the ground trajectory of Jason-2 satellite changes. Therefore, the GDR data of Jason-3's 001–023 cycle and Jason-2's 281–303 data of the same period are used in this paper. Jason-3 Cycle 001 corresponds to Jason-2 Cycle 281 (the number of cycles appearing below is Jason-3 Cycle), from 17 February to 2 October 2016. Jason-3's data products can be downloaded free of charge through the FTP download address available on the AVISO website. In this paper, the quality of Jason-3 satellite sea level height data is evaluated by comparing and analyzing the main geophysical parameters of Jason-2 and Jason-3, sea level height discrepancy at crossover points and along-trace sea level anomaly.

Geophysical Parameters. In this paper, based on the published observation data, the four main geophysical parameters of Jason-2 and Jason-3 satellites will be compared and analyzed, and the accuracy of the main geophysical parameters of Jason-3 will be evaluated by the main geophysical parameters of Jason-2 whose observation accuracy has been determined.

Sea Level Height Discrepancy at Crossover Points. According to the sampling characteristics of the radar altimeters of Jason-3 and Jason-2 satellites, the repeated observation positions of the two satellites are the crossover points of their trajectories. The reasonable calculation and selection of the crossover points are the basis of the cross evaluation of the two satellites. The discrepancy of sea level height at the crossover points is the difference of sea level height at the crossover points of different satellite trajectories, which is the main means to analyze the overall performance of altimetry system. In order to reduce the impact of the non system performance on the final evaluation results, this paper adopts the methods of Jason-3 and Jason-2 self-crossover points sea level height discrepancy correlation analysis and two satellites crossover points sea level height discrepancy correlation analysis to improve the accuracy of the evaluation results.

Along-trace Sea Level Anomaly. The analysis of along-trace sea level anomaly is a means to evaluate the performance of altimetry system. By analyzing the mean value and standard deviation of along-trace sea level anomaly cycle by cycle, we can identify the stability, distribution of anomalies and their spatial distribution characteristics of altimetry system. Especially in the verification/calibration stage, Jason-2 and Jason-3 operate in the same orbit (± 1 km difference) at an interval of 1 min and 20 s, When they measure the same sea area at approximately the same time, it can be considered that their sea surface height (SSH) correction terms are the same. Therefore, directly comparing their sea level anomaly (SLA) differences without using the correction can more objectively compare the performance of the two satellites.

2.2   Data Preprocessing

The percentage of data loss in Jason-3 and Jason-2 cycles is shown in Fig. 1. It shows that the percentage of data loss in Jason-3 cycles of 000 and 003 is 45.35%, 21.07%, and that in other cycles is less than 0.3%. Jason-3 data loss is mainly due to GPS software upload (181–233 orbits of 003 period), Gyro calibration (018 orbits of 003 period), instrument correction (235 orbits of 005 period), GPS shutdown (65 and 113 orbits of 006 period), OPS error (17 orbits of 008 period) and Poseidon-3B restart (144 and 148 orbital partial data of 008 period) [13].

In order to obtain reliable altimetry data, it is necessary to eliminate contaminated data points, this paper refers to the criteria of threshold screening and identification screening in Jason-3 Data Manual [14]. In the global ocean region, the results of identification screening criteria and Jason-3's 20th cycle GDR data are as follows: Table 1, threshold screening criteria and 20th cycle GDR data are as follows: Table 2. Surface_type is used to screen sea area. In order to mitigate the impact of rainfall pollution, this paper adds rainfall screening conditions. Ice, rainfall and threshold screening and statistics are based on the global ocean area after land data are removed. The editing criteria of Jason-2 satellite data adopt the same processing method.

The percentage of data edited by Jason-2 and Jason-3 in 23 cycles after removing land data is shown in Fig. 2. Statistical data show that the threshold screening ratio of Jason-3 and Jason-2 is consistent and stable with the change of the cycle, with the removal ratio of 1.5%. The identification screening is both high and on the rise, which is due to the data selection from February to October, during which the global ice cover and rainfall gradually increase. The abnormal identification screening ratio of the 001–003 periodic data of Jason-3 satellite is higher than that of the same period data of Jason-2, and the abnormal screening ratio is higher than that of the adjacent periodic data of Jason-3. According to grouping statistics, it is found that the anomaly is caused by the abnormal screening of the rain_flag of the 001–003 cycle of Jason-3 satellite. Figure 3 shows that the screening ratios of ice_flag for the first four cycles of Jason-3 and Jason-2 satellites are the same, but the rain_flag Jason-3 is significantly higher than Jason-2. After the 004 cycle, it is equivalent to Jason-2. Therefore, it can be concluded that the rain_flag of Jason-3 satellite in the period 001–003 is abnormally higher than the normal value, and returns to normal after the period 004.

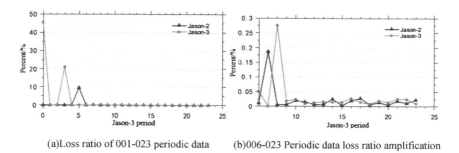

(a)Loss ratio of 001-023 periodic data   (b)006-023 Periodic data loss ratio amplification

Fig. 1. Percentage of data loss in Jason-3 and Jason-2 cycle-by-cycle ocean regions

Table 1. Jason-3 GDR data identification screening and 20th cycle screening statistics

Flag	Screening conditions	Editorial number/numbers	Editorial percentage/%
Surface_type	surface_type = 0	248543	29.48
Ice_flag	Ice_flag = 0	76635	12.89
Rain_flag	Rain_flag = 0	63577	10.69

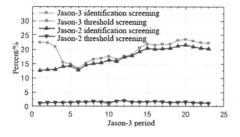

Fig. 2. Data edit percentage

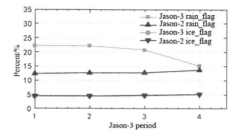

Fig. 3. Ice and rain flag edit percentage

Table 2. Jason-3 GDR data threshold screening and 20th cycle screening statistics

Parameters	Threshold		Editorial number/numbers	Editorial percentage/%
	Minimum	Maximum		
Ranging valid points/numbers	10	–	1698	0.29
RMS value of ranging/m	0	0.2	3649	0.61
Sea level height/m	−130	100	957	0.16
Dry tropospheric correction/m	−2.5	−1.9	0	0
Wet tropospheric correction/m	−0.5	−0.001	80	0.01
Ionospheric correction/m	−0.4	0.04	1978	0.33
Sea state deviation/m	−0.5	0	398	0.07
Tide/m	−5	5	0	0
Solid earth tides/m	−1	1	0	0
Polar tides/m	−0.15	0.15	0	0
Significant wave height/m	0	11	762	0.13
Backscattering coefficient/dB	7	30	438	0.07
Altimeter wind speed/m·s^{-1}	0	30	2081	0.35
Subsatellite deviation/deg^2	−0.2	0.64	502	0.08

(continued)

Table 2. (*continued*)

Parameters	Threshold		Editorial number/numbers	Editorial percentage/ %
	Minimum	Maximum		
RMS value of backscattering coefficient/dB	0	1	5221	0.88
Effective points of backscattering coefficient/numbers	10	–	2054	0.35
Total threshold editing	–	–	9658	1.62

3   Analysis of Geophysical Parameters

3.1   Significant Wave Height

Significant wave height (SWH) is one of the three basic parameters of altimeter measurement. It is used to study wind waves, surges and improve the accuracy of ocean prediction. In the statistical results of data analysis, the data rejection ratio of Jason-3 in the first three cycles is higher than that of Jason-2, this is due to differences in ground data processing strategies [14] and the use of rainfall markers.

Based on this situation, the daily mean values of significant wave heights were compared in the first three cycles without adding rain_flag screening. As shown in Fig. 4, the daily trend of Jason-3 and Jason-2 Ku band SWH shows strong consistency. The daily trend lines of Jason-3 and Jason-2 SWH almost coincide, but the value of Jason-3 is a little smaller, with a deviation of 0.3 cm. In conclusion, the daily average SWH values of Jason-3 and Jason-2 in Ku band have good consistency, it shows that Jason-3 inversion of SWH has high accuracy.

Figure 5 is the Ku-band SWH distribution curve of Jason-3 cycle 20 and Jason-2 cycle 20, The red solid line is Jason-3 and the blue dotted line is Jason-2. The figure shows that the distribution of SWH values is consistent. When SWH is less than 1.6 m, the percentage value of Jason-3 data is slightly higher. When SWH is greater than 1.6 m, the percentage value of Jason-2 data is slightly higher.

Figure 6 shows the global SWH distribution in Ku band of Jason-3 period 20 and Jason-2 period 300. It can be seen from the figure that the spatial distribution of SWH in both regions is highly consistent, especially in the region with large wave height in the Westerlies of the southern hemisphere.

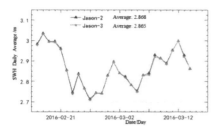

Fig. 4. Ku band SWH daily average

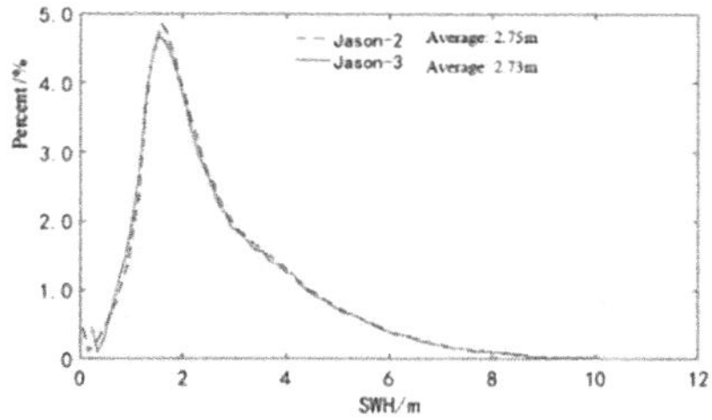

Fig. 5. Ku band SWH distribution curve

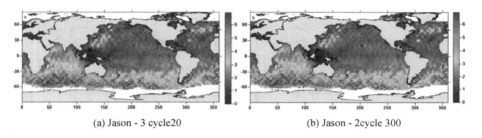

(a) Jason - 3 cycle20   (b) Jason - 2cycle 300

Fig. 6. Global significant wave height distribution in Jason 3 and Jason-2 Ku band

3.2   Backscattering Coefficient

Backscatter coefficient refers to radar scattering coefficient, which is one of the main observation parameters of radar altimeter. The daily mean values of backscattering coefficients in Jason-3 and Jason-2 Ku bands are shown in Fig. 7. The red and square markers are Jason-3, the blue and the upper triangle markers are Jason-2. The trend of daily variation of Jason-3 and Jason-2 is the same, but the average deviation of Jason-3 and Jason-2 is 0.2777 dB. Jason-3 is higher than Jason-2. The data distribution curve is shown in Fig. 8. The red solid line is Jason-3 and the blue dotted line is Jason-2. The distribution pattern of the two is consistent. When the backscattering coefficient is less than 13.5 dB, the percentage of Jason-2 data is higher, and when the backscattering coefficient is greater than 13.5 db, the percentage of Jason-3 data is higher.

Figure 9 shows the distribution of global backscatter coefficients in Ku band of Jason-3 and Jason-2 in the same period. It can be seen from the figure that they have good consistency in spatial distribution.

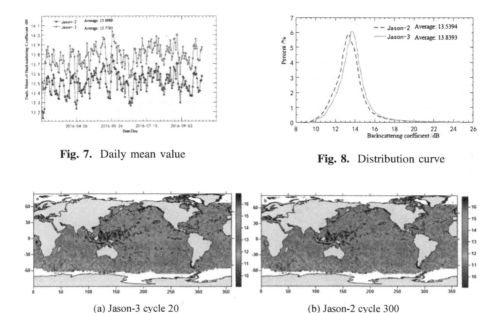

Fig. 7. Daily mean value

Fig. 8. Distribution curve

(a) Jason-3 cycle 20   (b) Jason-2 cycle 300

Fig. 9. Distribution of global backscatter coefficients in Jason 3 and Jason-2 Ku band

3.3   Ionospheric Delay

The radar pulse is also affected by the free electrons in the ionosphere. The higher the density of the free electrons, the lower the pulse propagation speed. The propagation effect of ionospheric free electrons on radar altimeter is called ionospheric delay. Because the ionospheric delay is inversely proportional to the square of the pulse frequency, the ionospheric delay is usually estimated by dual frequency observation. The daily average of ionospheric delays of Jason-3 and Jason-2 are shown in Fig. 10. The red and square markers are Jason-3, the blue and the upper triangle markers are Jason-2. The daily variation trend of Jason-3 and Jason-2 is the same, but the average difference between them is 0.5476 cm. Jason-3 is higher than Jason-2. The distribution curve is shown in Fig. 11. The red solid line is Jason-3 and the blue dotted line is Jason-2. The distribution pattern of the two lines is the same. When the ionospheric delay is less than −1.25 cm, the percentage of Jason-2 data is higher. When the ionospheric delay is greater than −1.25 cm, the percentage of Jason-3 data is higher.

Figure 12 shows the ionospheric delay distribution in the 20th and 300th periods of Jason-3. It can be seen from the figure that the spatial distribution of the two is basically the same, but there are some differences in the Westerlies of the southern hemi-sphere.

The above statistics show that the trend of dual frequency ionospheric delay of Jason-3 is consistent with that of Jason-2.

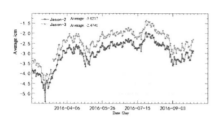

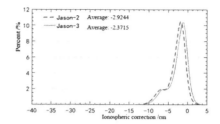

Fig. 10. Daily mean value

Fig. 11. Ionospheric correction distribution curve

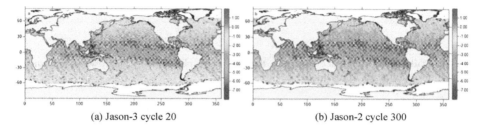

(a) Jason-3 cycle 20   (b) Jason-2 cycle 300

Fig. 12. Distribution of global ionospheric delay in Jason 3 and Jason-2 Ku band

3.4   Wet Tropospheric Delay

When the radar pulse passes through the atmosphere, it is affected by the attenuation of the atmosphere, resulting in the change of the propagation rate. Therefore, it is necessary to carry out the atmospheric delay correction for altimeter ranging. The atmospheric delay can be divided into dry delay and wet delay, in which the dry delay component is close to a constant, while the wet delay changes significantly. Wet tropospheric delays observed by Jason-3 and Jason-2 microwave radiometers are tested by using ECMWF model atmospheric wet delays. As shown in Fig. 13, the red and square markers are Jason-3, the blue and the upper triangle markers are Jason-2. The daily variation trend of the two satellites is the same, but the average difference between them is 0.142 cm. Jason-3 is higher than Jason-2. It can be seen from the figure that the values of Jason-3 in the first three cycles are abnormal, which is caused by the influence of abnormal rain threshold screening. The distribution curve of the difference between the observed values of microwave radiometer and ECMWF model of wet tropospheric delay is shown in Fig. 14, the red solid line is Jason-3 and the blue dotted line is Jason-2. The distribution pattern of the two is the same. When the value is less than 0.6 cm, the percentage of Jason-2 data is higher, and when the value is more than 0.6 cm, the percentage of Jason-3 data is higher.

Figure 15 shows the difference distribution of microwave radiometer observations and ECMWF model of wet tropospheric delay in Jason 3 period 20 and Jason-2 period 300. The spatial distribution of the two models is basically consistent in the global ocean.

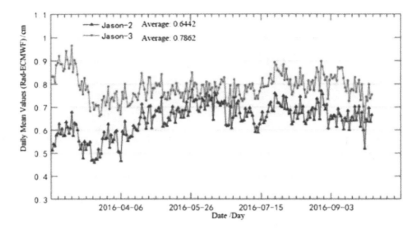

Fig. 13. Daily mean differences between the wet tropospheric delay microwave radiometer observations and the ECMWF model

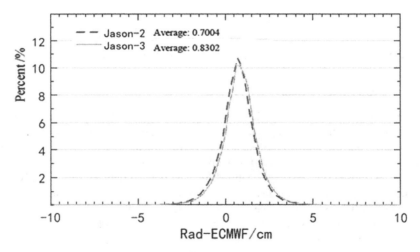

Fig. 14. Distribution curves of differential values between microwave radiometer observations with wet tropospheric delay and ECMWF models

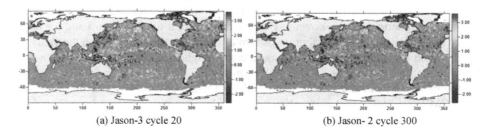

(a) Jason-3 cycle 20   (b) Jason- 2 cycle 300

Fig. 15. Distribution of global ionospheric delay in Jason 3 and Jason-2 Ku band

4 Analysis Discrepancy of Sea Level Height at Crossover Points

The discrepancy of sea level height at crossover points is the difference of sea level height at crossover points of different satellite trajectories, and is the main means to analyze the overall performance of altimetry system. The data quality and measurement accuracy of sea level height are evaluated by systematically analyzing the difference of sea level height at the crossover points. In order to reduce the difference of sea level height between crossover points caused by non-system performance, data editing constraints are set up when calculating crossover points, the maximum time interval of 10 days, sea level anomaly less than 20 cm, water depth less than 1000 m and geographic area selection of 50°N–50°S are used to weaken the impact of ocean change and reduce the impact of seasonal sea ice and geophysical correction model inaccuracy on data in high latitudes.

For the data that meet the above conditions, they are divided into six regions: 0–20° S, 20°S–40°S, 40°S–50°S, 0–20°N, 20°N–40°N and 40°N–50°N. The crossing points are obtained by using the method of regional quadratic polynomial fitting, and then the data are eliminated by using the 3δ standard to avoid the influence of residual suspicious values on the one-cycle overall data. Figure 16 shows the distribution of discrepancies before and after the elimination of Jason-3 19–21 cycle data. The total number of cross points removed in three cycles is 42, 34 and 25, accounting for 1.72%, 1.41% and 1.02% of the total data. After the elimination, the distribution of data is concentrated in (−15 cm, 15 cm). The data of Jason-2 and Jason-3 are edited and eliminated in one cycle respectively, and the average and standard deviation of discrepancy at crossover points are analyzed periodically.

4.1 Analysis Discrepancy of Sea Level Height at Self-crossover Points

After data screening and editing, the average value of sea level height discrepancy at Jason-2 and Jason-3 single periodic crossover points is shown in Fig. 17. The graph shows that the mean value of sea level height discrepancy at periodic crossover points fluctuates around 0 and most of the periods show negative values, the mean value of crossover points discrepancy of Jason-3 observation results is −0.1595 cm and the standard deviation is 0.7052 cm. The mean value of crossover points discrepancy of Jason-2 observation results in the same period is −0.079 cm and the standard deviation is 0.1618 cm. The results of data analysis report published by CNES are consistent with those obtained in this paper.

The standard deviation of sea level height discrepancy at Jason-1 and Jason-2 single-cycle crossover points is shown in Fig. 18, the two curves are similar. The mean of Jason-3 standard deviation is 4.989 cm and that of Jason-2 standard deviation is 4.979 cm. The results of data analysis report issued by CNES are consistent with those obtained in this paper, the two satellites have good performance and stability in the selection period. In the Jason-1/TOPEX verification stage, the mean standard deviation of sea level height discrepancies at intersections calculated from Jason-1 GDR-A data is 6.15 cm [15]. In the Jason-2/Jason-1 verification stage, the mean standard deviation

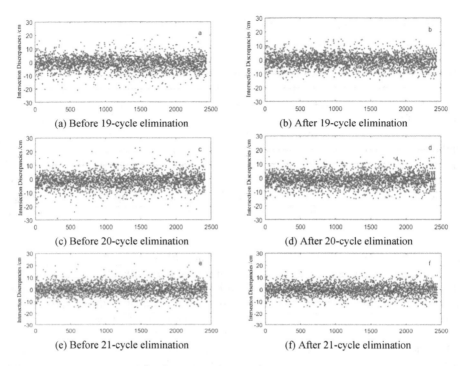

(a) Before 19-cycle elimination

(b) After 19-cycle elimination

(c) Before 20-cycle elimination

(d) After 20-cycle elimination

(e) Before 21-cycle elimination

(f) After 21-cycle elimination

Fig. 16. Jason-3 19-21 period 3δ standard data elimination before and after crossover points mismatch distribution

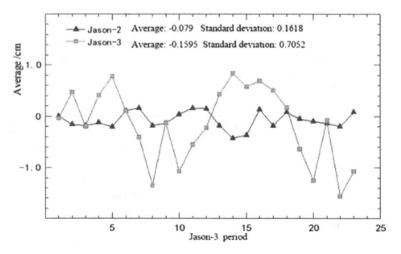

Fig. 17. Mean sea level discrepancy at a single periodic crossover points

of sea level height discrepancies at intersections calculated by Jason-2 GDRs 1–43 cycle is 5.07 cm [8]. In the Jason-3/Jason-2 verification stage, the mean value of the standard deviation of the sea level height discrepancy at the intersection calculated by Jason-3 GDR in the 001-023 period is 4.989 cm, which indicates that the accuracy of the altimetry series of satellites is continuously improved.

The existence of timescale deviation reduces the accuracy of altimetry satellite. Calculating timescale deviation is a means to evaluate the accuracy of altimetry, and appropriate correction method of timescale deviation can effectively improve the accuracy of altimetry. By choosing a suitable time window, it can be approximated that the crossover points discrepancy value is 0, so a simple least squares method is used to estimate the timescale deviation [16].

$$\tau = \frac{\sum_i P_i R_i}{\sum_i P_i^2} \tag{1}$$

τ. is time scale deviation, i stands for crossover points number, elevation difference of rail and lift orbit with Ri as crossover points i, Pi is the difference of the change rate of rail and lift orbit ranging.

Through self-crossover points data editing, using the above formulas, the mean deviations of the timescales for the 23 cycles selected by Jason-2 and Jason-3 are −0.03 ms and −0.035 ms, respectively. For the Jason-3 latitude (+40°C) area, the maximum range change rate is 15 m·s⁻¹, which will produce an error of 0.53 mm.

4.2 Analysis Discrepancy of Sea Level Height at Crossover Points

The data processing method of sea level height difference between Jason-3 and Jason-2 crossover points is the same as the above calculation method of sea level height difference from the crossover points. The spatial distribution of the crossover points meeting the requirements is shown in Fig. 19. It can be seen from the figure that the crossover points of Jason-3 and Jason-2 are evenly distributed in the global sea level, but they are more concentrated in 40S–50S and 40 N–50 N areas. Figure 20 shows the mean and standard deviation of sea level height difference at the crossover point of Jason-3 period 001-023 and same period Jason-2. The mean value and standard deviation of sea level height difference at the crossover points of 23 cycles have no abnormal fluctuation. The mean value of all cycles is 2.59 cm, and the average value of standard deviation is 5.742 cm. The results show that the difference of the crossover points between Jason-2 and Jason-3 is about (2.59 +5.742) cm in the deep-sea low ocean change area of 50°S–50°N.

The mean spatial distribution of sea level height discrepancies at the crossover points of Jason-3, 001-023 and Jason-2 in the same period is shown in Fig. 21. The results show that the sea level height measurements of the two altimeter satellites have good consistency, but the difference is relatively high (−3.5 cm) in the west of the Pacific Ocean, small in the east of the Pacific Ocean, and no significant deviation in other regions.

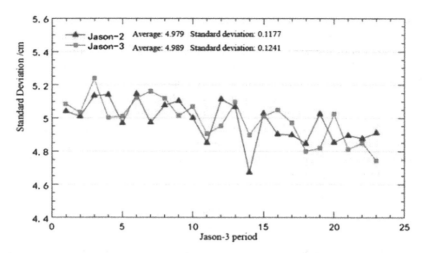

Fig. 18. Standard deviation of sea level height discrepancy at single periodic crossover points

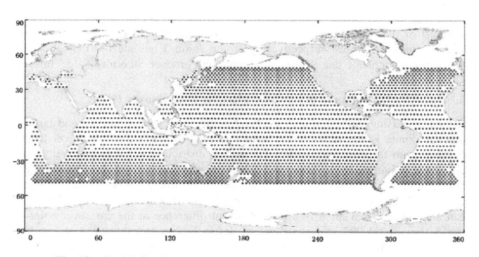

Fig. 19. Spatial distribution of the crossover points of Jason-3 and Jason-2

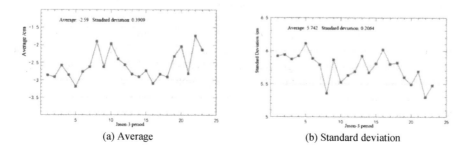

(a) Average   (b) Standard deviation

Fig. 20. Discrepancies of sea level height at the crossover points of Jason-3 and Jason-2

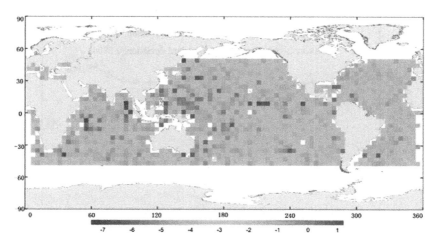

Fig. 21. Spatial distribution of mean sea level height discrepancies at the crossover points of Jason-3 and Jason-2

5   Analysis of Along-Trace Sea Level Anomaly

SLA is obtained by interpolating MSS of mean sea level model into SSH and subtracting it. Because the traces of Jason-2 and Jason-3 are not exactly the same, it is necessary to reduce the data to the same reference trace, and then calculate the difference between them to obtain the sea level anomaly along the trace.

5.1   Mean Value of Along-Trace Sea Level Anomaly

The periodic mean sea level anomaly of Jason-2 and Jason-3 are shown in Fig. 22, the characteristics of the two curves are similar and the time characteristics are stable, and there are no abnormal values. Figure 23 shows the periodic sea level anomaly difference between Jason-2 and Jason-3, red quadrate labeling with correction and blue upper triangle labeling without correction have similar characteristics of the two curves and stable time characteristics, with a deviation of 0.76 cm. The difference of global periodic sea level anomaly with correction is 2.9823 cm, and that without data correction is 2.2304 cm. The reason for the deviation is the difference of Jason-2 and Jason-3 correction, the difference of 2.2304 cm without correction reflects the difference of height measurement between Jason-2 and Jason-3, which is much smaller than the difference of 8.3 cm between Jason-1 and Jason-2. Figure 24 is the average sea level anomaly distribution of Jason-3 in the 20th cycle, from which the characteristics of ocean variations such as Antarctic circumpolar current can be well captured.

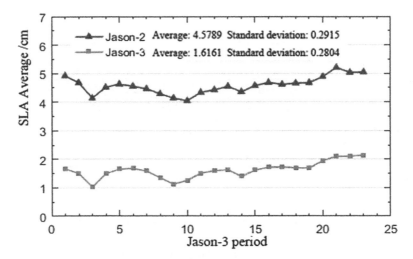

Fig. 22. Average sea level anomaly

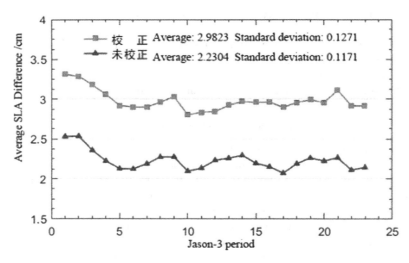

Fig. 23. Periodic sea level anomaly difference

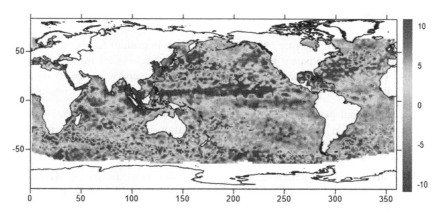

Fig. 24. Average sea level anomaly distribution in the 20th cycle of Jason-3

5.2   Standard Deviation of Along-Trace Sea Level Anomaly

Jason-2 and Jason-3 periodic along-trace sea level anomaly are shown in Fig. 25, the similarity between the two curves is very strong and the mean standard deviation is 10.64 cm. It can be concluded from the figure that the standard deviation of the along-trace sea level anomalies of Jason-2 and Jason-3 in the same period is approximately the same.

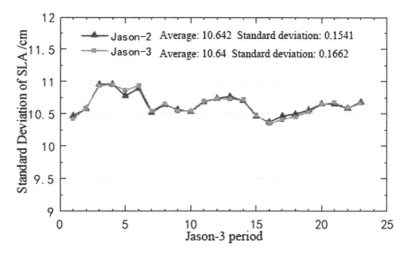

Fig. 25. Standard deviation of Jason-2 and Jason-3 cycle-by-cycle sea level anomaly

6   Conclusion

The quality evaluation of satellite altimeter data is very important for the application of data. By analyzing the statistical results of GDR data of 23 cycles in orbit in Jason-3 and Jason-2 verification and calibration phases, this paper carries out the evaluation of the data quality of Jason-3 satellite altimeter, and draws the following conclusions:

(1) The geophysical parameters of Jason-3 and Jason-2 satellite altimeter data have the same trend in the shape of value distribution curve and the trend of daily average change, but there are slight differences, the mean difference of significant wave height is 0.3 cm, the mean difference of backscattering coefficient is 0.2777 dB, mean difference of ionospheric delay correction is 0.5476 cm, the mean difference between radiometer observation data and ECMWF model values of wet tropospheric delay is 0.142 cm;

(2) The mean discrepancy of sea level height at self-crossover points of Jason-3 was −0.1595 cm, and the standard deviation was 4.989 cm. The mean discrepancy of sea level height at self-crossover points of Jason-2 in the same period was −0.079 cm, and the standard deviation was 4.979 cm. The mean and standard deviation of the crossover points difference between Jason-3 and Jason-2 were 2.59 cm and 5.742 cm respectively, it is shown that Jason-3 has reached the high precision of Jason-2 in the verification and calibration stage;

(3) The mean of along-trace sea level anomaly of Jason-3 is 1.6161 cm and the standard deviation is 10.64 cm. The mean of along-trace sea level anomaly of Jason-2 is 4.5789 cm, and the standard deviation is 10.642 cm. The mean difference of sea level anomaly was 2.9823 cm. Without any correction, the mean difference of sea level anomaly between Jason-3 and Jason-2 was 2.2304 cm.

From various research perspectives, the quality and performance of Jason-3 satellite altimeter have reached the level of sea level altitude data provided by Jason-2, which will have very strong scientific research and application value.

Acknowledgments. This work was supported in part by a grant from the Major Science and Technology Innovation Projects of Shandong Province (2019JZZY020103) and the National Science Foundation of China (41471330).

References

1. Jiang, X., Song, Q.: Satellite microwave measurements of the global oceans and future missions. Sci. Technol. Rev. (3), 105–111 (2010)
2. Jiang, X., Wang, X., Peng, H., et al.: The technology of precise orbit determination for HY-2A satellite. Strateg. Study CAE **15**(7), 19–24 (2013)
3. Chao, Y., Fu, L.L.: A comparison between the TOPEX/Poseidon data and a global ocean general circulation model during 1992–1993. J. Geophys. Res.: Oceans **100**(C12), 24965–24976 (1995)
4. Yang, L., Zhou, X., Xu, Q., et al.: Research status of satellite altimeter calibration. J. Remote Sens. **23**(3), 392–407 (2019)

5. Yang, G., Song, Q., Jiang, X., et al.: Preliminary assessment of the quality of HY-2A satellite sea surface height data. Acta Oceanol. Sinica **38**(11), 90–96 (2016)
6. Faugere, Y., Dorandeu, J., Lefevre, F., et al.: Envisat ocean altimetry performance assessment and cross-alibration. Sensors **6**(3), 100–130 (2006)
7. Ablain, M., Philipps, S., Picot, N., et al.: Jason-2 global statistical assessment and cross-alibration with Jason-1. Mar. Geodesy **33**(S1), 162–185 (2010)
8. Guo, J., Wang, J., Hu, Z., et al.: Temporal-spatial variations of sea level over China seas derived from altimeter data of TOPEX/Poseidon, Jason-1 and Jason-2 from 1993 to 2012. Chin. J. Geophys. **58**(9), 3103–3120 (2015)
9. Wen, J., Zhao, H., Jiang, Y., et al.: Research on the quality screening methods of satellite altimeter data—take Jason-3 data and Hongze Lake as an example. S. N. Water Transf. Water Sci. Technol. **16**(3), 194–200 (2018)
10. Xie, Y., Guo, J., Zhu, J., et al.: Simulation and analysis of significant wave height over Seas of Dongsha Island based on SWAN model. J. Shandong Univ. Sci. Technol. (Nat. Sci.) **35**(3), 17–24 (2016)
11. Peng, H., Lin, M., Mu, B., et al.: Global statistical evaluation and performance analysis of HY-2A satellite radar altimeter data. Acta Oceanol. Sinica **37**(7), 54–66 (2015)
12. Trano, P.Y.L., Stum, J., Dorandeu, J., et al.: Global statistical analysis of TOPEX and POSEIDON data. J. Geophys. Res. Oceans **99**(C12), 24619–24631 (1994)
13. Dumont, J.P., Rosmorduc, V., Carrere, L., et al.: Jason-3 Products Handbook. https://www.aviso.altimetry.fr/fileadmin/do-cuments/data/tools/hdbk_j3.pdf
14. Zhou, X., Miao, H., Wang, Y., et al.: Study on the determination of crossovers by piecewise fitting of satellite ground track. Acta Geod. Carto-gr. Sinica **41**(6), 811–815 (2012)
15. Dorandeu, J., Ablain, M., Faugere, Y., et al.: Jason-1 global statistical evaluation and performance assessment: calibration and cross-calibration results. Mar. Geodesy **27**(3–4), 345–372 (2004)
16. Prandi, P., Philipps, S., Pignot, V., et al.: SARAL/AltiKa global statistical assessment and cross-calibration with Jason-2. Mar. Geodesy **38**(sup1), 297–312 (2015)

Evaluating Spatial Details of Luojia-1 Night-Time Images Using Road Network Analysis

Huimin Xu[1] and Xi Li[2(✉)]

[1] Economics School, Wuhan Donghu University, Wuhan 430212, China
[2] State Key Laboratory of Information Engineering in Surveying, Mapping and Remote Sensing, Wuhan University, Wuhan 430079, China
li_rs@163.com

Abstract. Luojia-1 satellite is a new launched night-time light satellite providing 130 m resolution images. To evaluate the additional spatial details of Luojia-1 compared to VIIRS images, we employed road network analysis to compare the two kinds of images in Los Angeles, United States. In the road network analysis, we calculated the correlation coefficients between the distance to the primary road and the image radiance in 228 neighborhood areas, and we found that the average Spearman correlation coefficient is −0.3843 for Luojia-1 and −0.0974 for VIIRS, while those of the Pearson correlation coefficients are −0.3129 and −0.1370, respectively. In addition, we also calculated the Pearson correlation coefficients between the distance to the road intersections and the image radiance, and the average coefficient for Luojia-1 is −0.2967 and that of the VIIRS is −0.1100. The road network analysis suggests that the night-time light radiance in Luojia-1 is stronger correlated to the road network than VIIRS. All these findings show that Luojia-1 images provide richer information to reflect urban structures than VIIRS, indicating that Luojia-1 images have potential for studying urban socioeconomic parameters at a fine resolution.

Keywords: Night-time light · Luojia-1 · VIIRS · Spatial resolution · Road network

1 Introduction

In 1970s, the Defense Meteorological Satellite Program's Operational Linescan System (DMSP/OLS) had been proved effective to detect night-time light from oil fields [1] and human settlements [2]. Since then, the DMSP/OLS night-time light images had been widely applied in estimating energy consumption [3] and Gross Domestic Product (GDP) [4] as well as mapping population density [5] and urban extent [6, 7]. In 2011, the National Geophysical Data Center (NGDC) has released a time series dataset of DMSP/OLS image, and updated the dataset to 1992–2013. Since then, time series analysis of DMSP/OLS images have been employed for studies of urbanization [8–10], economic growth [11], humanitarian disasters [12] and light pollution [13]. Although NGDC provided time series DMSP/OLS between 1992 and 2013, it has stopped to update the recent products. Furthermore, the DMSP/OLS is lacking of

© Springer Nature Singapore Pte Ltd. 2020
Y. Xie et al. (Eds.): GSES 2019/GeoAI 2019, CCIS 1228, pp. 122–131, 2020.
https://doi.org/10.1007/978-981-15-6106-1_9

on-board radiometric calibration, and the spatial resolution is only 2.7 km. These problems have limited the usage of DMSP/OLS. In the end of 2011, the launch of Suomi National Polar-orbiting Partnership satellite's Visible Infrared Imaging Radiometer Suite (Suomi NPP/VIIRS) is a new milestone for the night-time light community. The day and night band (DNB) on VIIRS provides wide radiometric range and spatial resolution of 742 m, which are better than DMSP/OLS [14] and are able to provide rich information to model urban structures [15].

On June 2, 2018, at Jiuquan Satellite Launch Center, China's Wuhan University launched Luojia-1 01 satellite (abbreviated as Luojia-1 in the rest of this paper), which is a small satellite in 20 kg weight. One purpose of the satellite is to provide free global night-time light images at a finer resolution than VIIRS DNB. The spatial resolution of Luojia-1 is 130 m, with the local visiting time close to 9:30 PM. Luojia-1 satellite has a panchromatic band with wavelength between 460 nm–980 nm, which is wider than the VIIRS DNB wavelength (e.g. 500 nm–900 nm). The Luojia-1 data is now open to global users at Hubei Data and Application Network for High Resolution Earth Observation System (http://59.175.109.173:8888/app/login_en.html). A number of studies have been done to investigate the potential of Luojia-1 and its applications [16, 17], few analysis has been taken to explore the spatial information of Luojia-1 images. Although a previous study shows the Luojia-1 image has more spatial details than VIIRS based on the wavelet analysis [16], a more physical analysis is still needed to compare the two kinds of images. This paper aims to evaluate the spatial information provided by the Luojia-1 image, by comparing it with a VIIRS night-time light image, using road network analysis.

2 Materials and Methods

2.1 Data and Study Area

The study area is Los Angeles, the United States. The study data includes administrative boundaries, night-time light images and road network data. The administrative boundaries were downloaded from Los Angeles County Enterprise GIS (https://egis3. lacounty.gov/eGIS/) and shown in Fig. 1. The full dataset includes 272 neighborhood areas, and 265 neighborhood areas were extracted in this research since the rest of areas is not overlapped or only partly overlapped with the Luojia-1 image. However, some neighborhood areas in the 265 neighborhood areas will be discarded in part of this study because of sample size. The night-time light images include both Luojia-1 image and VIIRS images for Los Angeles. A Luojia-1 image (Fig. 2), which was captured on June 6, 2018, was downloaded from Hubei Data and Application Network for High Resolution Earth Observation System. The VIIRS image (Fig. 3) was extracted from monthly average composite for May 2018, downloaded from the National Geophysical Data Center (NGDC). The road network data for Los Angeles was downloaded from the Open Street Map (https://www.openstreetmap.org/).

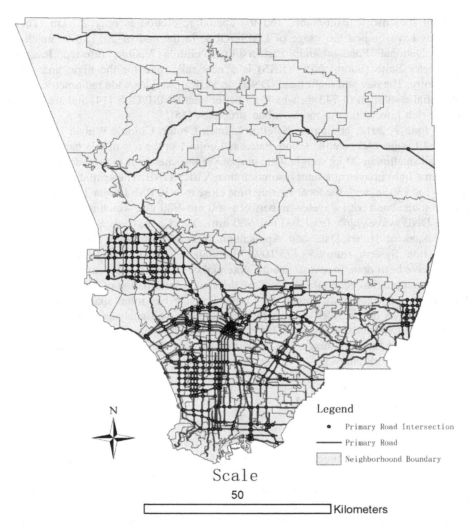

Fig. 1. The 265 neighborhood areas, primary roads and primary road intersections in Los Angeles, United States.

2.2 Preprocessing Data

All the geographic data were projected into coordinate system of WGS_1984_UTM_Zone_11N. For the Luojia-1 image and VIIRS image, they were resampled into 100 m resolution with bilinear spatial interpolation. For the Luojia-1 image, all the digital numbers (DNs) were transformed into radiance.

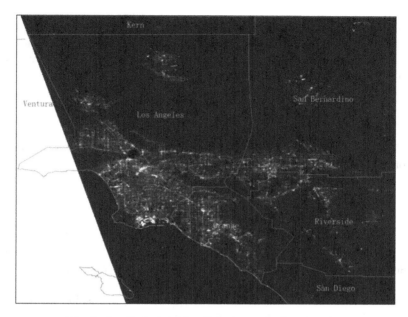

Fig. 2. Luojia-1 night-time light image for Los Angeles

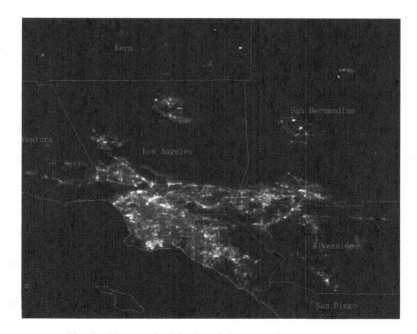

Fig. 3. The VIIRS night-time light image for Los Angeles

2.3 Statistical Analysis

To evaluate the relationship between different datasets, we used Pearson and Spearman's rank correlation analysis. The datasets include the remote sensing images and road network data. The Pearson coefficient is defined as

$$\rho_{XY} = \frac{\text{cov}(X, Y)}{\sigma_X \sigma_Y} \tag{1}$$

where X, Y are two observed vector data of two variable x, y, σ_X and σ_Y are standard deviation of X and Y, and cov(X,Y) is covariance between X and Y.

The Spearman's correlation coefficient is defined as:

$$r_{X,Y} = \frac{\text{cov}(rg_X, rg_Y)}{\sigma_{rg_X} \sigma_{rg_Y}} \tag{2}$$

where $r_{X,Y}$ is the Spearman's correlation coefficient between X and Y, rg_X, rg_Y are the rank number of X and Y, respectively, and σ_{rg_X} and σ_{rg_Y} are the standard variation of rg_X, rg_Y, respectively. While Pearson correlation reflects linear relationship between two variables, Spearman correlation reflects monotonic relationship, including both linear and nonlinear relationship, between two variables.

2.4 Road Network Analysis

It is known that the street light is a major contributor to the night-time light, and the street light has shown a linear feature in the night-time light images [18]. Normally, road intersections are always brighter than the surrounding areas since they have higher density traffic lights, so that the intersections can be viewed as point features in the night-time light images (Fig. 4). We will investigate how the night-time light images are responding to the linear features and point features, and the linear features are the primary road derived from the road network data, and the point features are the primary road intersections. Although there are several types of roads in Los Angeles, we only use the primary roads as our research material since this type of road has highest density lighting which can be effectively recorded by the satellite imagery.

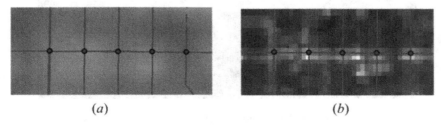

(a)   *(b)*

Fig. 4. A sample image patch overlapped with road network, showing the night-time light response to the primary road and primary road intersections: a) VIIRS and; b) Luojia-1

We employ buffer zone analysis to quantify how night-time light responds to the primary roads: 1) A buffer zone with distance of 500 m to the primary roads is generated. For each pixel in the buffer zone, we calculate its distance to the primary road and record its radiance value from night-time light images; 2) The whole buffer zone in the image is split into several subzones based on the number of the neighborhood areas; 3) For each buffer zone, there are several pixels, and the correlation between the distance and radiance is calculated. We record the correlation coefficient between road distance and Luojia-1 radiance, as well as the correlation coefficient between road distance and VIIRS radiance; 4) we compare the groups of correlation coefficients of both Spearman and Pearson for Luojia-1 and VIIRS.

In addition, we employ a similar way to quantify how night-time light responds to the primary road intersections: 1) Candidates of primary road intersections were extracted from the primary road network. We remove the fake intersections by hand since the topology of the road networks has some mistakes. For the intersections which fall into the same footprint of night-time light pixel (100 m resolution), we only remain one intersection for those intersections; 2) Buffers zones with distance of 500 m to the intersections are generated; 3) For each pixel in a buffer zone, we calculate its distance to the intersection and collect its radiance value from night-time light images of both Luojia-1 and VIIRS; 4) For each buffer zone, we record the correlation coefficient between distance to road intersection and Luojia-1 radiance, and the correlation coefficient between distance to intersection and VIIRS radiance; 5) We compare the groups of correlation coefficients of both Spearman and Pearson for Luojia-1 and VIIRS.

For both the primary roads and primary road intersections which are night-time light sources, the night-time light is regarded to be responding to the road network if the night-time light radiance is negatively correlated to the distance to the primary roads or primary road intersections, because the light sources have a distance-decay effect in spatial dimension.

3 Results

3.1 Visual Evaluation

When we compare the Luojia-1 and VIIRS images by visual evaluation (Figs. 2 and 3), the Luojia-1 image is much clearer than the VIIRS image. The linear features and point features, from the road networks, are apparent in the Luojia-1 image, while those in the VIIRS image are blurry although some of the features are still visible from the VIIRS image. For the linear features, only very bright roads in the Luojia-1 image can be seen in the VIIRS image, while the dim roads in the Luojia-1 image are invisible in the VIIRS image. However, this is only qualitative comparison, and the quantitative analysis will be carried out in rest of this section.

3.2 Night-Time Light Response to Primary Roads

We calculated the correlation coefficient between the image radiance and distance to the primary road in the buffer zone of each neighborhood area. For each neighborhood, we have a correlation coefficient value for each kind of image. Since some neighborhood areas have very little length of primary roads, they have very few pixels in the buffer zones. We excluded these neighborhood areas with area smaller than 600 m^2 (equal to 6 pixels at 100 m resolution), and finally we got 228 neighborhood areas for analysis. The Spearman correlation coefficients are shown in Fig. 5. For both the Luojia-1 image and VIIRS image, the image radiance is negatively correlated to the distance to the primary roads for most of the neighborhood areas ($p < 0.01$). In addition, the Luojia-1 image radiance is more negatively correlated to the distance to the primary roads than that of the VIIRS image, as shown in Fig. 5. These finding can be proved by quantitative evaluation in Tables 1 and 2.

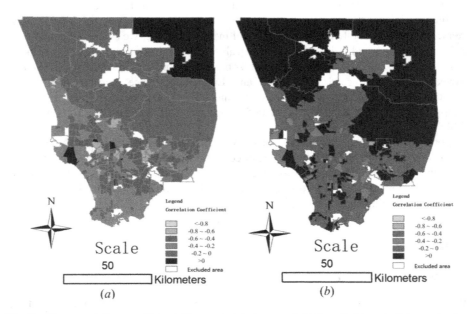

Fig. 5. The correlation coefficient (Spearman) between night-time light and distance to the primary road: (a) Luojia-1 image; (b) VIIRS image

For the Spearman correlation, 217 out of the 228 neighborhood areas have negative values of the correlation for Luojia-1 image ($p < 0.01$), while that of the VIIRS is only 161 ($p < 0.01$). As we change the threshold to smaller ones, from 0 to -0.5, we can see the contrast more obviously: 45 neighborhood areas have correlation coefficient smaller than -0.5 for Luojia-1 image, while that of the VIIRS image is only 8. The average correlation coefficient for Luojia-1 and VIIRS images are -0.3193 and -0.0915, respectively, also suggesting the correlation coefficient of Luojia-1 is much more negative than that of the VIIRS. In addition, we list the Pearson correlation results in

Table 1. Distribution of Spearman correlation coefficients (N = 228) between the radiance of night-time light images and the distance to the primary road

Sensor	Number of neighborhood areas	Number of neighborhood areas with correlation coefficient (c) less than a threshold				Mean value of the correlation coefficient
		$c < 0$	$c < -0.1$	$c < -0.3$	$c < -0.5$	
Luojia-1	228	220	212	152	70	−0.3843
VIIRS	228	157	117	27	10	−0.0974

Table 2. Distribution of Pearson correlation coefficients between the radiance of night-time light images and the distance to the primary road

Sensor	Number of neighborhood areas	Number of neighborhood areas with correlation coefficient (c) less than a threshold				Mean value of the correlation coefficient
		$c < 0$	$c < -0.1$	$c < -0.3$	$c < -0.5$	
Luojia-1	228	217	198	125	45	−0.3193
VIIRS	228	161	113	26	8	−0.0915

Table 2, which have the same pattern to the Table 1, and the only difference is the Pearson correlation is weaker than the Spearman correlation.

3.3   Night-Time Light Response to Primary Road Intersections

484 primary road intersections were selected. We calculated the correlation coefficients between image radiance and distance to the intersections in the buffer zones. Since some buffer zones are overlapped, it is complicated to show the results in a map, thus we only list the statistical results in Tables 3 and 4. From Table 3 of Spearman correlation, 431 out of 484 neighborhood areas have negative correlation coefficient for Luojia-1, and the number is 350 for VIIRS (p < 0.01). As we gradually move the threshold from 0 to −0.5, the contrast becomes larger: 111 out of 484 neighborhood areas have correlation coefficient less than −0.5 for Luojia-1, and the number is 34 for VIIRS. The average coefficient for Luojia-1 is −0.3129 and that of the VIIRS is −0.1370.

From Table 4 of Pearson correlation, 399 neighborhood areas have negative value of correlation coefficient for Luojia-1, while the number is 316 for VIIRS. The general pattern revealed by Table 4 is similar to that of Table 3. Both the Tables 3 and 4 indicate the image radiance of Luojia-1 is more strongly correlated to the road intersections than that of the VIIRS. In other words, the Luojia-1 image is more responsive to road intersection than the VIIRS image.

Table 3. Distribution of Spearman correlation coefficients between the radiance of night-time light images and the distance to the primary road intersections

Sensor	Number of neighborhood areas	Number of neighborhood areas with correlation coefficient (c) less than a threshold				Average correlation coefficient
		$c < 0$	$c < -0.1$	$c < -0.3$	$c < -0.5$	
Luojia-1	484	431	387	269	111	−0.3129
VIIRS	484	350	266	109	34	−0.1370

Table 4. Distribution of Pearson correlation coefficients between the radiance of night-time light images and the distance to the primary road intersections

Sensor	Number of neighborhood areas	Number of neighborhood areas with correlation coefficient (c) less than a threshold				Average correlation coefficient
		$c < 0$	$c < -0.1$	$c < -0.3$	$c < -0.5$	
Luojia-1	484	399	357	249	128	−0.2967
VIIRS	484	316	232	98	33	−0.1100

4 Conclusions

To evaluate the Luojia-1 image with some ground truth knowledge, the road network analysis was used. The basic assumption of the analysis is the night-time light along the road should have a distance-decay effect, so that the correlation analysis can help to quantify how much the night-time light responds to the night-time light sources. We found that the coefficient values from both the Spearman and Spearman correlation analysis for Luojia-1 image are not high, at an average value about 0.3, although much higher than those of the VIIRS image. This phenomenon occurs because there are many types of buildings such as shops, churches and gas stations along the roads, and these buildings also emitted light at night [18]. Therefore, the radiance does not well follow the distance-decay pattern, but is only correlated to the distance to the roads and road intersections. This study supports the previous study that Luojia-1 has more spatial details than VIIRS when the wavelet energy is used as the measurement. In summary, Luojia-1 images can provide more spatial details than the VIIRS images, suggesting that urban structures can be better analyzed with the Luojia-1 dataset.

Acknowledgements. This research was supported by Youth Fund of Wuhan Donghu University under grant no. 2018dhsk004 and Open Fund of State Laboratory of Information Engineering in Surveying, Mapping and Remote Sensing, Wuhan University, under grant no. 18T06.

References

1. Croft, T.A.: Burning waste gas in oil fields. Nature **245**(5425), 375–376 (1973)
2. Croft, T.A.: Nighttime images of the earth from space. Sci. Am. **239**, 86–98 (1978)
3. Welch, R.: Monitoring urban population and energy utilization patterns from satellite data. Remote Sens. Environ. **9**(1), 1–9 (1980)
4. Elvidge, C.D., Baugh, K.E., Kihn, E.A., Kroehl, H.W., Davis, E.R., Davis, C.W.: Relation between satellite observed visible-near infrared emissions, population, economic activity and electric power consumption. Int. J. Remote Sens. **18**(6), 1373–1379 (1997)
5. Sutton, P.: Modeling population density with night-time satellite imagery and GIS. Comput. Environ. Urban Syst. **21**(3), 227–244 (1997)
6. Henderson, M., Yeh, E.T., Gong, P., Elvidge, C., Baugh, K.: Validation of urban boundaries derived from global night-time satellite imagery. Int. J. Remote Sens. **24**(3), 595–609 (2003)
7. Small, C., Pozzi, F., Elvidge, C.D.: Spatial analysis of global urban extent from DMSP-OLS night lights. Remote Sens. Environ. **96**(3–4), 277–291 (2005)
8. Zhang, Q., Seto, K.: Mapping urbanization dynamics at regional and global scales using multi-temporal DMSP/OLS nighttime light data. Remote Sens. Environ. **115**(9), 2320–2329 (2011)
9. Huang, Q., et al.: Detecting the 20 year city-size dynamics in China with a rank clock approach and DMSP/OLS nighttime data. Landscape Urban Plan. **137**, 138–148 (2015)
10. Wei, Y., Liu, H., Song, W., Yu, B., Xiu, C.: Normalization of time series DMSP-OLS nighttime light images for urban growth analysis with Pseudo Invariant Features. Landscape Urban Plan. **128**, 1–13 (2014)
11. Henderson, J.V., Storeygard, A., Weil, D.N.: Measuring economic growth from outer space. Am. Econ. Rev. **102**(2), 994–1028 (2012)
12. Li, X., Li, D.: Can night-time light images play a role in evaluating the Syrian Crisis? Int. J. Remote Sens. **35**(18), 6648–6661 (2014)
13. Bennie, J., Davies, T.W., Duffy, J.P., Inger, R., Gaston, K.J.: Contrasting trends in light pollution across Europe based on satellite observed night time lights. Sci. Rep. **4** (2014). Article number: 3789
14. Elvidge, C.D., Baugh, K.E., Zhizhin, M., Hsu, F.C.: Why VIIRS data are superior to DMSP for mapping nighttime lights. Proc. Asia-Pacific Adv. Netw. **35**, 62–69 (2013)
15. Li, X., et al.: Anisotropic characteristic of artificial light at night – systematic investigation with VIIRS DNB multi-temporal observations. Remote Sens. Environ. **233**, 111357 (2019)
16. Li, X., Li, X., Li, D., He, X., Jendryke, M.: A preliminary investigation of Luojia-1 night-time light imagery. Remote Sens. Lett. **10**(6), 526–535 (2019)
17. Li, X., Zhao, L., Li, D., Xu, H.: Mapping urban extent using Luojia 1-01 nighttime light imagery. Sensors **18**(11), 3665 (2018)
18. Kuechly, H.U., et al.: Aerial survey and spatial analysis of sources of light pollution in Berlin, Germany. Remote Sens. Environ. **126**, 39–50 (2012)

Evaluation on Suitability of Oil and Gas Zoning in Nansha Sea

Min Ji[1], Xiaojia Liu[1(✉)], Yu Zhang[2], Fenzhen Su[2], and Yong Sun[1]

[1] College of Geomatics, Shandong University of Science and Technology,
Qingdao 266590, China
Liuxj199301@163.com
[2] State Key Laboratory of Resources and Environmental Information System,
Institute of Geographic Sciences and Natural Resources Research,
Chinese Academy of Sciences, Beijing 100101, China

Abstract. The Nansha Sea (NSS) is rich in oil and gas resources, which has caused great concern to neighboring countries and large-scale mining that brings a huge negative impact on the maintenance of China's marine right. In lack of marine functional zoning of the NSS and natural/social considerations in suitability evaluation of oil and gas zoning, this paper constructed a suitability evaluation model that involves resource conditions, topographic conditions, climate conditions, hydrological conditions, natural disasters, traffic conditions, social conditions. Based on the comprehensive score, the suitability was divided into four grades including the high suitable, the suitable, the less suitable and the unsuitable. Results showed that high suitable area are mainly distributed in Beikang basin; suitable area are mainly distributed in Lile basin, East Nanwei basin, Zhongjiannan basin and others in West Nanwei basin; less suitable area are mainly distributed in Nansha Trough basin and others are in West Nanwei basin. The paper would be helpful to the future exploration of oil and natural gas in the NSS.

Keywords: Nansha Sea · Oil and gas zoning · Suitability evaluation

1 Introduction

Natural resources are very important in securing China's socio-economic development. Our country's construction has a big demand for oil and gas resources. However, China's oil imports continue to grow and foreign dependence is as high as 67% [1]. The Nansha Sea (NSS) is rich in mineral resources especially oil and gas, which is a strong support and guarantee for China's sustainable development [2]. Nevertheless, as a cross-region of marine interests, the illegal exploitation has been carried out by countries around the NSS. It seriously affects China's marine energy security [3] and violates China's marine right. Therefore, there are demands for speeding up developments of resources mining in this area. In addition, oil and gas exploitation is not only affected by social factors but also natural factors. In consideration of the lack of strategic plans on the exploration, this paper made a comprehensive analysis of oil and

© Springer Nature Singapore Pte Ltd. 2020
Y. Xie et al. (Eds.): GSES 2019/GeoAI 2019, CCIS 1228, pp. 132–144, 2020.
https://doi.org/10.1007/978-981-15-6106-1_10

gas zoning in the field, so as to provide a reference for the exploration of oil and gas resources and the maintenance of marine right.

Regional suitability evaluation generally focus on specific sea/land use. It takes natural and social factors into consideration and establishes relevant index system. Then, by means of comprehensive evaluation, analyze degree of response in different spatial regions for specific uses. Currently, it is applied in tourism [4], fishery [5–7] and ecological protection [8–10]. Feng Youjian et al. [11] built index system for Ningbo marine tourism recreation to evaluate the suitability and thought that the marine functional zoning of Ningbo is basically based on the utilization of tourism resources. Yu Yonghai et al. [12] established the evaluation index system involving natural conditions, marine ecology, socioeconomics of coast and other factors and comprehensively applied analytic hierarchy process (AHP) and expert scoring method to evaluate the suitability of reclamation. According to the characteristics of seawater desalination, Huang Pengfei et al. [13] used fuzzy comprehensive evaluation model to evaluate the suitability of seawater desalination under current framework of marine functional zoning. Zhang Hexia et al. [3] comprehensively considered spatial location, present and potential resources, bidding and exploitation and used SAVEE method and GIS technology to quantitatively assess the strategic value of oil and gas resources exploitation in the central and southern South China Sea. Generally speaking, current suitability evaluation is based on natural zoning that takes spatial region as a whole and ignores the internal diversities, reducing the accuracy of evaluation results. Meanwhile, suitability evaluation indexes are closely related to the evaluation results. Furthermore, in lack of comprehensive considerations of natural/social influences, previous index system is difficult to analyse oil and gas zoning.

In order to evaluate the suitability of oil and gas zoning in the NSS, this paper comprehensively considered natural and social factors including resource conditions, topographic conditions, climatic conditions, hydrological conditions, natural disasters and traffic conditions to construct an evaluation model. At the same time, the suitability of oil and gas zoning was divided into four grades including the high suitable, the suitable, the less suitable and the unsuitable. Finally, through the grid, based on comprehensive score, every grids had own suitability that represents different degree of response of oil and gas zoning in the NSS.

2 Research Area

The NSS is rich in natural resources such as marine energy, oil and gas, islands and reefs, tourism and has unique tropical and subtropical ecosystem. On the remote sensing image, islands and reefs are dotted around the sea. There are more than 200 islands and reefs, which has became valuable marine resource in China. The NSS are pregnant with Wan'an basin, Zengmu basin, West Nanwei basin, Beikang basin, Nansha Trough basin, Brunei-Sabah basin, West Palawan basin, North Palawan basin, Li Le basin and so on [14]. There are abundant oil and gas resources in these sedimentary basins and the South China Sea is known as the "Second Persian Gulf". At present, the proven oil and gas reserves in the NSS are about 336×10^8t and account for 74.8% of the total in the South China Sea. The spatial distribution is shown in Fig. 1.

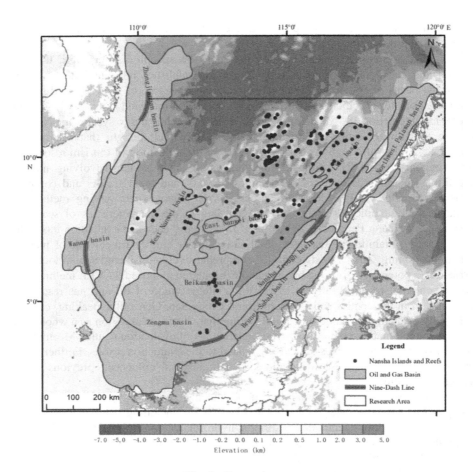

Fig. 1. Research area

3   Evaluation on Suitability of Oil and Gas Zoning

In order to evaluate the suitability of oil and gas zoning in the NSS, we built a suitability evaluation model, which is mainly made up of four aspects and involves natural and social factors. To start with, through literature review and expert consultation, built relevant index system. Then, collected, processed and managed data. And then, by means of AHP, determined weight of every indexes. At last, finished the comprehensive evaluation with weighted sum method. After we got the first edition map, result validation would be carried out. The model is shown in Fig. 2.

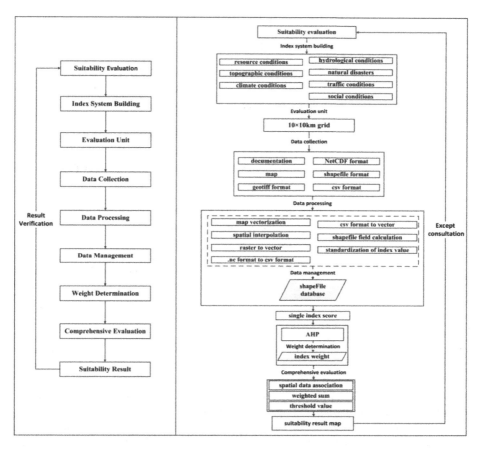

Fig. 2. Model of suitability evaluation of oil and gas zoning

3.1   Index System

Building a sound index system is the basis of objective evaluation and selected indexes should be comprehensive, representative, operable and independent [16]. By summarizing relevant academic literature and consulting expert, we built an index system of oil and gas zoning in the NSS, which includes three levels and involves natural and social considerations. The first level was also our evaluation goal that is evaluation on suitability of oil and gas zoning. There are natural and social aspects in the second level. Among natural factors, resource conditions, topographic conditions, climate conditions, hydrological conditions and natural disasters were selected. Among social factors, traffic conditions and social conditions were selected. Combined with accessibility of data, we refined the second level to set the third level. The index system is shown in Table 1.

Table 1. Index system of oil and gas zoning

First level	Second level	Third level
Evaluation on suitability of oil and gas zoning **B**	Resource conditions **B₁**	oil and gas basin area $\mathbf{B_{11}}$
		oil and gas reserve $\mathbf{B_{12}}$
		degree of exploitation $\mathbf{B_{13}}$
	Topographic conditions **B₂**	average water depth $\mathbf{B_{21}}$
	Climate conditions **B₃**	precipitation grade $\mathbf{B_{31}}$
		wind grade $\mathbf{B_{32}}$
	Hydrological conditions **B₄**	average wave height $\mathbf{B_{41}}$
	Natural disasters **B₅**	frequency of disaster $\mathbf{B_{51}}$
	Traffic conditions **B₆**	distance from islands and reefs $\mathbf{B_{61}}$
	Social conditions **B₇**	impact of enemy-occupied islands and reefs $\mathbf{B_{71}}$

3.2 Data Sources and Processing Methods

Index system normally consists of positive and negative indexes. The former has a positive correlation with evaluation object, which shows that when index value is higher, evaluation object is better. The latter has a negative correlation with evaluation object, which shows that when index value is higher, evaluation object is worse. In addition, different indexes have different dimensions and units which affects accuracy of evaluation result. In order to eliminate these adverse effects and make data normal, each index should be treated in a standardization manner [17]. In our index system, positive indexes included oil and gas basin area, oil and gas reserve and distance from islands and reefs. Aim at positive and negative indexes, there are different standardization formulas.

The standardization formula for positive index is [15]:

$$a_{ij}^{(k)} = 1 + 99 \frac{\left(b_{ij}^{(k)} - \mathrm{min}b_j\right)}{\left(\mathrm{max}b_{ij} - \mathrm{min}b_{ij}\right)} \tag{1}$$

In the formula, $b_{ij}^{(k)}$ is the k-th value of index B_{ij}; $a_{ij}^{(k)}$ is standardization value of $b_{ij}^{(k)}$; $\mathrm{max}b_{ij}$ is the maximum value in index B_{ij} and $\mathrm{min}b_{ij}$ is the minimum.

The standardization formula for negative index is [15]:

$$a_{ij}^{(k)} = 1 + 99 \frac{(\mathrm{max}b_{ij} - b_{ij}^{(k)})}{(\mathrm{max}b_{ij} - \mathrm{min}b_{ij})} \tag{2}$$

In the formula, $b_{ij}^{(k)}$ is the k-th value of index B_{ij}; $a_{ij}^{(k)}$ is standardization value of $b_{ij}^{(k)}$; $\max b_{ij}$ is the maximum value in index B_{ij} and $\min b_{ij}$ is the minimum value.

According to the distribution of oil and gas resources, the spatial scale of research area and expert opinions, evaluation unit in this paper is defined as 10×10 km. Combined with standardization formulas and the size of evaluation unit, we used some computing methods to process collected data and made each evaluation unit has own standardization value of each index for comprehensive evaluation. The data and processing of every indexes are as follows.

Oil and gas basin area: Through existing paper document that contains spatial distribution of oil and gas basins in the NSS and toolbox of ArcGIS for Desktop (version 10.4) software, vector data of the basins were obtained. If evaluation unit is completely covered by basin, the index value was given 100, otherwise, corresponding proportion.

Oil and gas reserve: The available oil/gas reserves are deemed as difference value between oil/gas reserves in the traditional territory and the reserves mined and tendered by surrounding countries [15] (Table 2). Because of different units of measurement, we respectively standardized oil and gas reserves and assigned the mean of standardization value to each evaluation unit. Oil/gas total reserves in each basin calculated by formula three. Oil/gas reserves in each evaluation unit calculated by formula four. The standardization value of the index of oil and gas reserve in each evaluation unit calculated by formula five.

$$m_i = S_i - p_i - q_i \qquad (3)$$

Table 2. Oil and gas reserves in each basin [15]

Basin name	Oil and gas reserves in traditional territory		Oil and gas reserves were mined		Oil or gas reserves were tendered	
	Oil/10^8t	Gas/10^{12} m^3	Oil/10^8t	Gas/10^{12} m^3	Oil/10^8t	Gas/10^{12} m^3
Wanan basin	16.3	0.96	2.2	0.13	14.1	0.83
Zhongjiannan basin	19.06	0.72	0	0	3.48	0.13
West Nanwei basin	8.43	0.3	0	0	2.64	0.09
East Nanwei basin	0.69	0.03	0	0	0	0
Lile basin	5.24	0.34	0	0	1.26	0.08
Northwest Palawan basin	4.42	0.41	0.02	0	2.95	0.28
Nansha Trough basin	1.53	0.09	0	0	0.68	0.04
Beikang basin	13.81	0.98	0	0	3.77	0.27
Zengmu basin	33.51	4.31	11.56	1.49	21.18	2.72
Brunei-Sabah basin	21.63	0.4	4.1	0.08	17.53	0.32

In the formula, m_i is the available reserves in i-th basin; S_i is the reserves of traditional territory in i-th basin; p_i and q_i respectively represent the reserves mined and tendered in i-th basin [15].

$$c_{ij} = \frac{d_{ij}}{D_i} \times m_i \qquad (4)$$

In the formula, c_{ij} is the reserves of the j-th evaluation unit of i-th basin; d_{ij} is area of the j-th evaluation unit in the i-th basin; D_i is the area of the i-th basin; m_i is same to formula three.

$$a_{ij} = \frac{(a_{ij}^1 + a_{ij}^2)}{2} \qquad (5)$$

In the formula, a_{ij} is standardization value of the index; a_{ij}^1 is standardization value of oil; a_{ij}^2 is standardization value of natural gas.

Degree of exploitation: We used the number of oil and gas platforms in each evaluation unit to represent degree of exploitation. Compared with interpretation result of GF-2 data, the spatial distribution of oil and gas platforms are derived from NPP-VIIRS night light data.

Average water depth: We downloaded the water depth data of geoTiff format from ETOP1 website located at https://www.ngdc.noaa.gov/mgg/global/ and its spatial resolution is approximately 2×2 km. Because of our evaluation unit size is 10×10 km grid, we used these methods that raster to point and spatial join to make each evaluation unit include only water depth value. Finally, through standardization formula, assigned standardization value to each evaluation unit.

Hydrological and climate elements: In this paper, we used precipitation, wind and height of wave to describe characteristics of climate and hydrological in the NSS. Relevant data provided by ECMWF website that is https://www.ecmwf.int/. These data has the time sequence of every day from 1 January, 2018 to 31 December, 2018 and the spatial resolution of $0.25 \times 0.25'$. First of all, through spatial interpolation, raster to point and spatial join, built hydrological and climate shapfile data. What's more, we calculated annual total precipitation, annual mean wind speed and annual mean wave height. Finally, by means of formula one or two, made these calculated values as input values of standardization formula to obtain standardization value of hydrological and climate indexes in each evaluation unit.

Frequency of disaster: The main disaster considered in this paper is typhoon. Through the website located at http://agora.ex.nii.ac.jp/digital-typhoon/, we downloaded table data of typhoon from 2008 to 2018, which includes time of occurrence, location, wind speed, radius of influence and so on. Then, based on the location and radius of influence, buffer was constructed to describe influence area in the NSS. Finally, through overlay analysis, we counted frequency in each evaluation unit and standardized the frequency to compute the standardization value of the index over 11 years.

Distance from islands and reefs: Zhang Hexia et al. [15] used gravity model to evaluate strategic value (Table 3) of islands and reefs in the NSS, which is based on natural value, economic value, shipping value, political and military value. Because of the NSS is far away from the mainland and considering safety, rest and replenishment during the exploitation of oil and gas, we selected distance from islands and reefs as the index in third level under traffic conditions. Currently, Taiping island, Nanxun reef, Yongshu reef, Huayang reef, Dongmen reef, Chigua reef, Zhubi reef, Meiji reef have been controlled by our country (Fig. 3). Meanwhile, in view of different strategic values of each island and reef, we calculated the mean of ratio between strategic value and the distance from midpoint of each evaluation unit to each island and reef (formula six). Then, through formula one, formula two and formula six, assigned standardization value to each evaluation unit.

$$d_i = \frac{\sum_{j=1}^{8} \left(\frac{v_j}{d_{ij}}\right)}{8} \tag{6}$$

In the formula, d_i is the mean, v_j is strategic value of j-th island and reef; d_{ij} is distance from midpoint of i-th evaluation unit to j-th island and reef.

Impact of the enemy-occupied islands and reefs: In order to successfully exploit oil and gas resources in the NSS, the impact of neighboring countries and their illegal occupation of islands and reefs (Fig. 3) need to be considered. We investigated the GDP of each country (https://data.worldbank.org/) to reflect each country's economic strength (Table 4). Then, we calculated the product between each country's GDP and the distance from midpoint of evaluation unit to the islands and reefs illegally occupied by each surrounding country and averaged the result (formula seven). Finally, through standardization formula, assigned the index value to each evaluation unit.

$$c_i = \frac{\sum_{j=1}^{44} \left(d_{ij}^k \times e_k\right)}{44} \tag{7}$$

In the formula, c_i is the averaged result in the i-th evaluation unit; d_{ij}^k represents the distance from the i-th evaluation unit to the j-th island and reef occupied by country k; e_k represents GDP of country k that represents Vietnam, Philippines, Malaysia and Brunei.

Table 3. Strategic value of islands and reefs controlled by China [15]

Islands and reefs	Strategic value
Chigua reef	2823.52
Dongmen reef	3687.13
Huayang reef	808.09
Meiji reef	3362.09
Nanxun reef	1882.42
Taiping island	7737.84
Yongshu reef	2672.24
Zhubi reef	2423.13

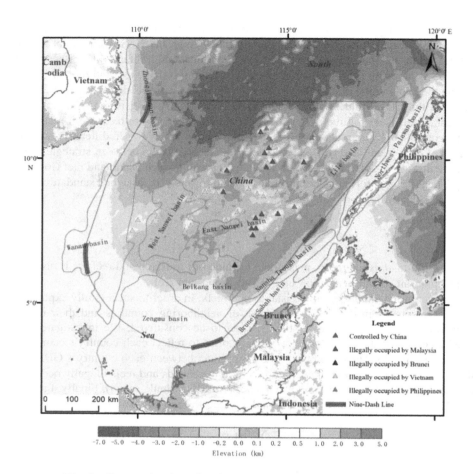

Fig. 3. Current situation of main Islands and Reefs in the Nansha Sea

Table 4. GDP of Malaysia, Philippines, Vietnam and Brunei from 2015 to 2018

Country	GDP/billion dollars				
	2015 year	2016 year	2017 year	2018 year	Mean
Malaysia	296.636	296.753	314.707	354.348	315.611
Philippines	292.774	304.898	313.62	330.91	310.5505
Vietnam	193.241	205.276	223.78	244.948	216.81125
Brunei	12.93	11.401	12.128	13.567	12.5065

3.3   Index Weight Determination

In order to reflect different importances of each evaluation index to evaluation goal, different weight coefficients must be assigned to each index after index system was built. It is worth mentioning that reasonable weight is very important on evaluation or decision-making [18]. Weight coefficient can be divided into subjective weight coefficient and objective weight coefficient. The former refers to that the weight is subjectively determined according to expert experience such as Delphi, AHP and expert scoring. The latter refers to that the weight is objectively determined according to actual data such as the entropy, standard deviation, and CRITIC [19]. We used AHP to determine the weight of each index (Table 5). Firstly, the judgment matrix of pairwise comparison was constructed by using the 1-9 scale according to expert experience. Secondly, calculated the maximum eigenvalue and corresponding eigenvector. Let p as the number of index level and q as the number of index in the i-th level (i = 1, 2, 3, ..., p). The calculation formula of the i-th index weight is:

$$w_j = \frac{w_j'}{\sum_{i=1}^{q} w_j'} \tag{8}$$

In the formula, w_j is the j-th index weight in i-th level; w_j' is the j-th eigenvector corresponding to the maximum eigenvalue of the judgment matrix.

Table 5. Weight of index system

First level	Second level	Weight	Third level	Weight
Evaluation on suitability of oil and gas zoning	Resource conditions	0.5204	oil and gas basin area	0.1546
			oil and gas reserves	0.2808
			degree of exploitation	0.0850
	Topographic conditions	0.0619	average water depth	0.0619
	Climate conditions	0.0277	precipitation grade	0.0092
			wind grade	0.0185
	Hydrological conditions	0.0399	average wave height	0.0399
	Natural disasters	0.0619	frequency of disaster	0.0619
	Traffic conditions	0.1307	distance from islands and reefs	0.1307
	Social conditions	0.1575	impact of enemy-occupied islands and reefs	0.1575

Thirdly, the consistency test was performed on the constructed judgment matrix. If the calculated value of consistency test is less than 0.10, the test is passed. Otherwise, the judgment matrix should be reconstructed. Finally calculate the total weight:

$$ws_j = w_j \times w_i \tag{9}$$

In the formula, ws_j is the weight of the j-th index to the root target; w_j is the j-th index weight in i-th level; w_i is the weight of the i-th index level to the root target.

3.4 Comprehensive Evaluation Result

We used weighted sum method to implement comprehensive evaluation and the weighted sum was computed by standardization index value and determinated weight (formula ten). Through expert advice, combined with the distribution of oil and gas

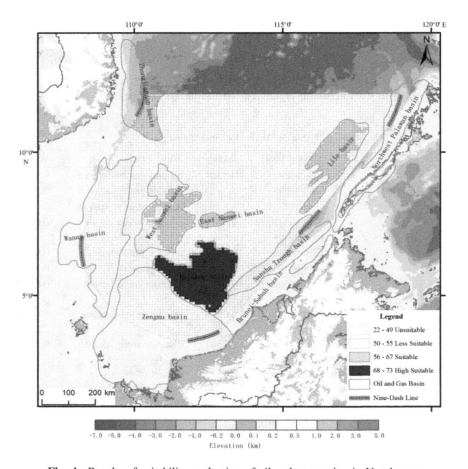

Fig. 4. Results of suitability evaluation of oil and gas zoning in Nansha sea

basins and oil and gas reserves, set the score of 22–49 as unsuitable area, 50–55 as less suitable area, 56–67 as suitable area and 68–73 as high suitable area (Fig. 4).

$$s_i = w_j \times v_{ij} \tag{10}$$

In the formula, s_i is the comprehensive score of i-th evaluation unit; w_j is the weight of the j-th index; v_{ij} is the score of the j-th index in i-th evaluation unit.

4 Conclusion and Discussion

The area of the Nansha Sea is about 734624.03 km^2 and we divided it into unsuitable area, less suitable area, suitable area and high suitable area with the goal of oil and gas zoning. Among them, The area of the unsuitable is 575,775.31 km^2 that accounts for 78.38% and these evaluation units are mainly distributed in the outside of oil and gas basin and these basins that Wanan, Zengmu, Brunei-Sabah and Northwest Palawan; the less suitable is 43969.78 km^2 that accounts for 5.98% and are mainly distributed in Nansha Trough basin and others in the western part of West Nanwei basin; the suitable is 69778.94 km^2 that accounts for 9.50% and are mainly distributed in Lile basin, East Nanwei Basin, Zhongjiannan basin and others in the eastern part of West Nanwei basin; the high suitable is 45100.00 km^2 that accounts for 6.14% and are mainly distributed in Beikang basin.

In view of the existing suitability evaluation treats the spatial area as a whole and its spatial scale limits the accuracy of the evaluation results. In addition, current index system lacks comprehensive consideration of natural and social influences for oil and gas zoning, which is difficult to obtain reasonable suitability evaluation results. This paper bulit a model of suitability evaluation for oil and gas zoning in the NSS and divided the NSS into grid units with the same size that 10 × 10 km. Meanwhile, we comprehensively utilized natural and social factors to optimize the suitability evaluation and obtained the spatial distribution of the suitable with four evaluation grades that the high suitable, the suitable, the less suitable and the unsuitable, based on the comprehensive score. Through the suitability evaluation modeling, the paper would be helpful to the future exploration of oil and natural gas in the NSS.

In order to adapt the oil and gas zoning to the national marine functional zoning, the comprehensive evaluation of suitability of oil and gas zoning in the NSS should be combined with the existing national marine functional zoning criterions so as to enhance the effectiveness and scientific value of the evaluation model in this paper.

Acknowledgments. This work was supported in part by a grant from the Strategic Priority Research Program of the Chinese Academy of Sciences (XDA13010401) and the National Science Foundation of China (41471330).

References

1. Qiu, Y.: Distribution Characteristics and Strategic Analysis of World Oil and Gas Resources, pp. 1–135. Chinese Academy of Geological Sciences, Beijing (2019)
2. Zhang, B., Chen, C.: Study on the characteristics and exploitation strategy of South China Sea oil and gas resources. Spec. Oil Gas Reservoirs (06), 5–8+108 (2004)
3. Zhang, H., Liu, Y., Li, M., et al.: Strategic value assessment of oil and gas exploitation in the Central and Southern South China Sea. Resour. Sci. **35**(11), 2142–2150 (2013)
4. Sun, J., Yang, J., Xi, J.C.: The comprehensive evaluation of suitability of marine tourism base in China. Resour. Sci. **38**(12), 2244–2255 (2016)
5. Wang, P., Lin, X., Fang, H., et al.: On present situation and comprehensive evaluation of mariculture environment suitability in Jinzhou City. Ocean Dev. Manag. **33**(5), 41–45 (2016)
6. Tang, L., Wang, Q., Liu, H., et al.: Habitat suitability of Stichopus japonicas, Scapharca broughtonii and Mytilus edulis in the shallow waters of Xiaoheishan Island. Acta Ecol. Sinica **37**(2), 668–682 (2017)
7. Zhang, Z., Zhou, J., Song, J., et al.: Habitat suitability index model of the sea cucumber Apostichopus japonicus (Selenka): a case study of Shandong Peninsula, China. Mar. Pollut. Bull. **122**(1–2), 65–76 (2017)
8. Welch, H., Pressey, R.L., Reside, A.E.: Using temporally explicit habitat suitability models to assess threats to mobile species and evaluate the effectiveness of marine protected areas. J. Nat. Conserv. **41**, 106–115 (2018)
9. Olsen, P.M., Kolden, C.A., Gadamus, L.: Developing theoretical marine habitat suitability models from remotely-sensed data and traditional ecological knowledge. Remote Sens. **7**(9), 11863–11886 (2015)
10. Rowden, A.A., Anderson, O.F., Georgina, S.E., et al.: High-resolution habitat suitability models for the conservation and management of vulnerable marine ecosystems on the Louisville Seamount Chain, South Pacific Ocean. Front. Mar. Sci. **4**, 335 (2017)
11. Feng, Y., Zheng, Z.: Suitability evaluation of ningbo marine tourism recreation and improvement of functional zoning. Ocean Dev. Manag. **34**(08), 92–96 (2017)
12. Yu, Y., Wang, Y., Zhang, Y., et al.: Research of evaluation methods for reclamation suitability. Mar. Sci. Bull. **30**(01), 81–87 (2011)
13. Huang, P., Yang, Z., Wang, R., et al.: Suitability assessment of sea utilization for seawater desalination indexes system and model construction. Environ. Sci. Manag. **40**(01), 554–558 (2015)
14. Liu, Z.: Distribution of sedmentary basins and petroleum potential in Southern South China Sea. Geotecton. Metallog. (03), 410–417 (2005)
15. Zhang, H.: Research on Strategic Value Evaluation of the Islands and Reefs in the Nansha Islands, pp. 1–116. Nanjing University, Nanjing (2014)
16. Li, X., Xing, Q., Wang, X., et al.: Construct principles and assessment method of index system. Math. Prac. Theory **42**(20), 69–74 (2012)
17. Ye, Z.: Selection of index forwarding and dimensionless methods in multi-index comprehensive evaluation. Zhejiang Stat. (04), 25–26 (2003)
18. Du, D., Pang, Q.: Modern Comprehensive Evaluation Method and Case Selection, pp. 1–207. Tsinghua University Press, Beijing (2005)
19. Wang, H., Chen, L., Chen, W., et al.: Multi-index comprehensive evaluation method and selection of weight coefficient. J. Guangdong Coll. Pharm. (05), 583–589 (2007)

Application of Sentinel-1A Data in Offshore Wind Field Retrieval Within Guangdong Province

Pinghao Wu[1,2,4], Kaiwen Zhong[2(✉)], Hongda Hu[2], Yi Zhao[1,2,4], Jianhui Xu[2,3], and Yunpeng Wang[1]

[1] Guangzhou Institute of Geochemistry, Chinese Academy of Sciences, Guangzhou 510640, China
[2] Key Lab of Guangdong for Utilization of Remote Sensing and Geographical Information System, Guangdong Open Laboratory of Geospatial Information Technology and Application, Guangzhou Institute of Geography, Guangzhou 510070, China
zkw@gdas.ac.cn
[3] Southern Marine Science and Engineering Guangdong Laboratory (Guangzhou), Guangzhou 511458, China
[4] University of Chinese Academy of Sciences, Beijing 100049, China

Abstract. Offshore wind is an important source of information for monitoring the interaction between fishery and marine water vapor environment. In this paper, the CMOD5 function was used to invert the wind field in the coastal waters of Guangdong province in March, May, July and December 2017. Compared with the measured wind speed, the reasonable result of the inversion wind speed is generally higher, with the mean absolute deviation of 1.95 m/s, the root mean square error of 2.7 m/s, and the correlation coefficient of 0.8. Due to the measured data not exactly matching the satellite transit time, the error in December is large. The inversion results of Sentinel-1A images are consistent with the measured data generally, which verifies that the COMD5 function is applicable to the inversion of offshore high-resolution marine wind field in Guangdong province, and provides a possibility for the next step of estimating the wind energy resources and reserves in Guangdong province.

Keywords: Offshore wind · Sentinel-1A · Wind stripe · Synthetic aperture radar · Remote sensing inversion

1 Introduction

Traditional methods of measuring wind are limited by cost and it is difficult to achieve continuous monitoring on a large scale. The low spatial resolution of scatterometer is easy to be affected by land echo in the coastal area, so it's unable to measure areas within a few tens of kilometers near the coast [1, 2]. SAR has the advantages of high spatial resolution, high precision, low cost, and sufficient data resources, and has become an important technical means for inverting the wind field.

© Springer Nature Singapore Pte Ltd. 2020
Y. Xie et al. (Eds.): GSES 2019/GeoAI 2019, CCIS 1228, pp. 145–150, 2020.
https://doi.org/10.1007/978-981-15-6106-1_11

The geophysical mode function (GMF) is a function describing the relationship between radar backscattering cross section and wind speed and wind direction. For the polarization mode of C-band VV, the CMOD5 function is a high wind speed inversion model improved by Hersbach [3] on the basis of CMOD4 function [4]. Based on the RISAT-1 images, Jagdish used the CMOD5.N, CMOD5 and CMOD_IFR2 functions to invert the wind field in the North Indian Ocean [5]. Qi Xianyun used the Yangtze River Port as a research area to compare the suitability of the wind field with CMOD4, CMOD-IFR2 and CMOD5 functions [6]. Based on Sentinel-1 images off the coast of the west coast of the United States, Zhang Kangyu compared several functions of CMOD4, CMOD-IFR2, CMOD5 and CMOD5.N, concluded that the CMOD5 function had higher accuracy in the process of high-precision wind field inversion [7]. The above scholars have used the CMOD5 function to invert the sea surface wind field and get good results. However, due to the particularity of the region, whether the function is suitable for high-precision wind field inversion of SAR images in the coastal waters of Guangdong Province remains to be verified.

With the Sentinel-1 satellite launched in 2014, the sea surface wind field of Guangdong Province was inversion based on CMOD5 function and wind stripe, to explore the Sentinel-1A images are applied to the possibility of the high-resolution wind retrieval in coastal areas of Guangdong Province, and provide a more accurate assessment of offshore wind energy resources in Guangdong Province.

2 Method

2.1 Data Pre-processing

The offshore wind measurement sites selected in this paper are distributed as evenly as possible in the coastal side of Guangdong Province. Sentinel-1A is one of the satellites in the binary system carrying a C-band sensor. The GRD images used in the paper have been processed multilooking, and the interference wide mode is IW, which data is 250 km in width, 20 m in distance resolution and 22 m in azimuth resolution.

Sentinel-1A image preprocessing can be seen in Fig. 1. Radiation calibration is to convert the gray value of Sentinel-1A image into normalized backscattering cross section (NRCS). Speckle filtering is to suppress speckle noise of remote sensing image and remove speckle noise of image itself. Terrain correction is to eliminate image distortion caused by terrain fluctuations.

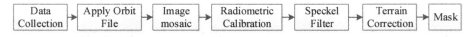

Fig. 1. The flow chart of image data preprocessing procedure

2.2 Wind Retrieval

The spiral of the atmospheric boundary layer is formed on the surface of the vortex, which result the periodic fringe in SAR image, which called the wind stripe [8]. This paper adopted the Fast Fourier Transform (FFT) method based on frequency domain extracted the sea wind direction of SAR image:

$$Y_{l,m} = \sum_{j=1}^{N} \sum_{k=1}^{N} X_{j,k} e^{-\frac{2\pi i(jl + km)}{N}} \tag{1}$$

Where Y is the low spectral number of the image, X is the grayscale value of the image after calibration, $l, m = 1, 2, \cdots, N$.

In this paper, the SAR data with VV polarization mode are adopted, and CMOD5 function was used to invert wind speed. The basic form of CMOD5 function is:

$$\sigma_{vv}^0 = B_0(1 + B_1 cos\phi + B_2 cos2\phi)^{1.6} \tag{2}$$

Among them, B_0, B_1, B_2 is a function of wind speed V and incident angle θ [9].

After data preprocessing Sentinel-1A images, we used FFT to extract the image wind stripe information, then contrasted CCMP data, and removed the 180° ambiguity of the wind direction (see Fig. 2). Finally, inverted the offshore wind speed based on CMOD5 function. The wind field inversion flow chart is shown in Fig. 3.

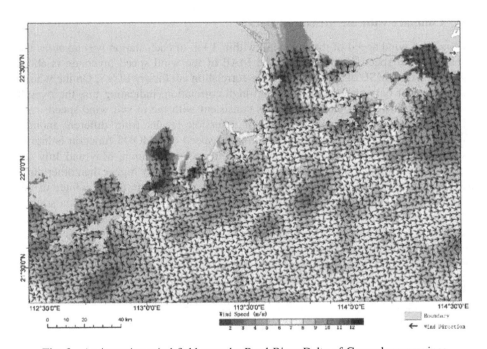

Fig. 2. An inversion wind field near the Pearl River Delta of Guangdong province

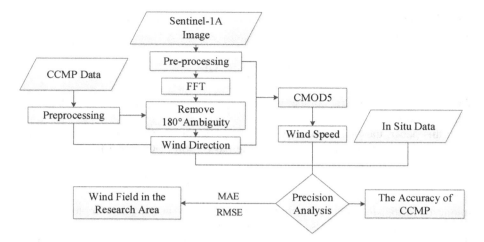

Fig. 3. The flow chart of wind field inversion

3   Results and Discussion

3.1   Compare with the Measured Data

The average wind speed of the inversion within 3 km of each station was taken as the inversion wind speed of the station. The MAE of the wind speed inversion is about 1.95 m/s, with RMSE of 2.70 m/s, and the correlation coefficient of 0.8. On the whole, the two sets of data have small error and high correlation, indicating that the overall inversion wind speed of these samples is consistent with the in situ wind speed.

From the comparison and analysis of inversion results from different months (Table 1), it can be seen that the inversion wind speed of CMOD5 function is higher than the measured wind speed, the average wind speed in March, May and July are relatively close, but the wind speed in December is significantly higher than other three months, with the average wind speed of 8.71 m/s. This is due to the winter in the coastal areas of Guangdong province, under the influence of cold air, the sea surface wind speed increasing significantly compared with other seasons. The MAE and RMSE in March, May and July between the inversion results and the measured data are both less than 2 m/s, and the correlation coefficient is also within the range of 0.7 to 0.9, indicating that there is a good consistency and correlation between the two sets of data in these three months, and the inversion results have a high accuracy. In December, the MAE is 3.64 m/s and the RMSE is 4.49 m/s, it can be seen that there was a large error between the retrieved wind speed and the in situ data. It can be seen that the inversion wind field in December has a large deviation from the measured data, most of which is due to the time difference between the image acquisition time and the measured data, in the case that the sea surface wind field changes greatly in winter, the wind speed is caused the difference between the two is also very large.

Table 1. Comparison of wind field inversion results by month.

Month	Number of samples	Average measured wind speed (m/s)	Average retrieval wind speed (m/s)	MAE (m/s)	RMSE (m/s)
March	9	6.99	7.16	1.57	1.74
May	15	4.02	5.17	1.47	1.93
July	18	5.97	6.10	1.23	1.57
December	14	8.71	10.46	3.64	4.49
All	56	6.30	7.11	1.95	2.70

3.2   Compare with the CCMP Data

In order to further discuss the reason for the large inversion error of wind speed in December, we compared the CCMP data of December with the data of the actual measurement station, and found out that MAE and RMSE are 2.96 m/s and 4.02 m/s respectively, with the correlation coefficient of 0.59 (Table 2). It can be seen that there are also some differences between CCMP data and in situ data in December, which is similar to the comparison between wind speed inversion result and in situ wind field data. In December of winter, the wind speed is the strongest, the number of windy days is the highest, and the number of windy days in the northeastern South China Sea can reach more than 20 days [10]. While the Sentinel-1A satellite transit time, the CCMP data time and actual measurement time points were not completely matched, resulting in large deviations between the retrieval wind field, the forecast wind field and the measured data in the weather with high wind speed.

Table 2. Comparison of CCMP wind fields in December

Station	Average measured wind speed/(m/s)	Average CCMP wind speed/(m/s)	MAE/ (m/s)	RMSE/ (m/s)
G3358	7.65	13.08	5.43	6.85
G7427	9.40	7.00	3.54	4.28
59490	1.7	7.61	5.80	5.80
G3704	2.87	4.66	1.85	2.24
G3597	14.35	13.31	1.30	1.66
G1368	13.05	8.60	4.45	4.48
G7526	11.40	11.46	0.27	0.28
Total	8.71	9.18	2.96	4.02

4 Conclusion

The reasonable inversion results in this paper prove the applicability of the COMD5 function in the research area, and provides the possibility for further estimation of wind energy resources in Guangdong province based on Sentinel-1A images. In addition, we conclude that the inversion wind speed of the CMOD5 function is mostly higher than the actual wind speed, and the characteristics of the offshore wind speed in Guangdong Province in December are verified to be larger than other seasons, leading to a large error between the inversion wind speed and the measured data. Further work like a large-scale wind farm retrieval or other methods will be applied.

Acknowledgment. This research is jointly supported by the Science and Technology Planning Project of Guangdong Province (2018B020207002, 2019B020208013, 2019B020208004), the GDAS' Project of Science and Technology Development (2018GDASCX-0902), and the Key Special Project for Introduced Talents Team of Southern Marine Science and Engineering Guangdong Laboratory (Guangzhou) (GML2019ZD0301).

References

1. Zhang, L., Shi, H., et al.: Overview of methods for satellite borne SAR image retrieval of sea surface wind field. Mar. Sci. Bull. **31**(6), 713–720 (2012)
2. Monaldo, F.M., Thompson, D.R., Beal, R.C., et al.: Comparison of SAR-derived wind speed with model predictions and ocean buoy measurements. Geosci. Remote Sens. IEEE Trans. **39**(12), 2587–2600 (2002)
3. Hersbach, H., Stoffelen, A., Haan, S.D.: An improved C-band scatterometer ocean geophysical model function: CMOD5. J. Geophys. Res. Oceans **112**(C3), 6–24 (2007)
4. Stoffelen, A., Anderson, D.: Scatterometer data interpretation: estimation and validation of the transfer function CMOD4. J. Geophys. Res. Oceans **102**(C3), 5767–5780 (1997)
5. Jagdish, K.S.A., Chakraborty, A., Kumar, R.: Validation of wind speed retrieval from RISAT-1 SAR images of the North Indian Ocean. Remote Sens. Lett. **9**(5), 421–428 (2018)
6. Qi, X., Zhou, Y., Tian, B., et al.: Inversion of the wind field near the Yangtze River Estuary based on Sentinel-1A. J. East China Norm. Univ. (Nat. Sci.) (6), 126–135 (2017)
7. Zhang, K., Huang, J., Xu, X., et al.: Spatial scale effect on wind speed retrieval accuracy using sentinel-1 copolarization SAR. IEEE Geosci. Remote Sens. Lett. **PP**(99), 1–5 (2018)
8. Chang, J.: Research on spaceborne SAR remote sensing technology and application of offshore offshore wind energy. Ocean University of China, Qingdao (2012)
9. Han, Y.: SAR and CALIPSO data inversion of sea surface wind speed, pp. 1–57, Ocean University of China, Qingdao (2010)
10. Wang, H., Sui, W.: Analysis of seasonal variation of offshore wind in 18 coastal regions of China based on CCMP wind field. Meteorol. Sci. Technol. **41**(4), 720–725 (2013)

Prospect of Power Inspection Using UAV Technology

Zhiyong Liu[1], Xiaodan Zhao[2(✉)], Hongchang Qi[1], Yanfei Li[1],
Gengbin Zhang[1], and Tao Zhang[1]

[1] China Southern Power Grid, Guangzhou Power Supply Bureau Co., Ltd.,
Power Transmission Management Station II, Guangzhou 510000, China
[2] Guangzhou Institute of Geography, Guangzhou 510070, China
dan_de@foxmail.com

Abstract. With the continuous iterative evolution and integration of new generation information technologies such as artificial intelligence, cloud computing, big data, Internet of Things (IoT), and mobile Internet, the Unmanned Aerial Vehicle (UAV) remote sensing technology will be qualitatively leap, and it will also drive the power industry into a new era of intelligence. This paper aims at reviewing the full-service process of UAV power inspection, expounds the application of new generation information technology in UAV power inspection, and forecasts the intelligent trend of power inspection. 1) Intelligent flight platform: the drone is closely integrated with 5G communication, gradually leading the 5G network UAV from network integration, real-time era to intelligent era; 2) intelligent patrol: the UAV intelligent control and other series of technologies Bottleneck will be overcome, networked "fixed/mobile" drone intelligent airport are developed, and UAV power inspection will be all-weather, unmanned and intelligent; 3) intelligent data analysis, introduction of artificial intelligence technology and continuous optimization of models will greatly improve fast and accurate inspection data intelligent analysis; 4) integration of IoT, big data, cloud computing will improve the multi-dimensional data integration, state monitoring full coverage, data stream and business flow integration coupling, and achieve intelligent equipment state evaluation and prediction; 5) Comprehensive application of the new generation of information technology: the construction of intelligent operation and maintenance system, and intelligent control platform for drone power inspection, will effectively improve management, and create a new situation of power inspection.

Keywords: Unmanned Aerial Vehicle (UAV) · Artificial intelligence (AI) · 5G communication · Internet of Things (IoT) · Power inspection

1 Introduction

During the "Thirteenth Five-Year Plan" period, China's power grid infrastructure construction has achieved rapid development. It is estimated that the total mileage of transmission lines will reach 1.59 million km in 2020, and will maintain a continuous growth of about 5% per year [1]. Inspection, care and maintenance put forward higher requirements. With the continuous advancement and development of the aircraft patrol

© Springer Nature Singapore Pte Ltd. 2020
Y. Xie et al. (Eds.): GSES 2019/GeoAI 2019, CCIS 1228, pp. 151–159, 2020.
https://doi.org/10.1007/978-981-15-6106-1_12

business, the Unmanned Aerial Vehicle (UAV) has been used as one of the important means of patrol of transmission lines, and normal operations have been carried out. The State Grid Corporation and the relevant departments of China Southern Grid Corporation are continuously deepening the construction of drone teams and improving them. Various types of support and support systems have gradually formed a new mode of inspection of transmission lines based on "machine inspection and personnel inspection" [2–4]. However, the contradiction between the increasing number of transmission equipment and the shortage of operation and maintenance personnel has become increasingly prominent. The transmission line inspection team faces a severe situation in which the total number of missing and structural defects are coexisting. The degree of intelligent inspection of drones is not high, and the application level is difficult to support the development requirements of transmission and inspection.

As the main feature of the modern era, the information technology revolution sweeping the world is evolving in the direction of integration, ubiquity, and intelligence, and has become the main driving force for social change. In this process, many new ideas, new concepts, new methods, and new technologies such as cloud computing, big data, Internet of Things, mobile Internet, and artificial intelligence are emerging. The new generation of information technology is gradually becoming a powerful engine for smart grids. At the same time, it has greatly promoted the innovation and development of drone technology.

The deep integration of the new generation of information technology and drones will inevitably continue to optimize and reconstruct the UAV power inspection technology system and framework, and push the UAV power inspection into a new stage of intelligence, with the "big cloud object shift". The concept of intelligent transportation supported by modern information technology came into being. This paper combines the application status of UAV grid inspection technology, starting from the technical development trend and industry application requirements, and imagining the future development direction and providing reference for the industry.

2   5G Network Connected Drones Lead the New Era of Power Inspection

Mobile communication technology is the key to improving real-time transmission of drone video, flight status monitoring, high-precision positioning and remote control. As a new generation of mobile communication technology, 5G's jump in bandwidth, delay, connection density, network performance, etc., will revolutionize the application of the UAV power industry. In the foreseeable future, the close integration of UAVs and 5G communication technology will lead a new era of power inspection with "network-connected drones" as the core.

Both the 5G network UAV terminal and the ground control terminal transmit data and control commands through the 5G network, and load various scenarios through the service server [5]. The future 5G network UAV system is composed as follows (see Fig. 1):

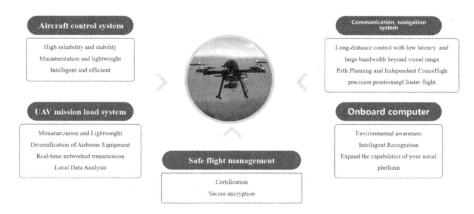

Fig. 1. 5G network connected drone platform composition

5th generation mobile networks has the characteristics of wide coverage area, low time delay, ultra-high bandwidth, large connection, etc. It can meet the requirements of UAV automatic driving and upgrade of obstacle avoidance technology, and will give UAV on-line real-time ultra-high-definition image transmission. State monitoring, ultra-long-range low-latency control, long-term stable online communication signals, high-precision positioning, secure network, autonomous obstacle avoidance and cluster control, etc., combined with network slicing and edge computing capabilities, will accelerate the innovation and development of UAV industry applications. At the same time, with the maturity of artificial intelligence, edge computing and other technologies, its in-depth integration with 5G technology will promote the application of networked UAV in power inspection from network to real-time, and realize the leap to unmanned intelligent inspection in the near future.

Networking stage: The UAV will be connected to the low-altitude cellular mobile communication network to realize UAV interconnection, over-the-horizon control, multi-machine coordinated flight, and data quasi-real-time return, etc. At present, the networking technology has realized some functions in the 4G network, but it is limited by the bandwidth, delay, interference coordination, etc. of the 4G network, it is imperative to combine 5G communication technology with UAV power patrol to meet the requirements of high speed and real-time.

Real-time phase: Real-time communication between UAV, ground stations, and dispatch management systems to realize real-time monitoring, real-time positioning, remote scheduling and control of UAV status. Based on the large-scale antenna array of 5G base station and single-station or multi-station co-localization, the positioning accuracy of the UAV is effectively improved, and the operation safety of over-the-horizon UAV is guaranteed. With the help of the characteristics of 5G network, such as large bandwidth transmission capability, end-to-end milliseconds delay and high reliability transmission, the existing UAV point-to-point communication technology is broken. The bottleneck of distance between data transmission and image transmission is the real-time return of high-definition images/videos, remote sharing of UAV shooting scenes, comprehensive control the operation site conditions; control the UAV flight path with ultra-long-range low-latency, carry out cluster cooperative operation, realize the ground station and management center to carry out internal and external field cooperation, and open up the information barriers of the inspection site and the management personnel behind the operation. At the same time, it supports flexible and efficient 5G network technology under ultra-high mobile speed, and combines scenarios such as dual connectivity and coordinated multi-point transmission to enhance the high mobility of the terminal, maintain the continuity of the inspection service and high system performance.

3 Artificial Intelligence Accelerates the Innovative Development of Electric Power Patrol Inspection

A new round of scientific and technological revolution and industrial transformation are emerging. The formation of big data, the innovation of theoretical algorithms, the improvement of computing power and the evolution of network facilities have driven the development of artificial intelligence into a new stage. Intelligentization has become an important direction for technology and industrial development. In July 2017, the State Council issued the "New Generation Artificial Intelligence Development Plan". On December 14, the Ministry of Industry and Information Technology issued the "Three-Year Action Plan for Promoting the Development of a New Generation of Artificial Intelligence Industry (2018–2020)", which raised artificial intelligence to the national strategic level. The deep integration of networked UAVs and the new generation of artificial intelligence technology promotes the UAV power inspection into a new stage of intelligence.

The development of UAV grid inspection technology has now passed the manual operation stage, entered the automatic inspection stage, and realized the automatic driving of the drone based on pre-programming methods such as manual teaching/three-dimensional route planning, as well as the auxiliary analysis of some typical defects and hidden dangers of power equipment, have been realized. With the continuous development of artificial intelligence, the power inspection of UAV will be driven from automation to intelligent leapfrog development. On the one hand, it promotes the intellectualization and operation autonomy of power inspection equipment. On the other hand, it effectively improves the intelligent processing level of patrol data of UAV, and creates a new situation of electric power patrol inspection of UAV.

3.1 Promote the Intelligence and Operation Autonomy of Power Inspection Equipment

With the continuous development and deep integration of information technology and sensor technology such as artificial intelligence, 5G communication, big data, etc., it will overcome a series of key technologies for UAV autonomous inspection, and comprehensively break through real-time perception and avoidance of complex scenes, real-time targets. Intelligent identification and tracking, intelligent path planning, intelligent flight control and self-determination, dynamic and precise positioning, environmental adaptive shooting, multi-machine multi-task collaborative control, collaborative semantic interaction, etc., a series of technical bottlenecks restricting UAV applications. Through the online environment awareness and information processing of the UAV system, the operating environment is fully perceived and obstacles are avoided, real-time intelligent obstacle avoidance and autonomous route planning, and the patrol inspection route and control strategy are generated independently according to the requirements of the inspection task.

To realize the intelligentization of the UAV power inspection and the intelligentization of multi-machine coordinated inspection under the environment of open, dynamic and complex transmission and distribution, intelligent, safe and efficient power inspection, and greatly enhance the intelligence of power inspection degree, inspection efficiency and quality, as well as power supply reliability, effectively solve structural problems such as lack of staff [6, 7]. And continue to deepen and expand the application of UAV power industry, explore the development of intelligent UAV foreign body removal, live water washing and other live operations, based on UAV's composite insulator hydrophobicity detection and other monitoring operations, and gradually promote the UAV detection and maintenance intelligence.

At the same time, through the development of an integrated UAV intelligent airport, the autonomous inspection technology of UAV based on "fixed platform" and "mobile platform" will be overcome to break through the limitation of the existing UAV's endurance capacity and form the continuous operation capability of the UAV.

The UAV intelligent airport is the infrastructure that guarantees the continuous operation of the drone, providing conditions for the landing site, storage, charging, and data transmission of the drone. The intelligent hangar of UAV can create an all-weather constant temperature and humidity storage space for UAV, which is equipped with precise landing guidance system, grabbing mechanism and selfcharging/automatic battery replacement system to ensure the endurance of UAV [8]. It has an independent environmental monitoring system to automatically judge the flight test conditions, which can support various power supply modes such as solar power supply and external power supply. Meanwhile, it is compatible with various UAV models. Through the deployment of networked fixed/mobile UAV intelligent airport, all-weather, all-day and full-autonomous multi-aircraft collaborative intelligent inspection can be realized, which greatly improves the inspection efficiency.

3.2   Improve the Intelligent Analysis Level of UAV Inspection Data

At present, the intelligent processing of drone inspection data is low, and the coupling with business data is not high, which cannot support the collaborative intelligent patrol mode based on data flow driven UAV.

With the continuous expansion of the inspection service of drones and the increasing number of machine inspection equipment, the processing and analysis of the data of transmission line machines will surely enter the era of "big data". At present, the amount of data on drone inspections has shown an exponential growth trend, which provides a large number of learning samples for artificial intelligence technology. By unifying and perfecting the image labeling rules, the artificial intelligence depth image recognition technology is used to construct and iterate the defect recognition algorithm to realize the rapid intelligence and standardization analysis of the hidden dangers of power equipment defects, and automatically generate defect hidden danger reports. At the same time, it explores the intelligent identification technology based on AI for airborne front-end defects, combined with interactive field operation technology such as line physical ID information identification, real-time intelligent diagnosis and identification of defects and hidden dangers can be carried out in the patrol inspection process to improve the timeliness of defect identification. Through the introduction of artificial intelligence technology, the efficiency and intelligence level of inspection data processing can be improved comprehensively, transmission line defects and external security risks can be effectively analyzed and mastered, line equipment operation status can be controlled in time, hidden dangers can be eliminated in time. Improve line running stability, safety, save human resources, decrease the cost of inspection.

4   Internet of Things, Big Data, and Cloud Computing Drive Global IoT and Situational Awareness

The Internet of Things (IoT) is an emerging technology involving multiple fields of information technology and has become one of the symbols of the global information age. It is called the third wave of electronic information technology after computers and the Internet. The Internet of Things has become an important part of the construction and operation of smart grids. With the full deployment of the State Grid Corporation of China and China Southern Power Grid Corporation in the Internet of Things, the power system is entering a new era of the IoT power system [8]. As an emerging computing model, cloud computing technology can effectively solve the storage of massive data and parallel computing of big data through virtualization, massive distributed data storage, parallel programming model and other technologies, which is the basis for supporting the application of intelligent technology in the field of power production [9]. The introduction of big data analysis technology can be used for in-depth mining for massive production and maintenance data, carry out situational awareness and global analysis, which has very important guiding significance for power production operation and maintenance management and decision-making. These emerging information technologies are profoundly changing the current way of power operation and maintenance.

Utilize Internet of Things technology, cooperate with multiple monitoring methods, break the barriers of data sharing, and build an all-round intelligent sensing monitoring system to realize online monitoring systems, UAV platforms, ground unmanned inspection platforms, satellite remote sensing platforms, as well as meteorological, geological, hydrological environmental monitoring and other massive multi-dimensional data fusion, full coverage of state monitoring, realizing the global inter-connection of the power grid, deep integration of data flow and business flow, complete acquisition of equipment life cycle data, and perceptual measurement of operating parameters of all working conditions, information exchange of influential factors in all scenarios provide data basis for lean management and control of power grid equipment.

Based on the entire IoT data of the power grid, we can control the operation and maintenance status of power lines in a timely and comprehensive manner. By using big data and cloud computing technology, systemic global analysis methods are used to construct equipment state evaluation and trend prediction models, and the mass production operation and maintenance data is deeply mined and multi-dimensional analysis, carry out situational awareness, real-time comprehensive evaluation and evaluation of equipment status, closed-loop management of pre-failure and safety risk prediction and early warning, real-time monitoring in the event, and comprehensive analysis after the event, comprehensively improve the status of transmission line state diagnosis, improve equipment status evaluation and the level of intelligence in trend forecasting. At the same time, based on the situational awareness results, according to the health status of different power equipment such as lines, poles and towers, a scientific differentiated inspection strategy is developed to assist the differentiated inspection of UAV, reduce the operation and maintenance costs, improve the inspection efficiency, and promote the intelligent upgrading of power grid operation and maintenance management.

5 Global Integrated Intelligent Management and Control Platform to Improve Lean Management Level

At present, the operation level and management level of UAV in various regions are uneven, and the requirements of lean management of equipment do not match the level of business development. The inspection operation lacks effective supervision, and the inspection equipment has device information management that relies on formalization. The inspection data storage and management are scattered, lack of effective means of integration and sharing, internal and external industries are out of line, and unified application and closed-loop management have not yet been formed, resulting in the inability to carry out multi-dimensional analysis and comprehensive application of inspection data, and can not provide more accurate decision-making basis for transmission line operation and maintenance departments [10, 11].

Utilizing the technological advantages of artificial intelligence, big data, Internet of Things, mobile internet, etc., we will build a network-wide integrated UAV intelligent management and control platform to promote business standardization, management and control informationization, operation intelligentization, and management leanness [12]. As the data center and intelligent production monitoring command center of transmission operation and inspection management the management and control platform integrate power inspection, online monitoring, power grid resource data, equipment and personnel information, and equipment operation status, meteorological/micro-meteorological data and other multi-source operation and maintenance data, and realizes deep integration of massive inspection data and share the whole network. By means of global analysis, multi-dimensional intelligent analysis and precise positioning of transmission line operation status are carried out, clearly and completely display the operation status, inspection status, elimination of transmission equipment and the implementation of the cycle, etc., provides a comprehensive, real-time and accurate decision-making basis for the intensive command of the transmission inspection.

It has powerful visualization functions, realizing real-time monitoring of UAV patrol and human patrol, recording and playback, grid resources, and 2D/3D visualization display of inspection results; intelligent three-dimensional coordination of intelligent inspection equipment and inspection personnel of drones Control and control, establish a comprehensive centralized management and control capability for the whole business process of "human patrol + aircraft patrol", realize global visual controllability of inspection, and effectively link all aspects of data collection, data processing, results management and elimination management, effectively improve the inspection effectiveness. Realize the interconnection and interoperability of internal and external data, front and back end data, form a complete closed loop of transmission operation and inspection work, comprehensively standardize the inspection of drones, ensure the safe and controllable operation, and promote the intelligent transmission and maintenance operation safely, efficiently and pragmatically. Improve the efficiency of the inspection and the level of lean management, so that the transmission inspection will be transformed from the original extensive management mode to the informatization and refined, and the production command and decision-making will be highly intelligent and intensive.

6 Conclusion

Smart grid is the inevitable trend of the development of the power industry. With the continuous iterative evolution and integration innovation of the new generation of technical information such as artificial intelligence, Internet of Things, big data, cloud computing, mobile communications, it will provide a powerful drive for the intelligent development of the power industry and reshape the new industrial era of power inspection.

This paper introduces the application of new generation of information technology in UAV power inspection from aspects of flight platform intelligence, data analysis intelligence, data application intelligence, etc. And looking forward to the intelligent trend of power inspection.

In the future, we will overcome a series of technical bottlenecks related to intelligent inspection of UAV. By deploying networked mobile/fixed-UAV nests, we using 5G network to connect UAV, ultra-long-distance and low-delay control UAV or multiple UAV cooperate to carry out all-weather, all-autonomous and unmanned intelligent patrol inspection, and simultaneously carry out intelligent analysis of defects. Through the global integrated management and control platform, UAV centralized management and control, unified management of data, deep mining, intelligent analysis and comprehensive application are realized, and multi-dimensional intelligent analysis and early warning are carried out on the operating state of the transmission line. At the same time, we should promote standardization of business, informatization of management and control, intellectualization of operation and lean management. Change the traditional operation and maintenance methods, improve the safety of operations, inspection efficiency and operation quality, continuously improve the intelligent transmission inspection level and the lean management level of transmission industry, and create a new situation of transmission and transportation inspection.

Acknowledgements. The research was jointly supported by China Southern Power Grid Guangzhou Power Supply Bureau Co., Ltd. Key Technology Project (080000KK52190001); Guangdong Provincial Science and Technology Program (2017B010117008); Guangzhou Science and Technology Program (201806010106, 201902010033); the National Natural Science Foundation of China (41976189, 41976190); the Guangdong Innovative and Entrepreneurial Research Team Program (2016ZT06D336); the Southern Marine Science and Engineering Guangdong Laboratory (Guangzhou) (GML2019ZD0301); the GDAS's Project of Science and Technology Development (2016GDASRC-0211, 2018GDASCX-0403, 2019GDASYL-0301001, 2017GDASCX-0101, 2018GDASCX-0101).

References

1. Tu, J., Feng, Z.H., Liang, W.Y., et al.: Application analysis of small Unmanned Aerial Vehicles in power line patrol inspection. Electr. Age (11), 75–77 (2016)
2. Chen, T.T.: Application analysis of UAV inspection transmission line technology. Sci. Technol. Innov. (4), 154–155 (2016)
3. Wei, S.T., Li, L., Yue, L.P., et al.: Research on human-UAV collaborative patrol mode of power transmission corridor. Zhejiang Electr. Power **35**(3), 10–13 (2016)
4. Tao, C.Z., Huang, Y.M., Li, Y.C., et al.: Application of Unmanned Aerial Vehicles in transmission line inspection. China Sci. Technol. Inf. (16), 36–37 (2016)
5. Tong, W.: Research on application scene of 5G new air interface technology in networked UAV. Telecom Power Technol. **36**(04), 212–213 (2019)
6. Fan, B.K., Zhang, R.Y.: UAV system and artificial intelligence. Geomatics Inf. Sci. Wuhan Univ. **42**(11), 1523–1529 (2017)
7. Fan, B.K.: Academician Fan Bangkui: six directions, knowing the future of drones. Robot Industry (01), 59–64 (2017)
8. Hu, Z.M., Li, K., Tang, G.F., et al.: A new mode of "nest-nest" inspection of UAVs for transmission lines. Jiangxi Electr. Power **42**(12), 13–15 (2018). +25
9. WANG, K.: The application of the Internet of Things in the 5G era in the power system. Telecom Power Technol. **35**(05), 187–188 (2018)
10. Song, Q.Y.: Research on Storage Optimization and Parallel Processing of Power Equipment Monitoring Big Data Based on Cloud Platform. School of Electrical and Electronic Engineering, Beijing (2016)
11. Wang, Y.: Design of UAV power line inspection system based on ArcGIS. Tianjin University (2016)
12. Feng, S.: Design of UAV power line inspection support system. Tianjin University (2016)

Identification, Extraction and Three-Dimensional Building Model Reconstruction Though Faster R-CNN of Architectural Plans

Kunyang Ma[1,2]([✉]), Yi Cheng[1,2], Wen Ge[1,2], Yao Zhao[2],
and Zhang Qi[1]

[1] Information Engineering University, Zhengzhou 450000, China
394018925@qq.com
[2] Collaborative Innovation Center of Geo-Information Technology for Smart
Central Plains, Zhengzhou 450000, China

Abstract. High-precision three-dimensional model is a high value information nowadays. Especially the 3D model of inner space of buildings can be used widely. However, it is not quite easy to acquire it. Building Information Model (BIM) provides a way to get detailed information of building to be build. But for most buildings already there the only thing that recording their internal structure we have are architectural plans. In order to get high precision model of building inner space, people scanning architectural plans into computer and vectoring every elements of building and reconstruction it by 3D building software as 3DMax and so on. Although this method is time consuming and high cost, it is the general method used in 3D building fields in recent years. In this study, we tried to speed up the modeling efficiency by identify and extraction information of building components automatically by deep learning algorithms. Faster R-CNN object-detection model, which was proved to be efficiency in image recognition is used in the paper to identity and extract independent building functional components such as tables, beds, cabinets and toilets and other equipment. This components are elements of plans but not exists actually in buildings. By deleting these "useless" components we can get a "clean" map of architecture and the vector data of building inner space can be obtained by image recognition algorithms as refine, extract center line and other operations. This paper proposed a method also to building the topology relationships of the vector data to drawn into room models in the buildings. And then, using room models to form building models. In order to verify the availability and generality of the model, we designed a method to transform the model we build to IFC, which is the famous BIM Standard model. Experiments shows that Faster R-CNN algorithm can provide high accuracy identification results, with the help of it the automation degree and efficiency of vectorization can be improved. The topology relationships defined to describe relations inside and between rooms are effective to form an entire building model. And the transform from 3D building model extracted from architectural plan in this way to IFC are feasible.

Keywords: Three-dimensional building model · Inner space · Faster R-CNN · Three-dimensional topology

© Springer Nature Singapore Pte Ltd. 2020
Y. Xie et al. (Eds.): GSES 2019/GeoAI 2019, CCIS 1228, pp. 160–176, 2020.
https://doi.org/10.1007/978-981-15-6106-1_13

1   Introduction

Researches show that 90% of people's daily life is related to indoor space [1]. As the primary choice to express and reconstruct the indoor scene, three-dimensional model has the better user experience and the more realistic spatial analysis effect [2]. It is also the basis of realizing various indoor location services and the key integrating indoor and outdoor. At present, the mainstream methods of indoor three-dimensional modeling cannot meet the requirements of large-scale three-dimensional modeling rapidly [3–6]. Architectural plans can be used for indoor three-dimensional modeling, because of its low cost and rich content including geometry, size, material and the structure of building. However, many interference factors such as tables and beds are also in the plan. At the same time, most of the building models now available are ideographic [7], which lack the accurate description of spatial location and effective expression of the relationship between entities and is difficult to apply effectively to indoor location services and building management. Therefore, how to solve the problems mentioned above is the key issues of high-precision indoor three-dimensional modeling.

Guannan Li [8] and his group make up an integrated three-dimensional model in 3ds Max by MaxScript. They extract the information of layer objects, with the hierarchical relationship between them and main components retained, using the LISP command invoked from a secondary development interface of Tangent CAD.

After scanning and digitizing the architectural drawings, Yuanshu Li [9] builds an ideographic model with the walls extracted by using the sparse pixel vectorization algorithm, and building entities such as doors and windows which identified on the wall with the algorithm of LDA. The poor efficiency and accuracy of extraction and the lack of topological information in three-dimensional models make it difficult for the method in above studies, which extracting building structure information directly, to be widely used.

2   Methods

Based on the experience of previous studies and the idea of "reverse extraction", this paper proposes an efficient method which can identify and eliminate decorative components using object-detection technology, then establish the three-dimensional topological relationship among the entities in the building, which is the foundation for three-dimensional modeling and indoor location service. The diagram of main idea is shown in Fig. 1.

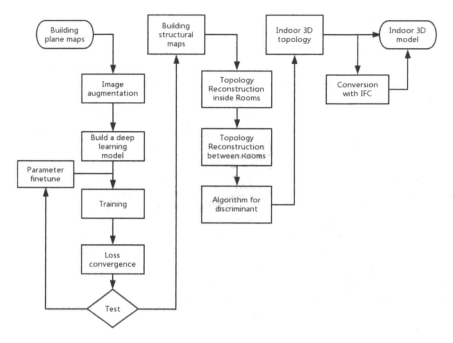

Fig. 1. Main idea of this paper

2.1   Geometric Extraction Based on Faster R-CNN

Characters of Architectural Plans

Compared with other data, architectural entities with unclear boundaries are dispersed in abstract forms in the architectural plan [10]. The abstraction, decentralization and overlap of information are important reasons for the difficulty of identification. Figure 2 shows the building plane map.

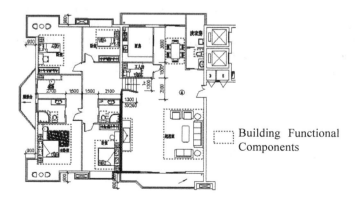

┌┈┈┈┐ Building Functional
└┈┈┈┘ Components

Fig. 2. Building plane map

Focusing on the research objectives of this study, building components in the architectural plan will be divided into two categories: architectural structural components and architectural functional components, according to the *Standard for Terminology of Civil Architectural Design* [11] and IFC (Industry Foundation Classes) standard.

Definition 1: Building structural components, which support the roof and floor and connect the rooms and floors in the building, includes walls, columns, doors, windows, railings and stairs.

Definition 2: Building functional components, which do not affect the internal structure of building, includes tables, beds, cabinets and toilets and other equipment.

Principle

Faster R-CNN model was proposed by Shaoqing Ren [12], where Region Proposal Network (RPN) is used to replace the Selective Search [13] method for rapid Region Proposal extraction. This model can predict what and where is the target through the training method of end-to-end. The anchor mechanism and border regression algorithm in the model make the recognition and prediction of mass data faster and more efficient.

(1) Anchor Mechanism

After receiving the feature map, a convolution layer with a convolution kernel of 3×3 will be used to convolute the feature map into feature vectors, at the same time, nine anchors according to three different areas and three different lengths are defined at the center of convolution core.

The feature vectors with anchors will be passed into a classified convolution layer with a convolution kernel of 1×1 to determine the scores of fore-ground (target) and background (non-target) in the anchors by Softmax classification function. Formulas are as follows:

$$s_i = \frac{e^i}{\sum_j e^j} \tag{1}$$

In this formula, i denotes the first element in a j-dimensional vector. As can be seen from the formula, the Softmax function maps a vector to a value on $(0,1)$ and the sum is 1.

(2) Bounding-Box Regression

The feature vectors with anchors are simultaneously fed into a regression convolution layer with a convolution core of 1×1, simulating a fully connected layer, to predict the modified parameters including the displacement of the centers and the change of the edges of anchors, t_x, t_y, t_w, t_h. The predictive formulas are as given follows:

$$t_x = \frac{(G_x - P_x)}{P_w} \tag{2}$$

$$t_y = \frac{(G_y - P_y)}{P_h} \tag{3}$$

$$t_w = \log\frac{G_w}{P_w} \tag{4}$$

$$t_h = \log\frac{G_h}{P_h} \tag{5}$$

$$x = w_a t_x + x_a \tag{6}$$

$$y = h_a t_y + y \tag{7}$$

$$w = w_a e^{t_w} \tag{8}$$

$$h = h_a e^{t_h} \tag{9}$$

In the formulas, x and y indicate the coordinate of the center point of proposal region, w and h represent width and height of proposal region, G_i represents the true value, P_i represents the predicted value, $(i = x, y, w, h)$.

Procedures of Method

Based on the basic principle of Faster R-CNN, the main steps of geometric extraction model are shown as follows:

Step 1, image preprocessing. In order to remove the noise and highlight the main information of the architectural plan with different formats and wide sources, we removed the annotated axes and text descriptions first, then binary processing is done to obtain a black-and-white dot matrix.

Step 2, extract feature map. After the pretreatment, VGG-16 [14], a convolution neural network consisting of 13 convolution layers and 3 full connection layers are used as a feature extractor to generate the feature map for the follow-up Neural Networks. Figure 3 shows the structure of VGG-16 network.

Step 3, generate proposal regions. RPN, which plays a role in target location and image rough detection by anchor mechanism and bounding-box regression algorithm, is the most nuclear network of Faster R-CNN. When the feature map is transferred into the RPN network, the preliminary prediction of proposal regions (where and what are the targets) are obtained. The structure of RPN network is shown in Fig. 4.

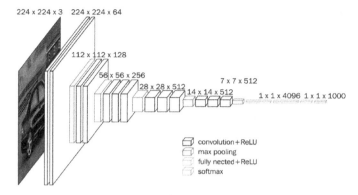

Fig. 3. The structure of VGG-16

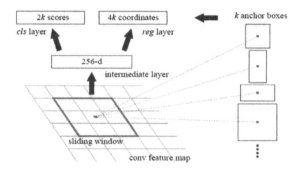

Fig. 4. The structure of RPN

Step 4, determine the target. The feature map in step1 and the proposal regions generated by RPN are simultaneously introduced into the network in this step to make secondary classification and regression. First, 300 target boxes are screened out from thousands of proposal regions by Non-Maximum Suppression (NMS) algorithm. Then, the target boxes are mapped to the feature maps and pooled into the same size and shape through the ROI pooling layer. Finally, the maps are transferred respectively into classification and regression network, which has the same structure and algorithm as RPN network, but the convolution layer is replaced by fully connected layer to abandon the weight sharing in different positions of convolution layer in order to realize further classification and better correction of target boxes.

Step 5, derive the results which contain the label and coordinate of targets that disturb indoor three-dimensional modeling from Faster R-CNN.

Step 6, remove the disturbances. A python script is used to automatically read the maps and delete the functional components by the graphical method according to the results obtained in the previous step to get structural maps of building internal.

The flow chart is designed as Fig. 5.

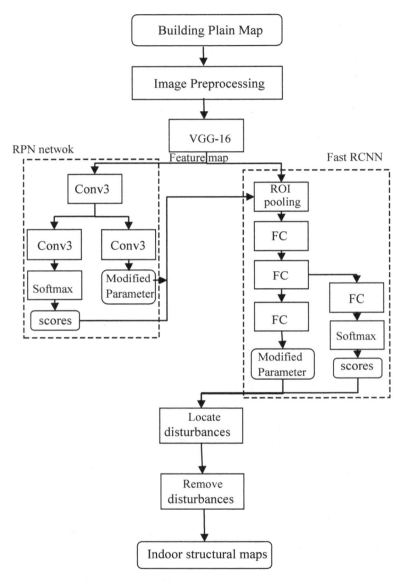

Fig. 5. The flow chart of geometric extraction

2.2   Building the Topological Model of Indoor Space

To build up a topological relationship of indoor space is essential for effective management of large-scale building [15]. Before that, we should transfer the 'clean' map obtained above to a vectorgraph by extracting center line. The topology reconstruction of indoor three-dimensional model is divided into two parts: inside rooms and between rooms. The topological from point to building in three-dimensional is established as shown in Fig. 6.

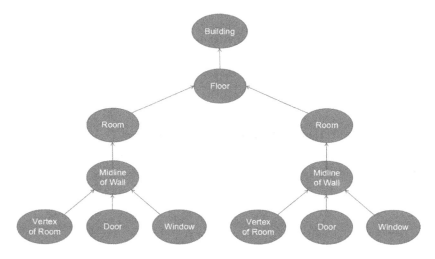

Fig. 6. Hierarchical structure of indoor topological model

Topology Reconstruction Inside Rooms

Rooms, corridors and stairwells that are considered as special rooms, are divided into vertex points, doors, windows, midline of walls and bottom of rooms, according to the different attribute types and functions. The attribute information of each element and the topological relationship between each other are established as Table 1 and Table 2.

Table 1. Attribute information of elements

Name	Attribute name
Vertex	Num
	Coordinate
	Floor
Door	Num
	Coordinate
	Width
	Height
Window	Num
	Coordinate
	Width
	Windowsill
	Top

Table 2. Relationships between elements

Relationship	Elements	Topological information
Component	Midline of walls and Vertexes	Num
		Start
		End
		Width
		Floor
	Midline of walls and bottom of rooms	Num
		Function
		Boundary
		Floor
Inclusion	Midline of walls and doors, windows	Num of midline of walls
		Num of doors
		Num of windows

Topology Reconstruction Between Rooms

Because there are many common parts in the boundary elements between rooms, which results in data redundancy and low efficiency of data management, it is necessary to establish three-dimensional topological relationship to uniquely duplicate elements. In the three-dimensional model of building interior, there are only two spatial relationships between rooms: adjacency and disjoint. In order to express the topological relationship between rooms more clearly, there are three kinds of situations in which the boundary elements exist in the common part.

(1) The two elements are completely common.
(2) The two elements are partially public, and there is a relationship of intersection or coverage between them.
(3) One element is completely contained in another.

According to the three categories above, spatial topological relationships between rooms can be classified into the following eight categories (Fig. 7).

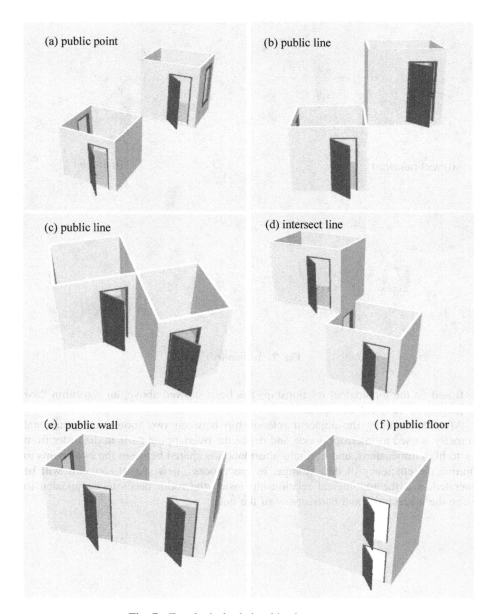

Fig. 7. Topological relationships between rooms

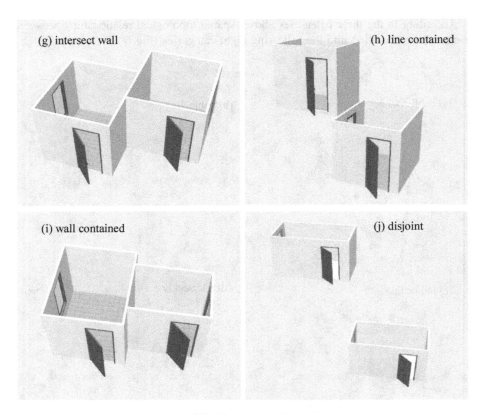

Fig. 7. (*continued*)

Based on the topological relationships has been showed above, an algorithm flow for judging is designed as Fig. 8:

After determining the adjacent relationship between two rooms, computational geometry is used to interrupt divide and delete the overlapping parts in the order from low to high dimensions, and attribute information is shared between the two rooms to improve the efficiency of data storage. In this process, new spatial elements will be generated, and the topological relationship inside the room needs to be updated to ensure the correctness and consistency of the data.

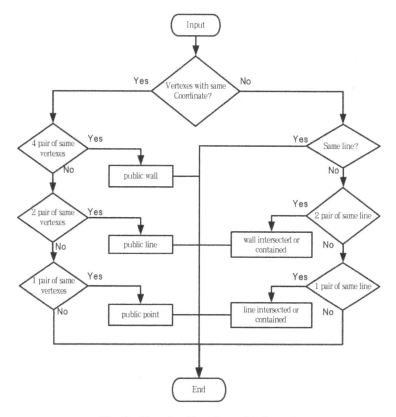

Fig. 8. The algorithm flow of judgment

2.3 Mapping to IFC

IFC (Industry Foundation Classes) which contains about 800 entities, 358 attribute sets and 121 data types in the latest version, defines a unified data format for building information to facilitate the interaction of data among applications in various industries, and it was first proposed by IAI (International Alliance for Interoperability) in 1997. Based on the indoor spatial topology model mentioned above, a transformation mapping between indoor spatial topology model and IFC standard (version 2×4) is established after analyzing the rules of how entity information and spatial relationship is described in IFC (see Table 3 and Table 4).

Table 3. Mapping rules of entities

Indoor spatial topology model	IFC
Vertex	IFCCartesianPoint
Door	IFCDoor
Window	IFCWindow
Midline of wall	IFCBoundedCurve
Bottom of room	IFCFaceOuterBound

Table 4. Mapping rules of relationships

Indoor spatial topology model	IFC
Vertexes to midline of wall	IFCRelAggregates
Doors to midline of wall	
Windows to midline of wall	
Midline of wall to bottom of room	
Public point	IFCRelConnects
Public line	
Public wall	
Components to room	IFCRelContainedInSpatialStructure
Rooms to floor	
Floors to building	

3   Experiment

3.1   Experimental Date

The data set contains 800 different building plans with no size requirement, each of which contains at least one object to be identified. The data sets are randomly divided into training sets used to fit models, cross-validation sets used to adjust model parameters and test sets for evaluating model performance in the ratio of 0.6:0.2:0.2. The experiment focused on the building component of beds and was carried out for better analyze the detection effect of the model, and data sets were divided into two categories: only objects and objects with interference in figure. Organization of data set is shown in Table 5.

Table 5. Organization of data set

	Only objects	Objects with interference	Total
Number of figures	314	486	800
Number of objects	345	1259	1603

3.2   Model Training

In this experiment, the parameters of the Faster R-CNN model are changed as follows: the number of candidate areas reserved for maximum suppression is set from 300 to 100 to improve training efficiency, the batch_size is set to 24 in order to take account of hardware computing ability and training efficiency, training times is set to 200,000. Training would be stopped when the model converges. The training set and verification set are used to train the model, and LabelImg is used to unify labeling of experimental objects on turning pictures firstly. After labeling, the training begins with about 1 s per step. 60 000 steps later, the model converges gradually and the loss value oscillates around 1. Trends of value are shown in Fig. 9.

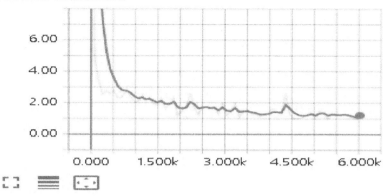

Fig. 9. Trends of value

After training, deriving the parameters of the model and testing the recognition effect of the model on test set. Precision means that how many of the detected items are right, recall means how many of the exact items are detected and mAP shows the comprehensive performance of model. The test results are shown in Table 6.

Table 6. Test results of model

Classification	Precision	Recall	mAP
Only objects	98%	91%	84%
Objects with interference	92%	83%	69%
Total	95%	87%	76%

3.3 Model Tests

After training, the plane maps of a two-storey building is used for testing. Firstly, the data is imported into the detection model to identify and delete the building's functional components, as shown in Fig. 10.

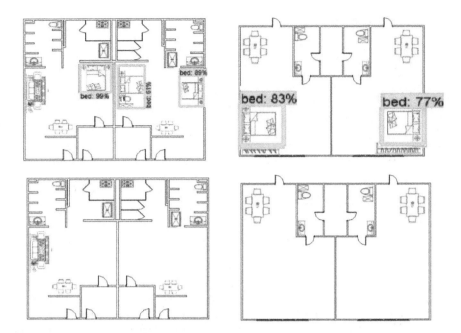

Fig. 10. Identify and delete building functional components

According to the method described in the third section, the indoor 3D topological relationship of buildings is built and converted to IFC standard file based on mapping rules. The contents of the document are shown in Fig. 11.

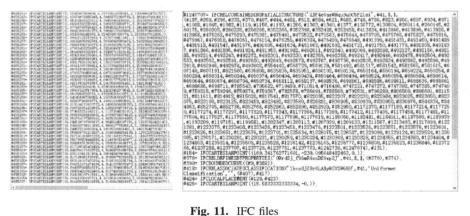

Fig. 11. IFC files

Finally, build up the three-dimensional model of building interior as Fig. 12.

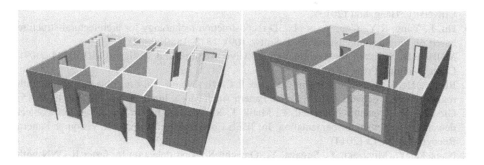

Fig. 12. Three-dimensional model of building interior

4   Summary

High-precision three-dimensional model is an important database for realizing integration of indoor and outdoor. On account of the idea of "reverse extraction method", this paper designs a detection model based on Faster R-CNN by using building plan for its accurate and cheap, which can automatically and efficiently obtain the internal structure map of building. At the same time, topology model of indoor space is established under the guidance of IFC, which can be used to accelerate three-dimensional modeling. Experiments show this method is suitable for three-dimensional reconstruction because of better accuracy and efficiency.

References

1. Worboys, M.: Modeling indoor space. In: ACM SIGSPATIAL International Workshop on Indoor Spatial Awareness, Chicago, p. 7 (2011)
2. Li, D.R., Liu, Q., Zhu, Q.: An integrated representation of outdoor and indoor in CyberCity GIS. Geomat. Inf. Sci. Wuhan University **28**, 253 (2003)
3. Isikdag, U., Zlatanova, S., Underwood, J.: A BIM-oriented model for supporting indoor navigation requirements. Comput. Environ. Urban Syst. **41**, 112 (2013)
4. Xue, M., Feng, L.I.: A framework for CAD/GIS/BIM online integration in project life-cycle management. Geogr. Geo-Inf. Sci. **31**(06), 30–34 (2015)
5. Cheng, L., Gong, J., Li, M., Liu, Y.: 3D building model reconstruction from multi-view aerial imagery and Lidar data. Acta Geodaetica Cartogr. Sin. **38**(06), 494–501 (2009)
6. Allen, P.K., Stamos, I., Troccoli, A., Smith, B.: 3D modeling of historic sites using range and image data. In: Proceedings of the IEEE International Conference on Robotics and Automation, ICRA, pp. 145–150. Taipei (2003)
7. Wang, X.F., Wang, Y.J.: Organization and scheduling of indoor three-dimensional geometric model based on spatial topological. Geomat. Inf. Sci. Wuhan Univ. **42**(01), 35–42 (2017)

8. Li, G.N., Wang, X.F., Zhang, C., et al.: Information extraction method for 3D building component based on T-arch. Geospat. Inf. **15**(07), 109–111 (2017)
9. Li, Y.S.: Fast 3D Building Modeling Based on Sparse Pixel Vectorization. Zhejiang University, Hangzhou (2013)
10. Hu, J., Yang, R.Y., Cao, Y., et al.: 3D reconstruction technology for architectural structure based on graphics understanding. J. Softw. (09), 1873–1880 (2002)
11. MOHURD. Standard for Terminology of Civil Architectural Design (GB/T50504-2009). China Planning Press (2009)
12. Ren, S., He, K., Girshick, R., Sun, J.: Faster R-CNN: towards real-time object detection with region proposal networks. IEEE Trans. Pattern Anal. Mach. Intell. **39**(6), 1137–1149 (2015)
13. Girshick, R., Donahue, J., Darrell, T., Malik, J.: Rich feature hierarchies for accurate object detection and semantic segmentation. In: IEEE Conference on Computer Vision & Pattern Recognition, p. 13 (2014)
14. Nagaoka, Y., Miyazaki, T., Sugaya, Y., Omachi, S.: Text detection by faster R-CNN with multiple region proposal networks. In: IAPR International Conference on Document Analysis and Recognition. ICDAR, Kyoto, p 15 (2018)
15. Zhu, Q., Li, D.R., Gong, J.Y., et al.: The design and implementation of CyberCity GIS. Geomat. Inf. Sci. Wuhan Univ. **01**, 8–11 (2001)

Satellite-Based Water Depth Estimation: A Review

Kaixiang Wen[1,2], Yong Li[1,2(✉)], Hua Wang[1], Wenlong Jing[2],
Ji Yang[2], Chen Zhang[2], and Zhou Wang[3]

[1] Guangdong University of Technology, Guangzhou 510070, China
5952546@qq.com
[2] Key Lab of Guangdong for Utilization of Remote Sensing and Geographical
Information System, Guangdong Open Laboratory of Geospatial Information
Technology and Application, Guangzhou Institute of Geography,
Guangzhou 510070, China
[3] State Grid Panzhihua Power Supply Company, Panzhihua 617000, China

Abstract. This paper reviews the advances in water depth estimations from
satellite images. According to the sensor type, current satellite-based water
depth estimations are divided into two categories: optical-based and SAR-based
(synthetic aperture radar) methods. By analyzing and discussing on the
advantages and disadvantages of various methods. The following points can be
summarized: First, the accuracy of optical remote sensing water depth detection
method is higher than that of SAR-based models. Besides, optical-based method
is simpler and less restricted than the SAR-based methods. Second,
hyperspectral-based approach performed better than the multispectral optical
remote sensing approaches. Since hyperspectral images provide richer spectral
information than multispectral, which significantly matter in the water depth
detection. Third, remote sensing water depth detection method is the best way to
obtain the underwater topographic features of the inaccessible waters. Finally,
on the basis of summarizing the previous water depth estimation methods, the
development prospects of shallow seawater deep inversion technology are discussed and analyzed. For example, the establishment of an inversion model by
fusing multiple remote sensing data, and the use of artificial intelligence technology for water depth inversion.

Keywords: Water depth inversion · Synthetic aperture radar · Hyperspectral ·
Multispectral remote sensing

1 Introduction

The water depth data is an important factor for obtaining and sensing the underwater
topography. It is also an important project construction reference data for the development of the coastal marine economy [1, 2]. Traditional methods for measuring water
depth include sounding rods, sounding line hammers, sonar sounding, etc. [1–3].
Although the accuracy is high, manual intervention, repeated operations, and high
measurement cost are not met. Especially in the water depth measurement of some
disputed islands and reefs, is impossible to carry out by relying on traditional artificial

© Springer Nature Singapore Pte Ltd. 2020
Y. Xie et al. (Eds.): GSES 2019/GeoAI 2019, CCIS 1228, pp. 177–195, 2020.
https://doi.org/10.1007/978-981-15-6106-1_14

measurements. The development of remote sensing technology has solved the problem of traditional sounding surveying in large-area, all-weather, low-cost, disputed sea areas [3, 4].

Satellite remote sensing technology has just emerged in the 1960s and remote sensing water depth inversion has attracted attention. With the increasing types of remote sensing satellite sensors, the model research of water depth inversion using remote sensing images has been increasingly enriched. Water depth inversion based on multispectral optical images is particularly prominent. including theoretical analytical methods, semi-theoretical semi-empirical methods, and statistical methods [3–5]. There are also water depth inversion methods based on hyperspectral imagery such as Artificial Neural Network (ANN), principal component analysis, Bierwirth algorithm, linear unmixing method, look-up table method and spectral differential statistical method [6–9]. In addition, marine satellites such as Synthetic Aperture Radar (SAR) technology also play an important role in shallow water depth detection, such as Wave Number Spectrum Balance Equation (WNSBE), direct water depth Method and Bathymetry Assessment System (BAS) [10, 11]. Throughout the development process of remote sensing water depth inversion. Water depth inversion methods can be roughly divided into two categories: optical remote sensing inversion and Synthetic Aperture Radar (SAR) inversion. But there are many types of optical remote sensing sensors, rich data sources, and simple inversion methods, so they are the main methods of remote sensing water depth inversion. Although SAR remote sensing can be free from cloud and weather conditions, it needs to consider the influence of wind direction, wind speed, ocean wave and underwater topography in the process of water depth inversion. Therefore, these factors limit the wide range of applications of SAR water depth inversion technology.

This article has compiled a large number of methods for water depth inversion. A systematic review of the current theories and methods of water depth inversion from two aspects: optical remote sensing and synthetic aperture radar (SAR) inversion. What's more, the development prospects of current remote sensing water depth inversion methods are analyzed and prospected.

2   Water Depth Inversion of Optical Remote Sensing Imagery

2.1   Water Depth Inversion Method of Multispectral Imagery

The principle of optical remote sensing for detecting water depth is mainly based on the ability of light to transmit to water, and the transmission ability of visible light is particularly prominent in many spectral bands. The smaller the attenuation coefficient of water to visible light, the better its penetration into water. The value of visible light attenuation coefficient determines the measurable depth of multi-spectral remote sensing in water detection. When remote sensing of water depth, electromagnetic waves must be through the atmosphere and water. Because the electromagnetic wave in the visible light band has the strongest atmospheric transmittance and the smallest water body attenuation. So, it is the main relying band in optical water depth remote sensing measurement [3, 12–14]. Optical remote sensing water depth inversion models can be divided into: theoretical analytical models, semi-theoretical semi-empirical models and statistical models.

Theoretical Analytical Method. The theoretical analysis method establishes a mathematical expression based on the amplitude value received by the satellite sensor, the water body reflectance value and the water depth value. Obtain the water depth value by solving the expression. The more common one is the simplified classical radiation equation, which is based on the Two-Stream Approximation Model. The Two-Stream Approximation Model divides the water depth into two part at any depth D: Dup and Ddw, so that the radiance value can be divided into two values. The change of radiance value is estimated by studying the relationship between radiance values of different water depth. Therefore, the distribution of water depth can be obtained by the water depth variable D of the analysis process [3, 15–17].

Radiative transfer equation:

$$\frac{dL(D, \varphi, \theta)}{dD} = -KL(D, \varphi, \theta) + L_P(D, \varphi, \theta) \tag{1}$$

$dL(D, \varphi, \theta)$ is the plane radiance from the water surface D to the θ angle and φ azimuth angle of the propagation direction; $L(D, \varphi, \theta)$ is the stroke function, representing the scattering gain; KLD, φ, θ is the attenuation loss function.

Lyzenga et al. [18] simplified the radiative transfer equation using a two-layer flow approximation model and abandons the reflection effect in the water to obtain the mathematical expressions of water surface reflection and water depth D. The expression is as follows:

$$R' = \frac{R_b\sqrt{1 - X^2}\cosh(KD) + (X - R_b)\sinh(KD)}{R_b\sqrt{1 - X^2}\cosh(KD) + (1 - XR_b)\sinh(KD)} \tag{2}$$

$X = \frac{\beta_b}{\alpha + \beta_b}, K = (\alpha^2 + 2\alpha\beta_b)^{\frac{1}{2}}$ is water body attenuation coefficient; R' is effective scattering of water; R_b is the bottom reflectance of the water body; α is absorption coefficient; β_b is backscatter coefficient of water; D is indicate the water depth value.

The theoretical analytical method mainly obtains the water depth value by solving the physical parameters of each water body through the radiation transfer equation. Zhongliang Ping [16] used MSS-4 photos in the Jiaozhou Bay to derive a physical model of shallow sea water depth inversion based on the relationship between sea water transmittance, backscatter coefficient, sea floor reflectance and sea surface reflectance. Qidong Chen et al. [19] He derived a physical model of shallow sea depth inversion based on the radiative transfer equation and the optical thickness of the water body. In addition, he also considered the effects of chlorophyll and suspended sediment. Their experiments were applied in the Feilaixia Reservoir area in Guangdong and achieved good inversion results. Although this method has higher versatility, the calculation process requires a lot of physical and optical parameters, which is cumbersome and difficult to obtain. Therefore, it has not been widely promoted in practical applications.

Semi-theoretical Semi-empirical Method. The theoretical basis of the semi-theoretical semi-empirical method is the radiation attenuation of light in water. It is a method to achieve water depth inversion based on a combination of theoretical models and empirical parameters. This method mainly includes single-band linear regression method, dual-band linear regression method, multi-band linear regression method and logarithmic conversion ratio [3, 15–17].

1. Single-band linear regression method

The basis of the single-band linear regression method is the basement reflection water depth inversion model. Polcyn et al. [20] and Tanis et al. [21, 22] first applied this method:

$$L_i = L_{i\infty} + C_i R_B(\gamma) e^{-k_i f Z} \tag{3}$$

L_i is the amount of radiation in the i-band recorded by the satellite sensor; C_i is the constant of solar irradiance, atmospheric transmittance, water surface transmittance, and water surface refraction; $L_{i\infty}$ is the amount of radiation in the deep-water area; f is the length of the water path.

Although the theoretical inversion accuracy is high, the model relies on the conditions that are difficult to satisfy in practical applications. Therefore, scholars have improved the model to a certain extent to make it more suitable for water depth inversion in different waters. A water depth inversion method based on water body backscatter, as proposed by Wei [23]:

$$D = -\frac{1}{k} \ln(1 - (L_i - B)/C) \tag{4}$$

B, C is the coefficient of the equation. Determined by the water depth value D and the radiation value L_i.

Qing Hang et al. [24] used the single-band model to perform water depth inversion on the Dongjiang River channel in the Pearl River Delta region based on the Landsat TM3 and TM2 bands and obtain the river topographic map. Based on the single-band model, Shufang Tian et al. [25] used correlation analysis to make a quantitative analysis of water depth and established a remote sensing model of salt-lake water depth. Xiaolei Guo et al. [26] compared single-band, dual-band, and multi-band water depth inversion methods at Longwan Port on Hainan Island using WorldView-2 images. They concluded that although the single-band linear regression method is simple to calculate, has few parameters, and is easy to implement, the accuracy of the obtained water depth value is poor (see Table 1).

2. Dual-band linear regression method

The dual-band linear regression method is based on the single-band linear regression method. Its purpose is to weaken the influence of the water body attenuation coefficient and the absolute value of the water body bottom reflectivity in different water types and water body sediments. It can suppress errors caused by surface fluctuations, satellite scanning angles and solar elevation angles. The dual-band linear regression equation can be expressed as follow:

Table 1. Accuracy analysis of water depth inversion in linear regression model [26]

Linear regression model	R^2	Fitting formula	MAE/m	MRE (%)	RMSE/m
Single band	0.434	Z = −35.810 * (X_{B3}) + 218.361	2.20	88.46	2.66
Dual band ratio	0.669	Z = −56.552 * (X_{B3}/X_{B1}) + 256.246	2.05	69.74	2.43
Multi-band	0.611	Z = −7.404 * (X_{B3}) − 49.337 * (X_{B4}) + 12.113 * (X_{B5}) + 12.542 * (X_{B6}) + 198.585	2.15	58.17	2.55

Suppose there are any two bands that have a constant reflectivity on different substrates:

$$\frac{(R_{a1})^{c1}}{(R_{a2})^{c2}} = \frac{(R_{a1})^{c1}}{(R_{a2})^{c2}} = C \tag{5}$$

$$D = \frac{w_1}{2K_1}(lnR_{b1} + lnk_1 - X_1) + \frac{w_2}{2k_2}(lnR_{b2} + lnk_2 - X_2) \tag{6}$$

$X_1 = \ln[R_{E1} - R_{W1}]; X_2 = \ln[R_{E2} - R_{W2}]; w_1, w_2$ is dual-band weighting factor.

When $w_1 = \frac{C_1K_1}{C_1K_1 + C_2K_2}, w_2 = \frac{-C_2K_2}{C_1K_1 + C_2K_2}, d = \frac{w_1}{2K_1}lnk_1 - \frac{w_2}{2K_2}lnk_2$, we can obtain the final water depth expression D:

$$D = aX_1 + bX_2 + c \tag{7}$$

Because the dual-band linear regression method can overcome the shortcomings of the single-band linear regression method, it is more widely used. Qingjiu Tian et al. [27] used the TM3 and TM4 bands, which are sensitive to water depth, to perform water depth inversion in the offshore waters of Jiangsu. The predictions obtained from their experiments fit well with the measured water depth. Dong Zhang et al. [28] used the TM4 or TM2 band to establish a linear regression relationship to perform water depth inversion in the Yangtze Estuary waters also obtained good results. Ming Chen et al. [29] performed the ratio processing on the TM2 and TM4 band. What's more, they made 16 sections at 1 km intervals according to the size of the study area. They compared according to the water depth of each section and got a good result with an average agreement rate of 82%.

3. Multi-band linear regression method

The multi-band linear regression method is extended to N bands and the reflectivity of N different water bodies based on the dual-band linear regression method. John et al. [30] derived multi-band linear regression water depth inversion method based on e-exponential decay optical mode:

$$D = a_1X_1 + a_2X_2 + \ldots + c \tag{8}$$

Multi-band comprehensive consideration of all aspects of water depth information has higher water depth inversion accuracy than single-band and dual-band. So it is more widely used. Clark et al. [31] used multi-band linear regression and Landsat4 imagery to perform water depth inversion on islands in the southeastern coast of Puerto Rico in the Caribbean. They obtained the root mean square error between the predicted water depth and the measured water depth is 1.86 m. Kaichang Di et al. [32] introduced piecewise linear regression and data normalization techniques to improve the multi-band linear regression method. Therefore, this method is more suitable for large-scale sea water depth inversion with improved accuracy. Dianyuan Xu [33] uses the image gray values of MSS4, 5, and 7 to establish a relationship with the measured water depth. Then, based on this relationship, the predicted water depth distribution is obtained and the Yellow River waters are interpreted.

4. Logarithmic conversion ratio method

When using the linear regression method for water depth inversion, the difference between the radiance value received by the satellite sensor and the radiance value in the deep water region may be negative. Therefore, it may not be used in complex sea areas with large range and low radiance value. For this reason, Stumpf et al. [34] proposed a log-transformation ratio method by correcting the linear model:

$$D = m_1 \frac{\ln(nR_w(\lambda_i))}{\ln(nR_w(\lambda_j))} - m_0 \tag{9}$$

D is the water depth; m_1, n is a constant used to adjust the water depth ratio; m_0 is used to compensate for zero meter depth; $R_w(\lambda_i)$, $R_w(\lambda_j)$ is the reflectance of the corresponding band.

In the study of later water depth inversion, Zhen Tian [8] changed the adjustment factor of Stumpf model to two and compensated the water depth inversion results of different water bodies to some extent. And the inversion accuracy obtained by studying the waters around the East Island is higher than that of the Stumpf model. The improved model is as follows:

$$D = m_1 \frac{\ln(nR_w(\lambda_i))}{\ln(mR_w(\lambda_j))} + m_0 \tag{10}$$

Xiaolei Guo et al. [35] built a geo-adaptive model based on the log-ratio model to perform water depth inversion in the eight ports of Hainan. This model improves the problem of uneven substrate and water body in the global model inversion process. Therefore, the accuracy of water depth inversion can be improved and the error can be controlled within 1 m. Anna Chen et al. [36] based on GeoEye-1 multispectral imagery used log-linear models, log-transformation ratio models, and improved log-transformation ratio models to perform water depth inversion for Wuzhizhou Island and Penang. They concluded that the improved logarithmic conversion ratio model has the highest inversion accuracy and stability in different water quality and depth. but the logarithmic conversion ratio model is not good, while the logarithmic linear model is susceptible to seawater environment and not stable enough.

Statistical Methods. The statistical method is to obtain water depth data by analyzing the correlation between the gray value of the multispectral remote sensing image and the measured water depth value. Its mathematical expression is as follows:

$$D = f(X_1, X_2 \ldots X_n) \tag{11}$$

D is the water depth value; $X_i(i = 1, 2, \ldots n)$ is the band of the image.

Lyzenga et al. [18, 37] used principal component analysis to assume that the gray value of the image is linearly related to the water depth value:

$$X_i = ln(L_i - L_{si}) \tag{12}$$

Generate new variables Y_i by rotating the coordinate system on X_i;

$$Y_i = \sum_{j=1}^{N} A_{ij}X_j \tag{13}$$

If this is a pure rotation, then only Y_i is a variable related to water depth, and the other $i - 1$ variables are only related to the bottom reflectance; the water depth variable can be expressed as:

$$Y_N = B_m - C_z \tag{14}$$

B_m is the bottom composition function, C_z determined by the water attenuation coefficient. This method provides a more accurate water depth value by selecting the appropriate band in a clear ocean with a depth of no more than 15 m.

Artificial neural networks [38] is also a type of statistical method often used in water depth inversion. Because of its self-learning, self-organizing, adaptive and nonlinear dynamic processing, it has more powerful approximation ability than the traditional statistical method in simulating nonlinear change systems. The network of error reverse propagation algorithm, referred to as BP neural network [39]. It takes the spectral characteristic value of the image as input and the water depth value as the desired output. The simplest structure is usually three layers of an input layer, an intermediate layer (hidden layer) and an output layer (see Fig. 1).

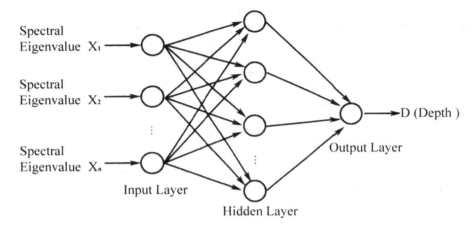

Fig. 1. Schematic diagram of BP neural network structure [40]

BP network algorithm weight formula is as follows:

$$E = \frac{1}{2} \sum_{j=1}^{m} (Y_j - \hat{Y}_j)^2 \tag{15}$$

Y_j is the target value, $\hat{Y}_j$ is the expected output value, m is the number of neurons in the output layer.

$$w_{ij}(p+1) = w_{ij}(p) + \alpha \Delta w_{ij}(p+1) - \mu \frac{\vartheta E(p)}{\vartheta w_{ij}(p)} \tag{16}$$

α is the momentum coefficient μ is the learning coefficient and the range of values is between 0 and 1.

Although BP neural network has powerful computing power, it is easy to cause over-fitting, slow convergence and sensitivity to initial weight. What's more, it requires a large amount of known water depth data as a training sample, which limits its wide application. Bin Cao et al. [40] based on the traditional BP neural network algorithm and overcome its shortcomings proposed an improved algorithm for water depth inversion. Cao used this algorithm to experiment on Ganquan Island in Sansha, Hainan. The results show that the improved BP network algorithm has relatively high accuracy and fast convergence speed. Yanjiao Wang et al. [3] used 3 layers of BP neural network and 600 water depth samples as training samples to better invert the water depth of the Nangang section of the Yangtze River estuary, with an average absolute error of 0.9 m.

2.2   Water Depth Inversion Method of Hyperspectral

The resolution of hyperspectral spectrum reaches nm level, and the number of bands is several dozen more than that of multi-spectrum. Therefore, there are many advantages to using water depth detection in its image. Hyperspectral spectrum can provide multiple bands for band combination. What's more, Hyperspectral spectrum use its rich image band information to identifies waterborne sediments and water types, reducing the impact of other factors in water depth inversion [8].

The Artificial Neural Network Method. The mathematical model is the same as the BP neural network in the statistical method. Yingni Shi [9] uses the principal component analysis of the input variables and normalizes the peaks to improve the learning speed of the neural network. And then, perform water depth inversion experiments in the study area. They concluded that the water depth based on the inversion of hyperspectral image neural network algorithm is better than the semi-analytical model. Sandidge et al. [41] used the BP neural network algorithm to perform water depth inversion in the waters around Florida based on the correlation established between the hyperspectral image and the measured water depth. The results show that the artificial neural network predicts the water depth is consistent with the measured water depth, which is superior to the traditional statistical methods.

Semi-analytical Model Method. Hyperspectral Optimization Process Exemplar (HOPE) is the main method for current water depth inversion of hyperspectral remote sensing. The core of the algorithm is to establish the functional relationship between remote sensing image reflectivity, water body, water body sediment and water depth. This method was first proposed by Lee et al. [42, 43] and its expression is as follow:

$$R_{rs}(\lambda) = f[a(\lambda), b_b(\lambda), \rho(\lambda), H, \theta_\omega, \theta_v, \varphi] \tag{17}$$

$a(\lambda)$ is the water absorption coefficient; $b_b(\lambda)$ is the backscatter coefficient of water; $\rho(\lambda)$ is the water body reflectivity; H is the water depth; θ_ω is the lower surface solar zenith angle; θ_v is the lower surface nadir angle of view; φ is the field of view azimuth.

This model has introduced large number of water and physical parameters, so it's comprehensive calculation with small error has been widely used. Lee et al. [44] used this model algorithms and AVIRIS hyperspectral images to perform experiments in Tampa Bay, Florida has obtain better water depth inversion results. Mcintyre M L et al. [45] used a semi-analytical model and hyperspectral imagery for water depth inversion in the 10-15 m waters of Florida. Obtained the average deviation of the predicted water depth is 4.9% and the RMS is 7.83%. Lee et al. [45] Based on the water depth inversion model and use the NASA EO-1 hyperspectral image with a resolution of 30 m has obtained an average error of only 11% compared to LIDAR. Zhen Liu et al. [7] used the HOPE algorithm and semi-analytical model perform water depth and optical parameters inversion of the South Island reef. Their experiment obtained a good result. For other methods, Zhang et al. [46] used a lookup table approach to reverse the water depths of the waters around Peter Island in the Virgin Islands. Zhishen Liu [47] inverted the water depth of the Smith Island Bay using Principal Component Analysis and obtained the error in the deep-water area is 1 m and the shallow water zone error is 0.2 m.

3   The Method of Synthetic Aperture Radar (SAR) Water Depth Inversion

3.1   The Principle of Synthetic Aperture Radar (SAR) Water Depth Inversion

The working band of SAR does not directly penetrate seawater to detect underwater terrain. It indirectly measures underwater terrain by receiving backscattering from the sea surface, because the undulations of the underwater terrain change the backscattering intensity of the sea surface as a change of bright and dark stripes on the radar image.

3.2   The Inversion Method Water Depth Estimation System

The water depth estimation system [11, 48] is currently a highly accurate algorithm for shallow seawater deep inversion using radar images. This algorithm was developed by the Dutch company ARGOSS. It consists of a SAR image simulation model and a data assimilation model. The schematic of the BAS system is as Fig. 2:

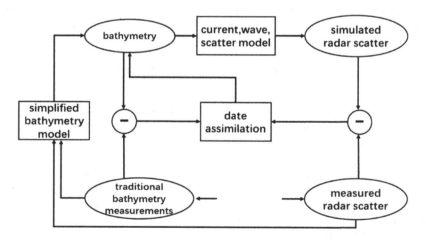

Fig. 2. BAS water depth estimation system [49]

Calkoen et al. [49–51] used the BAS technique to conduct a water depth estimation test at the tidal inlet between the small islands of Ameland and Schiermonnikoog in Waddenzee, northern Netherlands. The mean square error of water depth obtained by this work is 0.36 m. However, the shortcoming is that the method is more troublesome to measure in shallow waters of 30 m and requires intensive measurement of water depth sampling spacing.

3.3 The Inversion Method of Direct Water Depth

Direct water depth inversion is based on the interaction of three models of radar underwater terrain imaging and reuse simulation model to obtain water depth data. Among them, the simulation model includes the Nevi-Stokes equation, the continuity equation, the spectral action balance equation, and the radar backscatter model [1, 2]. The specific calculation steps are as Fig. 3:

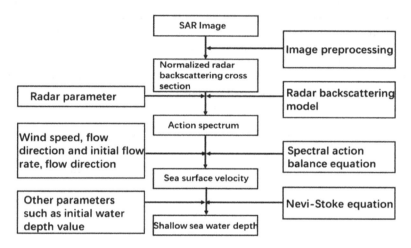

Fig. 3. Direct water depth inversion

$$\frac{\partial V_x}{\partial_t} + V_x\frac{\partial V_x}{\partial_x} + V_y\frac{\partial V_x}{\partial_y} - FV_y + g\frac{\partial \xi}{\partial_x} + g\frac{V_x\sqrt{V_x^2 + V_y^2}}{c^2(h+\xi)} - \frac{\tau_x}{\rho(h+\xi)} = 0 \qquad (18)$$

$$\frac{\partial V_y}{\partial_t} + V_x\frac{\partial V_y}{\partial_x} + V_y\frac{\partial V_y}{\partial_y} - FV_x + g\frac{\partial \xi}{\partial_y} + g\frac{V_y\sqrt{V_x^2 + V_y^2}}{c^2(h+\xi)} - \frac{\tau_y}{\rho(h+\xi)} = 0 \qquad (19)$$

$$\frac{\partial \xi}{\partial_t} + \frac{\partial[(h+\xi)V_x]}{\partial_x} + \frac{\partial[(h+\xi)V_y]}{\partial_y} = 0 \qquad (20)$$

The above Eq. (18) is the Nevi-Stokes equation and the (19) continuity equation. V_x, V_y is the water flow velocity in the X and Y directions, ξ is the relative horizontal potential height, h. The distance between the bottom of the sea and the horizontal potential surface, F is the Coriolis parameter, c is XieCai parameter, τ_x, τ_y is the wind stress in the X and Y directions, ρ is the wind stress in the X and Y directions, g is gravitational acceleration [1, 2, 10].

$$\frac{dA\left(\vec{r},\vec{k},t\right)}{d_t} = \left(\frac{\partial}{\partial_t} + \frac{d\vec{r}}{d_t}\frac{\partial}{\partial\vec{r}} + \frac{d\vec{k}}{d_t}\frac{\partial}{\partial\vec{k}}\right)A\left(\vec{r},\vec{k},t\right) = S_r\left(\vec{r},\vec{k},t\right) \qquad (21)$$

The above formula (21) is the spectral action balance equation; $A\left(\vec{r},\vec{k},t\right)$ is the action spectrum, $\vec{r} = (x,y)$ representing spatial variables, $\vec{k}$ is the representative wave number, $S_r\left(\vec{r},\vec{k},t\right)$ is the original function representing the wave composition wave.

The radar backscattering model [1, 2, 11] has a two-scale model, Bragg model and Kirchoff scattering model. The two-scale model is mainly used for rough sea surfaces that cannot clearly distinguish the undulating scale. The kirchhoff scattering model is mainly used for large-scale undulating rough sea surface and the situation of 0°–20° satellite incident angle. However, The Bragg model is mainly used for small-scale undulating rough sea surface and the situation of 20°–70° satellite incident angle. Because the incident angle of spaceborne synthetic aperture radar is mostly 20°–70°, most researchers use this model as the radar backscattering model.

$$\sigma_{pq}^0(\theta) = 16\pi k^4 cos^4(\theta)\left|G_{pq}(\theta)\right|^2 \varphi(k_B,0) \qquad (22)$$

p, q related to polarization, wave number, angle of incidence, and function G_{pq}, $k_B = 2ksin\theta$ is the Bragg wave number, G_{pq} function expression is as follow:

$$G_{HH}(\theta) = \frac{\varepsilon_r - 1}{\left(cos\theta + \sqrt{\varepsilon_r - sin^2\theta}\right)^2} \qquad (23)$$

$$G_{VV}(\theta) = \frac{\varepsilon_r - 1[\varepsilon_r(1 + sin^2\theta) - sin^2\theta]}{\left(\varepsilon_r cos\theta + \sqrt{\varepsilon_r - sin^2\theta}\right)^2} \qquad (24)$$

ε_r Representing the complex permittivity of seawater.

Bin Fu [1, 2] inversion of the Xiaoyinsha sea area in Keelung Island and Jiangsu offshore by direct inversion method. The root mean square error of the retrieved water depth value and chart water depth value in the Xiaoyinsha area is 0.42 m. Xiaolei Bi [52] used the direct inversion method to carry out the water depth test in the Taiwan Strait and classified according to radar polarization. Draw the best conclusion from the results of the full-polarization SAR inversion.

3.4   The Inversion Method of Wavenumber Spectrum Balance Equation (WSBE)

Based on the Bragg model, large-scale background flow field model, and two-dimensional shallow water dynamics equations, the steepest descent method is used to iteratively obtain the shallow sea depth [11]. The numerical model is as follows:

$$G(V_x, V_y, \theta_1) = cos^2\theta_1 \frac{\partial V_x}{\partial x} + cos\theta_1 sin\theta_1 \times \left(\frac{\partial V_x}{\partial y} + \frac{\partial V_y}{\partial x}\right) + \frac{\partial V_y}{\partial y} sin^2\theta_1 \quad (25)$$

(25) is a first-order approximate analytical expression of the gradation value of the SAR image; wherein, θ_1 is the sea surface wind direction.

$$\frac{\partial V_x}{\partial t} + V_x \frac{\partial V_x}{\partial x} + V \frac{\partial V_x}{\partial y} - fV_y = -g\frac{\partial \xi}{\partial x} + \frac{cV_x}{(h+\xi)} \quad (26)$$

$$\frac{\partial V_y}{\partial t} + V_x \frac{\partial V_y}{\partial x} + V \frac{\partial V_y}{\partial y} - fV_x = -g\frac{\partial \xi}{\partial y} + \frac{cV_y}{(h+\xi)} \quad (27)$$

$$\frac{\partial \xi}{\partial t} + \frac{\partial((h+\xi)V_x)}{\partial x} + \frac{\partial((h+\xi)V_y)}{\partial y} = 0 \quad (28)$$

(26) (27) (28) is the two-dimensional shallow hydrodynamic equation.

$$J(V_x, V_y, h, \xi) = \int_{D_0} (f_1^2 + f_2^2 + f_3^2 + f_4^2) dxdy \quad (29)$$

(29) is targeted function $J(V_x, V_y, h, \xi)$, wherein:

$$f_1 = \alpha_1 \left(-fV_x + g\xi_x + \frac{cV_x}{h+\xi} + V_x V_{xx} + V_x V_{xy} + V_{xt}\right) \quad (30)$$

$$f_2 = \alpha_2 \left(fV_x + g\xi_y + \frac{cV_x}{h+\xi} + V_y V_{yy} + V_x V_{xy} + V_{yt}\right) \quad (31)$$

$$f_3 = \alpha_3 \left[\left(l_m l_n + \frac{1}{2} l_\alpha l_\beta \xi_{\alpha\beta} L_m L_n\right) \frac{\partial u_m}{\partial x_n} - G\right] \quad (32)$$

$$f_4 = \alpha_4 \left(V_x(h+\xi)_x + V_y(h+\xi)_y + V_{xx}(h+\xi) + V_{yy}(h+\xi) + \xi_t\right) \quad (33)$$

Changshui Xia [53] used the wave number balance equation to invert the water depth in the Tanggu sea area. They found that the inversion water depth and the actual water depth are in good agreement. The average relative error of the two sections taken in the image is 8.9% and 9.2%, respectively. Jungang Yang et al. [54] used WNSBE technology and two ERS-2 images imaged at different time to invert the water depth near the Taiwan Strait. Experiments show that the smallest inversion error can reach 2.23 m (Fig. 4).

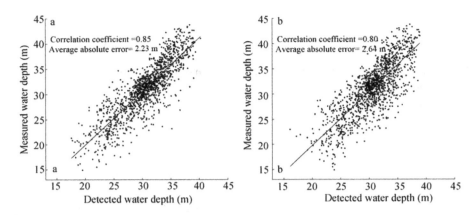

Fig. 4. Water depth inversion comparison chart [54]

Zejun Li et al. [55] improved the method on the basis of Jungang Yang scholars. He used this method and ERS-2 and ENVISAT data to conduct water depth inversion experiments to obtain ideal inversion results (Table 2):

Table 2. Water depth inversion results [55]

Name	Absolute variation error	Relative variation error
ERS-2 data	3.8 m (8.9%)	1.18 m (27%)
ENVISAT data	4.5 m (10.8%)	2.94 m (69%)

The method of retrieving the water depth based on the synthetic aperture radar image is not directly obtained water depth information on image. So, the water depth obtained by the interaction calculation of shallow sea topography, sea surface current, wind speed and the like. The BAS method requires more measured water depth information when calculating. The more accurate the water depth data is, the higher accuracy the experimental results will be. Submarine depth inversion is required in areas with complex underwater terrain or large height differences. Otherwise, the error is large. The direct water depth inversion method has higher requirements on the quality of SAR images. What's more, the image noise contributes a lot to the water depth error. The image gray scale simulation model of the wave number balance equation has higher requirements on wind direction. So, it's calculation of water depth is more complicated than the former two.

4 Prospect of Remote Sensing Water Depth Inversion Technology

The remote sensing water depth inversion technology has formed a relatively complete theoretical and practical basis from the 1960s to the present. This paper expounds the research results of remote sensing water depth inversion from the aspects of optical remote sensing and synthetic aperture radar remote sensing. Inductively analyze its advantages and disadvantages in order to present a rich and comprehensive review article for readers. As we all know, water depth inversion still has large uncertainties, such as remote sensing image deformation caused by atmospheric and topographical effects, low resolution of image, weak water permeability of water band, complex water reflection mechanism, and uncertainties in water body. Therefore, improving the accuracy of water depth inversion is a systematic problem. It is necessary to consider not only the optimization of inversion methods, but also the means of obtaining data and the method of improving data preprocessing.

From the previous review, we can see that the main problems of water depth inversion technology are as follows:

1. Remote sensing image preprocessing

Since the imaging process of satellites is affected by sensors, atmosphere and terrain. If we want to obtain a real image, it is necessary to perform atmospheric correction, geometric correction, and noise filtering on the original image. The image cannot be realistically restored in various pre-processing, and other errors may be introduced. For example, atmospheric correction processing needs to use different models for different remote sensing images to correct the effects of atmospheric molecules such as aerosols. In addition, the image resolution and the measured water depth are not an exact match relationship, which is a pixel value corresponding to a range of water depths. That is, the water depth data we actually measured is the precise water depth of a certain point. But the image such as a pixel of the Sentinel_2 green band corresponds to a range of $10 \text{ m} \times 10 \text{ m}$ on land. What's more, for complex seabed topography, this range of water depth changes will be large, which undoubtedly affects the accuracy of water depth inversion.

2. Nautical water depth

Water depth inversion is mainly applied to uninhabited sea areas. So, the measured water depth often depends on the water depth data on the chart. The water depth data on the chart is the reference surface depth at a certain point, but the instantaneous water depth at a certain point is recorded during image imaging. Therefore, it is necessary to change the chart water depth to the synchronized instantaneous water depth based on the time of image formation and the tide data. This correction process will undoubtedly introduce errors and lead to low accuracy of later water depth inversion.

3. Water depth division

The water depth inversion technology is not only regional, but also has different water depth intervals. This is mainly because different wave bands penetrate water bodies at different depths and carry different amounts of water depth information. Therefore, for the same inversion method, it is also necessary to perform inter-zone inversion of water depth to ensure better inversion results.

4. Water quality

Most of the water depth inversion are clear Class I water bodies. These water suspensions are less, not turbid, and have a homogeneous bottom material, so the water depth inversion is not suitable for all water bodies. If the water bodies is not clear, the sun's light waves will be reflected and scattered by the suspended sediment in the water after entering the water body, which weakens the emissivity of the water. On the other hand, if the water contains a lot of impurities such as chlorophyll and plankton, these impurities will also reflect and absorb sunlight. Only a small amount of light waves are absorbed and reflected by the water bottom and then captured by the receiver. However, if the substrate is not uniform, the reflected light wave will be further weakened. What's more, The weakly reflected light received by the receiver will also contain spectral information of impurities such as chlorophyll. If these problems are not dealt with in later water depth inversion models, this will cause large errors in experimental results.

5. Synthetic Aperture Radar Inversion Difficulties

Although Synthetic Aperture Radar has the advantage of being weather-free throughout the day, the inversion method is computationally intensive. This method is greatly affected by terrain slope, sea surface waves, and wind speed, so that its inversion accuracy is poor. This is why it is difficult to be widely used.

Nevertheless, the water depth inversion technology is still an important water depth detection technology for current sea chart sounding and marine engineering safety. With the launch of high-resolution satellites, hyperspectral satellites, and ocean sounding radars, researchers will explore more accurate water depth inversion techniques in this area, such as:

- Combined inversion of multiple data sources: Hyperspectral satellite data combined with Multi-spectral satellite data. Inversion of Airborne Lidar Data with Multispectral and Hyperspectral Data. Long-term sequence satellite data inversion. Synthetic Aperture Radar combined with Optical image inversion.
- Classification of water quality parameters: Classification of suspended sediment, chlorophyll, and bottom sediment in water. Then use it as an influence factor to establish a relevant model for water depth inversion.
- Water body inversion by area and depth by depth: Proper partitioning of a wide range of water bodies and then selecting a matching model for water depth inversion. So, we can get more accurate water depth data. Different depths of depth are also used for segmentation inversion. Selecting appropriate water depth inversion models for different water depths makes the fitting effect better and reduces the inversion error.
- Artificial intelligence technology water depth inversion: Autonomous training of large amounts of water depth data and remote sensing data using algorithms of machine learning and deep learning. Use its excellent computing power and fitting

accuracy to obtain accurate inversion water depth and achieve automatic inversion. Of course, this also needs to solve the problem of various data source fusion, time series problems [3, 14, 17].

Remote sensing water depth inversion has the advantages of low cost, large range and real-time. It complements traditional water depth measurement methods and has broad development prospects. If we can make a major breakthrough in data acquisition and inversion methods, there will be a wider range of practical applications.

Acknowledgments. This study was jointly supported by the National Natural Science Foundation of China (41976189, 41976190); Special Fund for Development of Guangdong Academy of Sciences (2019GDASYL-0502001); China Southern Power Grid Guangzhou Power Supply Bureau Co., Ltd. Key Technology Project (0877002018030101SRJS 00002); Guangdong Provincial Science and Technology Program (2017B010117008); Guangzhou Science and Technology Program (201806010106, 201902010033); the Guangdong Innovative and Entrepreneurial Research Team Program (2016ZT06D336); the Southern Marine Science and Engineering Guangdong Laboratory (Guangzhou) (GML2019ZD0301); the GDAS's Project of Science and Technology Development (2016GDASRC-0211, 2018GDASCX-0403, 2019GDASYL-0301001, 2017GDASCX-0101, 2018GDASCX-0101).

References

1. Bin, F.: Shallow Sea Bottom Topography Mapping by SAR. Ocean University of China, Qingdao (2005)
2. Cao, B.: A Study of Remotely-Sensed Data Processing in Bathymetry. Information Engineering University, Zhenzhou (2017)
3. Wang, Y., Dong, W., Zhang, P.: Advances in water depth visible light sensing methods. Marine Notif. **26**(5), 92–101 (2007)
4. Ma, Y., Zhang, J., Zhang, J., Zhang, Z., Wang, J.: Progress in shallow water depth mapping from optical remote sensing. Adv. Marine Sci. **36**(3), 331–351 (2018)
5. Huang, W., Wu, D., Yang, Y., et al.: Shallow sea multi-spectral remote sensing water depth inversion technique. J. Marine Technol. **32**(2), 43–46 (2013)
6. Bierwirth, P.N., Lee, T.J., Burne, R.V.: Shallow sea-floor reflectance and water depth derived by unmixing multispectral imagery. PE & RS **59**(3), 331–338 (1993)
7. Liu, Z., Hu, L., He, M.: Inversion of shallow water depth and optical parameters around Islands and reefs in the South China Sea by EO-1/Hypeion data. Period. Ocean Univ. China **44**(5), 101–108 (2014)
8. Tian, Z.: Research on Deep Multi/High Spectral Remote Sensing Model and Water Depth Topographic Map Production Technology in Shallow Seawater. Shandong University of Science and Technology, Qingdao (2015)
9. Shi, Y.: Ultra-spectral Remote Sensing Shallow Seawater Inversion Based on Artificial Neural Network Technology. Ocean University of China, Qingdao (2005)
10. Wang, X.: Remote Sensing Imaging Mechanism and Inversion of Typical Underwater Terrain by SAR in Shallow Sea. Zhejiang University, Zhejiang (2018)
11. Fan, K., Huang, W., He, M., et al.: Overview of remote sensing technology for shallow seawater topography by SAR. Progr. Geophys. **24**(2), 714–720 (2009)

12. Eugenio, F., Marcello, J., Martin, J.: High-resolution maps of bathymetry and benthic habitats in shallow-water environments using multispectral remote sensing imagery. IEEE Trans. Geosci. Remote Sens. **53**(7), 3539–3549 (2015)
13. Jiran, L.I., Zhang, H., Hou, P., et al.: Mapping the bathymetry of shallow coastal water using single-frame fine-resolution optical remote sensing imagery. Acta Oceanologica Sinica **35** (1), 60–66 (2016)
14. Zhao, Y.: Principles and Methods of Remote Sensing Application Analysis. Science Press, Beijing (2013)
15. Zhang, Y.: Study of fathoming method by RS technology. J. Hohai Univ. **26**(6), 68–72 (1998)
16. Ping, Z.: Mathematical model of visible light remote sensing sounding. Ocean Lake Marsh **13**(3), 225–230 (1982)
17. Ye, M., Li, R., Xu, G.: Multi-spectral water depth remote sensing methods and research progress. World Sci. Technol. Res. Dev. **29**(2), 76–79 (2007)
18. Lyzenga, D.R.: Passive remote sensing techniques for mapping water depth and bottom features. Appl. Opt. **17**(3), 379 (1978)
19. Chen, Q., Deng, R., Qin, Y., et al.: Remote sensing of water depth in Feilaixia Reservoir Area, Guangdong Province. J. Sun Yat-Sen Univ. (Nat. Sci. Ed.) **51**(1), 122–127 (2012)
20. Polcyn, F.C., Sattinger, I.J.: Water depth determinations using remote sensing techniques. Remote Sens. Environ. **II**, 13–16 (1969)
21. Tanis, F.J., Hallada, W.A.: Evaluation of landsat thematic mapper data for shallow water bathymetry. In: Proceeding of 18th International Symposium on Remote Sensing of Environment, Ann Arbor, Michigan, pp. 629–643 (1984)
22. Tanis, F.J., Byrne, H.J.: Optimization of multispectral sensors for bathymetry applications. In: Proceeding of 19th International Symposium on Remote Sensing of Environment, Ann Arbor, Michigan, pp. 865–874 (1985)
23. Wei, J., Daniel, L.C., William, C.K.: Satellite remote bathymetry: a new mechanism for modeling. Photogram. Eng. Remote Sens. **58**(5), 545–549 (1992)
24. Hang, Q., Wang, X.: Remote sensing research methods and applications of river evolution. J. Sun Yat-Sen Univ. Nat. Sci. Ed. **38**(5), 109–113 (1999)
25. Tian, S., Hong, Y., Qin, X.: Remote sensing of water depth in high-concentration salt lakes. Remote Sens. Land Resour. **18**(1), 26–30 (2006)
26. Guo, X., Qiu, Z., Shen, W., et al.: Shallow water depth inversion in Longwan port based on WorldView-2 remote sensing image. J. Marine Sci. **35**(3), 27–33 (2017)
27. Tian, Q., Wang, J., Du, X.: Remote sensing study of Jiangsu coastal water depth. J. Remote Sens. **11**(3), 373–379 (2007)
28. Zhang, D., Wang, W., Zhang, Y.: Remote sensing of water depth in the Yangtze River estuary. J. Hohai Univ.: Nat. Sci. Ed. **26**(6), 86–90 (1998)
29. Chen, M., Li, S., Kong, Q.: Satellite remote sensing water depth in the Yangtze River estuary. J. Water Resour. Water Eng. **2003**(2), 61–64 (2003)
30. John, M.P., Robert, E.S.: Water depth mapping from passive remote sensing data under a generalized ratio assumption. Appl. Opt. **22**(8), 1134–1135 (1983)
31. Clark, R.K., Fay, T.H., Walker, C.L.: Bathymetry calculations with Landsat 4 TM imagery under a generalized ratio assumption. Appl. Opt. **26**(19), 4036–4038 (1987)
32. Di, K., Qian, D., Wei, C., et al.: Shallow water depth extraction and chart production from TM image in Nansha Islands and nearby sea area. Remote Sensing of Land and Resources **41**(3), 59–64 (1999)
33. Xu, D.: Remote sensing study on water depth distribution in the Yellow River estuary. Remote Sens. Technol. Appl. **7**(3), 17–23 (1992)

34. Stumpf, R.P., Holderied, K., Sinclair, M.: Determination of water depth with high-resolution satellite imagery over variable bottom types. Limnol. Oceanogr. **48**(1), 547–556 (2003)
35. Guo, X., Qiu, Z., Tan, Q., et al.: Shallow water depth extraction from OLI remote sensing image in Basuo port. Hydrogr. Surv. Charting **37**(6), 54–57 (2017)
36. Chen, A., Ma, Y., Zhang, J.: Applicability of water depth passive optical remote sensing inversion model. Marine Environ. Sci. **37**(6), 953–960 (2018)
37. Lyzenga, D.R.: Remote sensing of bottom reflectance and water attenuation parameters in shallow water using aircraft and Landsat data. Int. J. Remote Sens. **2**(1), 71–82 (1981)
38. Li, X., Xu, Y., Wang, Y., et al.: Establishment and application of BP artificial neural network adaptive learning algorithm. Syst. Eng. Theory Pract. **24**(5), 1–8 (2004)
39. Wang, Y., Zhang, Y.: Study on remote sensing of water depth based on BP artificial neural networks. Ocean Eng. **23**(4), 33–38 (2005)
40. Cao, B., Qiu, Z., Zhu, S., et al.: Improvement of BP neural network remote sensing water depth inversion algorithm. Bull. Surv. Mapp. **2017**(02), 40–44 (2017)
41. Sandidge, J.C., Holyer, R.J.: Coastal bathymetry from hyperspectral observations of water radiance. Remote Sens. Environ. **65**(3), 341–352 (1998)
42. Lee, Z., Carder, K.L., Mobley, C.D., et al.: Hyperspectral remote sensing for shallow waters. 2. Deriving bottom depths and water properties by optimization. Appl. Opt. **38**(18), 3831–3843 (1999)
43. Lee, Z., Carder, K.L., Chen, R.F., et al.: Properties of the water column and bottom derived from airborne visible infrared imaging spectrometer (AVIRIS) data. J. Geophys. Res. Oceans **106**(C6), 11639–11651 (2001)
44. Mcintyre, M.L., Naar, D.F., Carder, K.L., et al.: Coastal bathymetry from hyperspectral remote sensing data: comparisons with high resolution multibeam bathymetry. Mar. Geophys. Res. **27**(2), 129–136 (2006)
45. Lee, Z.P., Casey, B., Arnone, R.A., et al.: Water and bottom properties of a coastal environment derived from hyperion data measured from the EO-1 spacecraft platform. J. Appl. Remote Sens. **1**(1), 011502 (2007)
46. Zhang, L., Teng, H., Meng, C., et al.: Hyperspectral remote sensing water depth detection method based on semi-analytical model. Ocean Mapp. **31**(4), 17–21 (2011)
47. Liu, Z., Zhou, Y.: Direct inversion of shallow-water bathymetry from EO-1 hyperspectral remote sensing data. Chin. Opt. Lett. **9**(6), 060102 (2011)
48. Huang, W., Fu, B., Zhou, C., et al.: Research on remote sensing method of spaceborne synthetic aperture radar in shallow seawater. In: National Remote Sensing Technology Academic Exchange Conference (2001)
49. Calkoen, C.J., Hesselmans, G.H.F.M., Wensink, G.J., et al.: The bathymetry assessment system: efficient depth mapping in shallow seas using radar images. Int. J. Remote Sens. **22**(15), 2973–2998 (2001)
50. Wensink, G.J., Hesselmans, G.H.F.M., Calkoen, C.J., et al.: The bathymetry assessment system. Oceanography **62**(97), 214–223 (1997)
51. Hennings, I., Metzner, M., Calkoen, C.J.: Island connected sea bed signatures observed by multi-frequency synthetic aperture radar. Int. J. Remote Sens. **19**(10), 1933–1951 (1998)
52. Bi, X.: SAR Detection Model for Shallow Seawater Terrain Based on Polarization Information. China University of Petroleum, Qingdao (2010)
53. Xia, C., Yuan, Y.: SAR image simulation and inversion of seabed topography in Tangguhai area. Adv. Marine Sci. **21**(4), 437–445 (2003)
54. Yang, J., Zhang, J., Meng, J.: Using SAR images to detect underwater Terrain in shallow seas of Taiwan. Chin. J. Ocean Lakes **28**(3), 636–642 (2010)
55. Li, Z., Wang, X., Yu, X., et al.: An improved method for SAR image inversion of shallow sea topography. Electron. Meas. Technol. **2012**(4), 86–89 (2012)

Remote Sensing Monitoring of Resources and Environment and Intelligent Analysis

Research on Rapid Extraction Method of Urban Built-up Area with Multiple Remote Sensing Indexes

Yuntao Ma[1], Zhiwei Xie[1(⊠)], Yingchun You[1], and Xin Xuan[2]

[1] School of Transportation Engineering, Shenyang Jianzhu University,
Shenyang 100168, China
zwxrs@sina.com
[2] College of Geomatics, Shandong University of Science and Technology,
Qingdao 266590, China

Abstract. The urban built-up area is a key monitoring and statistics data of national and provincial statistical departments, and has become an important scale to reflect the level of urban development and predict the potential of urban development. Based on the Landsat 8 images, by using the way of spectral signature analysis, the study selected four indices, Modified Normalized Difference Built-up Index (MNDBI), Normalized Difference Vegetation Index (NDVI), Modified Normalized Difference Water Index (MNDWI) and a set of mineral index (Clay, Iron, Ferrous), and combined with the band operation and K-means unsupervised classification algorithm to extract the range of urban built-up area. Taking Shenyang as an example, the method proposed in this paper was used to extract the built-up area of Shenyang and the Kappa coefficient of the extracted built-up area range was 0.9091 after the post-classification treatment and refinement treatment of the built-up area.

Keywords: Landsat · Built-up area · MNDBI · NDVI · MNDWI

1 Introduction

The urban built-up area refers to the area within the urban administrative region that has actually been developed and constructed, been provided with the municipal public utilities and public facilities basically. Accurate acquisition of urban built-up area has important guiding significance for urban construction, management and research, and is also one of the important indicators reflecting the comprehensive economic strength and urbanization level of a city [1, 2]. However, urban built-up area is a complex land use type, which cannot be accurately extracted from remote sensing images by simple methods [3]. DMSP/OLS nighttime light data are widely used at present [4], but it is susceptible to the problems of saturation, diffusion and low resolution of nighttime light data, so it is still a great challenge to extract urban built-up area by DMSP/OLS data alone [5]. Object-oriented image segmentation method requires multiple experiments to select appropriate threshold, segmentation method and rule set, which takes a lot of time [6, 7]. And the index-based information extraction of built-up area has been increasingly applied [3, 8] due to its simplicity and ease of use. The paper based on

© Springer Nature Singapore Pte Ltd. 2020
Y. Xie et al. (Eds.): GSES 2019/GeoAI 2019, CCIS 1228, pp. 199–209, 2020.
https://doi.org/10.1007/978-981-15-6106-1_15

Modified Normalized Difference Built-up Index (MNDBI) [9], Modified Normalized Difference Water Index (MNDWI) [10], and Normalized Difference Vegetation Index (NDVI) [11], combined with a group of mineral indexes and unsupervised classification methods [9, 12], proposed a rapid extraction method of urban built-up area based on multiple indexes. The overall accuracy of extraction results after post classification and refinement of built-up area can reach 97.14%, and the Kappa coefficient is 0.9091.

2   Extraction and Post Classification of Built-up Area

2.1   Extraction of Built-up Area

Ground objects in the urban built-up area of Shenyang can be divided into five types: built-up area, forest land, arable land, bare soil and water [9, 10]. This paper randomly selected 2,500 pixel points as samples, 500 pixel points for each type, and the reflection spectrum characteristic curve of ground objects is shown in Fig. 1. The spectral signatures of five types of ground objects was analyzed through the reflection spectrum characteristic curves, which shows the differences between bare soil and built-up area in Swir1 was very small. So if the Normalized Difference Built-up Index (NDBI) [8] (Formula (1)) was selected, the bare soil would be strengthened at the same time as the built area, which is not conducive to the accurate extraction of the built area. This study selected Modified Normalized Difference Built-up Index (MNDBI) proposed by

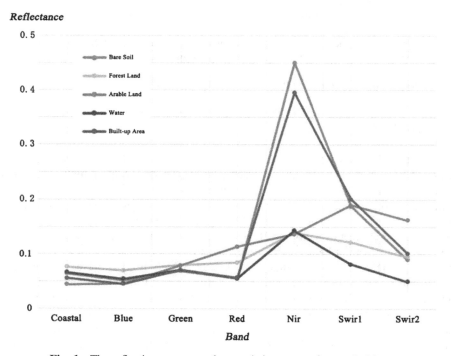

Fig. 1. The reflection spectrum characteristic curves of ground objects

Yueyao Hu to preliminary extract the built-up area of Shenyang. MNDBI used Swir2 instead of Swir1 band of NDBI, enhanced the difference between the built-up area and the bare soil, and effectively reduced the problem of miscellany between bare soil and built-up area [9]. The MNDBI formula is shown in Formula (2).

$$NDBI = \frac{Swir1 - Nir}{Swir1 + Nir} \tag{1}$$

Where $Swir1$ is the reflectance of band 6 Shortwave Infrared 1 (1.57–1.65 μm), Nir is the reflectance of band 5 Near-Infrared (0.85–0.88 μm).

$$MNDBI = \frac{Swir2 - Nir}{Swir2 + Nir} \tag{2}$$

Where $Swir2$ is the reflectance of band 7 Shortwave Infrared 2 (2.11–2.29 μm).

Using MNDBI alone to extract the urban built-up area cannot avoid the interference of water and vegetation in and around the urban built-up area, so vegetation index and water index were introduced to reduce the influence caused by the wrong classification of vegetation and water into built-up area. The Normalized Difference Vegetation Index (NDVI) proposed by Rouse was selected, as shown in Formula (3). The Modified Normalized Difference Water Index (MNDWI) was selected because it still has good extraction effect in the case that the background of the buildings is more [10]. This paper used Swir2 band instead of Mir [13] because the Landsat 8 OLI sensor does not contain Mir band. The formula is shown in (4).

$$NDVI = \frac{Nir - R}{Nir + R} \tag{3}$$

Where R is the reflectance of band 4 Red (0.64–0.67 μm).

$$MNDWI = \frac{G - Mir}{G + Mir} \tag{4}$$

Where G is the reflectance of band 3 Green (0.53–0.59 μm), Mir is the reflectance of band 7 Mid-Infrared (2.08–2.35 μm) of TM.

After index calculation, the threshold models of built-up area, water and vegetation were constructed according to the statistical results, and binary classification was carried out to eliminate the mixed water and vegetation in the built-up area, so as to improve the classification accuracy.

Due to urban development and irregular distribution of bare soil, there were still a lot of bare soil in the built-up area and surrounding towns after removing the disturbance of misclassified water and vegetation, which would affect the extraction accuracy of the built-up area. In this study, the method of combining a group of mineral indexes [12] with k-means unsupervised classification proposed by Ward in 2000 was used to eliminate the bare soil mixed in urban built-up area. This method can effectively reduce the wrong sample points which were misclassified into the built-up area, reduce the discrete points outside the built-up area significantly, reduce the false alarm rate, and improve the overall accuracy and Kappa coefficient [9]. This group of mineral indexes includes:

Iron Oxide Index:

$$Iron = \frac{Red}{Blue} \tag{5}$$

Where *Blue* is the reflectance of band 2 Blue (0.45–0.51 μm).

Clay Mineral Index:

$$Clay = \frac{Swir1}{Swir2} \tag{6}$$

Ferrous Metal Index:

$$Ferrous = \frac{Swir1}{Nir} \tag{7}$$

K-means clustering algorithm is an unsupervised real-time clustering algorithm advanced by Mac Queen [14]. On the basis of minimizing the error function, the data is divided into a predetermined number of clusters. The principle of the algorithm is simple and it is easy to process a large amount of data [15].

2.2 Data Post-classification and Construction Area Refinement

Because the spectral characteristics of buildings in urban villages and urban built-up areas are similar, many small patches were formed around the built-up area, which affected the extraction accuracy. Within the built-up area, due to the existence of large lakes, rivers and parks, the spectral characteristics of them are quite different from those of the most widely distributed buildings in the built-up area, which often be missed and form voids. In order to reduce the impact of small patches and voids and further meet the needs of classification, the necessary post-classification and the refinement steps of build-up area were needed. Majority Analysis and Clump were used to post process the classification results.

Majority Analysis is a form of convolution filtering for smoothing and sharpening purposes actually. It replaces the central pixel category with the pixel category which occupies the most important position (the largest number of pixels) in the transform kernel.

For the voids within the built-up area, although the low-pass filtering can smooth the image well, it is affected by the coding of surrounding objects easily. Clump processing solved this problem well. Clump processing is an algorithm that uses mathematical morphological operators (erosion and dilation) to cluster and merge adjacent similar classification regions.

The extraction results of built-up area after post-classification still had some voids that cannot be filled and small patches that cannot be removed. And it did not conform to the definition of built-up area and was not conducive to data statistics, so built-up area needed to be refined. The range of built-up area was transformed into vector and imported into ArcMap 10.2. The fragmented map patches outside the built-up area were deleted and the voids in the built-up area were filled, and the built-up area became a whole map patch.

3   Experiment and Analysis

In this paper, Landsat 8 OLI image data in Shenyang on August 31, 2017 were adopted, including 8 multispectral bands with resolution of 30 meters and a panchromatic band with resolution of 15 meters. In order to obtain surface albedo, radiometric calibration and atmospheric correction were carried out. In this paper, the processing of remote sensing image data also includes image fusion and clipping. The technical route of this experiment is as Fig. 2.

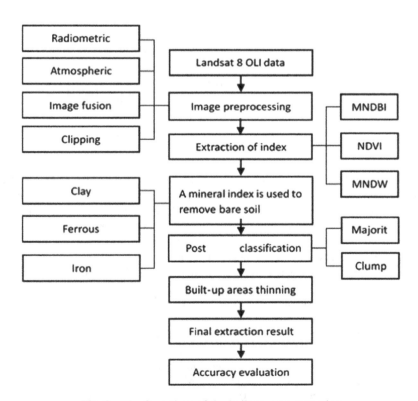

Fig. 2. The flow chart of the built-up area extraction

3.1   Extraction of Index and Establishment of Threshold Model

Figure 2 is the flow chart of the built-up area extraction. After many experiments and statistical analysis, it is found that the brightness of the ground object corresponding to each index is enhanced after calculation, while other objects are suppressed. Therefore, it is easy to achieve the goal of binary classification by establishing a threshold model, so as to extract the required ground objects for the next built-up area extraction and vegetation and water mixed removal. The calculated value range of each classification is shown in Table 1.

Table 1. The classification value range

Index	Features	The minimum	The maximum
MNDBI	Bare soil	−0.3576	1.300
	Forest land	−0.6664	−0.3820
	Arable land	−0.7255	−0.4907
	Water	−0.6695	−0.3004
	Built-up area	−0.5032	0.5248
NDVI	Bare soil	−1.2022	0.5553
	Forest land	0.5698	0.8286
	Arable land	0.6659	0.8137
	Water	0.1328	0.8106
	Built-up area	−0.4083	0.5391
MNDWI	Bare soil	−0.5151	−0.2320
	Forest land	−0.2709	−0.0692
	Arable land	−0.2020	−0.0546
	Water	−0.1240	0.4789
	Built-up area	−0.5122	0.1162

According to the value range of five types of ground objects after calculation and combining with the actual situation, MNDBI threshold was chosen as −0.50 in order to retain the scope of urban built-up area to the greatest extent. At this time, the arable land was well removed, and would not be considered in the next extraction.

After visual inspection, part of woodland, water and bare soil were still mixed in the extracted built area. These ground objects in the binary images obtained only through MNDBI calculation and classification were difficult to be distinguished, so NDVI and MNDWI was introduced to extract and remove them in the next step.

According to the statistical results after NDVI calculation, 0.56 was selected as the threshold value, as a result, not only was the built-up area preserved completely, but also the forest land was removed well, the situation that part of the non-built-up area of the forest land was misclassified into the built-up area was reduced effectively, and the fine patches were reduced. Through the calculation of NDVI, the vegetation mixed in urban built-up area was removed well. After MNDWI calculation, according to the statistical results, a threshold value of 0.10 was selected to conduct binary classification

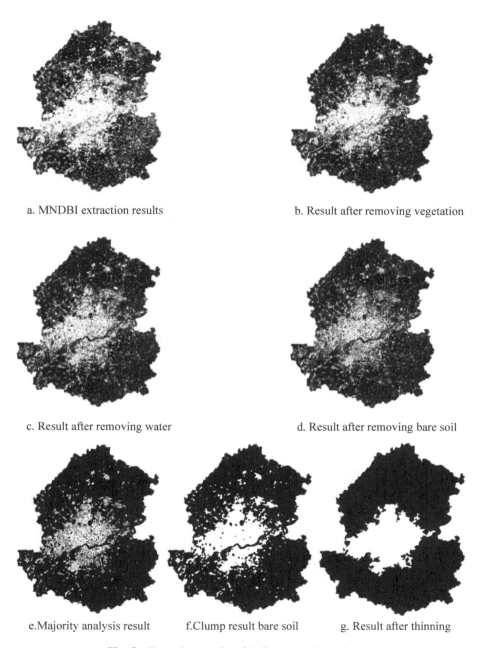

a. MNDBI extraction results   b. Result after removing vegetation

c. Result after removing water   d. Result after removing bare soil

e.Majority analysis result   f.Clump result bare soil   g. Result after thinning

Fig. 3. Extraction results of built-up area in each step

of the calculated results. The index could well remove the water mixed in the built-up area and the small patches outside the built-up area, while retaining the urban built-up area to the greatest extent.

3.2   A Group of Mineral Indexes Are Used to Remove Bare

There was still a situation that bare soil was misclassified into built-up area after exponential extraction, which was not conducive to the post-processing of data and the evaluation of the final accuracy. In this paper, a group of mineral indexes were introduced to identify and eliminate the misclassified bare soil in the urban built-up area. Firstly, Iron, Clay and Ferrous were calculated, and the calculated results were superimposed. K-means unsupervised classification algorithm was used to classify the mineral indexes. The classification results were visually identified, and the bare soil part was selected to be removed from the results obtained by index extraction. The comparison of extraction results of each step is shown in Fig. 3.

4   Conclusion and Analysis

According to the definition of urban built-up area, the built-up area includes the areas that have been developed and constructed in the urban administrative area, the municipal public facilities and public facilities have been fully equipped in the urban

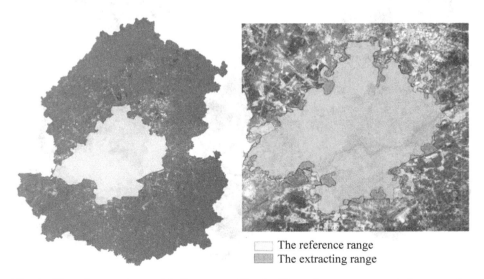

☐ The reference range
▨ The extracting range

Fig. 4. The reference range of built-up area (hand-extracting result (blue)) (Color figure online)

Fig. 5. The reference range and the extracting range of built-up area

administrative area, includes the buildings, highways, bridges, lakes, rivers and other infrastructures that have been built and improved in the urban area. Because of the variety and complexity of the ground objects species, it was not typical to use the Confusion Matrix Using Ground Truth ROIs function of ENVI 5.1 to evaluate the accuracy of ground objects by using pixels based on their spectral characteristics. This paper used area as evaluation criterion to establish confusion matrix to evaluate the classification results, which could minimize subjectivity and reduce the impact of other objects within the built-up area.

In 2017, the built-up area of Shenyang was 633.8 km^2 [16]. Taking the pre-processed remote sensing data of Shenyang in August 2017 as the base map, referring to the remote sensing images of the same period in 2015 and 2016, and combining the image data and information provided by Google Earth and Auto Navi Map, a map patch representing the urban built-up area of Shenyang was plotted. As shown in Fig. 4, its area is 633.8 km^2. Figure 5 shows the comparison between accuracy assessment area and built-up area. The "combined" superposition analysis tool in ArcMap 10.2 was used to combine the experimental results with the property sheet of map patches in the built-up area of Shenyang. By comparison, the correct classification, misclassification and missing area of built and non-built were calculated. Confusion matrix was made and the User's Accuracy, Producer's Accuracy, Overall Accuracy and Kappa coefficient of each step were calculated and were used to evaluate the Accuracy, the results are shown in Table 2.

Table 2. The extracting accuracy chart of the built-up area in each step (User's accuracy, Producer's accuracy, Overall accuracy and Kappa coefficient)

Precision index	MNDBI extraction	Mineral index	Post classification	Proper refining
User's accuracy	89.04%	66.96%	81.43%	92.46%
Producer's accuracy	26.49%	45.46%	68.64%	91.91%
Overall accuracy	52.845	79.28%	89.55%	97.14%
Kappa coefficient	0.1180	0.4140	0.6818	0.9091

The results show that the extraction effect was poor only through MNDBI, and only the area of the User's Accuracy meets the needs. However, many ground objects in the built-up area were misclassified into non-built-up areas, resulting in low overall accuracy and poor reliability. The results showed that it was not nearly enough to extract the built-up area only through MNDBI because of the complexity of the ground objects included in the urban built-up area, and the elimination and post classification steps of other interfering ground objects must be carried out.

Eliminating the influence of water, vegetation, bare soil and other ground objects could reduce the area of non-built area misclassified in the built-up area significantly, but it also increased the voids in the built-up area, reduced the area, and reduced the user accuracy. However, the overall classification accuracy and reliability were improved significantly, but the available state has not been reached yet. Further post classification was needed to increase the built-up area, and improve user's accuracy and overall accuracy.

The post classification filled the voids existing in the built-up area, effectively increased the area of built-up areas, and significantly improved the user's accuracy. The area of the non-built area which was misclassified into the built-up area and the area of the built-up area which was misclassified into the non-built area were also significantly reduced. The overall accuracy and Kappa coefficient were improved, which basically achieved the desired results of the experiment.

The refinement of built-up area effectively removed the hard-to-remove villages and large areas of land to be developed in the periphery of the city, and filled in the large and hard-to-fill voids within the built-up area. Compared with the results before refinement, the accuracy of the built-up area has been improved, the accuracy index has reached more than 90%, and the extracted data has reached the available state, which also proved the necessity of refinement of the built-up area.

The various remote sensing indexes adopted in this paper have the advantages of convenient calculation, low learning cost, easy realization and the built-up area can be extract fast. In addition, combined with the processing steps of removing vegetation, water and bare soil mixtures, post-classification and refinement of built-up areas, the final extraction results have high classification accuracy and location accuracy, which all meet the statistical requirements. This method is suitable for rapid extraction and long-term change monitoring of built-up area, and can play a very good role in the monitoring of geographical conditions.

Acknowledgments. This study was supported by Natural Science Foundation of Liaoning Province of China: Study on Remote Sensing Monitoring Method for Maize Planting Area in Liaoning Province (No. 20180550479).

References

1. Li, G.D., Fang, C.L., Wang, S.J., Zhang, Q.: Progress in remote sensing recognition and spatio-temporal changes study of urban and rural land use. J. Nat. Resour. **31**(04), 703–718 (2016)
2. Xu, Z.N., Gao, X.: A novel method for identifying the boundary of urban built-up areas with POI data. Acta Geogr. Sin. **71**(06), 928–939 (2016)
3. Xu, H.Q., Du, L.P., Sun, X.D.: Index-based definition and auto-extraction of the urban built-up region from remote sensing imagery. J. Fuzhou Univ. (Nat. Sci. Ed.) **39**(05), 707–712 (2011)
4. Song, J.C., et al.: A method of extracting urban built-up area based on DMSP/OLS Nighttime data and Google Earth. J. Geo-Inf. Sci. **17**(06), 750–756 (2015)
5. Li, Z., Yang, X.M., Meng, F., Chen, X., Yang, F.S.: The method of multi-source remote sensing synergy extraction in urban built-up area. J. Geo-Inf. Sci. **19**(11), 1522–1529 (2017)
6. Zhong, S.Y., Li, X.X., Bai, Y.H., Feng, J.: The method of extracting built-up areas based on multi-scale segmentation and spectral features. Softw. Guide **17**(09), 180–184 (2018)
7. Tong, B., Shen, W.: Object-oriented Landsat 8 image of the city proper extraction method research. J. Liaoning Prov. Coll. Commun. **19**(02), 21–25 (2017)

8. Zha, Y., Gao, J.Q.: Use of normalized difference built-up index in automatically mapping urban areas from TM imagery. Int. J. Remote Sens. **24**(3), 583–594 (2003)
9. Hu, Y.Y.: Investigation of a modified normalized built-up index and a post processing scheme for BUILT-UP extraction in urban area. Geomat. Sci. Technol. **5**(3), 83–92 (2017)
10. Xu, H.Q.: A study on information extraction of water body with the modified normalized difference water index (MNDWI). J. Remote Sens. **9**(5), 589–595 (2005)
11. Rouse, J.W., Haas, R.H., Schell, J.A., Deering, D.W.: Monitoring vegetation systems in the great plains with ERTS. NASA Spec. Publ. **351**, 309 (1973)
12. Douglas, W., Stuart, R.P., Alan, T.M.: Monitoring growth in rapidly urbanizing areas using remotely sensed data. Prof. Geogr. **52**(3), 371–386 (2000)
13. Xu, H.Q., Tang, F.: Analysis of new characteristics of the first Landsat 8 image and their eco-environmental significance. Acta Ecol. Sin. **33**(11), 3249–3257 (2013)
14. Mac, Q.J.: Some methods for classification and analysis of multivariable observation. Comput. Chem. **4**, 257–272 (1967)
15. Duan, M.X.: Research and Application of Hierarchical Clustering Algorithm. Master, Central South University (2009)
16. Liaoning Statistical Department: Liaoning Investigation Team of National Statistical Bureau: Liaoning Statistical Yearbook 2018. China Statistical Press, Shenyang (2018)

Granger Causality Analysis of Grass Response to Climatic Changes Over Tibetan Plateau

Hua Wang[1,2], Yuke Zhou[1(✉)], and Chenghu Zhou[1]

[1] Institute of Geographic Sciences and Nature Resources Research,
Chinese Academy of Sciences, Beijing 100101, China
zhouyk@igsnrr.ac.cn

[2] Chinese People's Liberation Army 31009, Beijing 100088, China

Abstract. Due to the complex natural environment, vegetation on the Tibetan Plateau (TP) has a sensitive response to climatic changes. Thus, it is of great importance to explore the cause effect of climatic shifts on vegetation. Based on the long-term satellite NDVI dataset during 1982–2012, we analysed the causal relationship of vegetation greenness with temperature and precipitation by using the Granger causality test at monthly and seasonal temporal scale for each pixel. The results show that (1) the proportion of pixels with stationary time series for NDVI vs. temperature and NDVI vs. precipitation is greater at month scale than at seasonal scale, which is 99% and 98% at monthly scale, 64% and 71% at seasonal scale, respectively. (2) At month scale, the lagging time period of the average temperature and total precipitation on NDVI is around 12–13 months at monthly scale and show similar temporal profile across various vegetation types. At seasonal scale, the lagging time period is mainly occurred in 3, 4 and 6 quarter and very different in desert steppe, steppe and meadow. (3) For 98% area of the TP, average temperature change is found to granger cause of NDVI. While for 89% of the TP, except for the south-eastern TP, NDVI is supposed to granger cause of average temperature change at month scale. At seasonal scale, average temperature change is granger cause of changes in NDVI approximately accounting for 92% area of TP, where the central part of TP is excluded. However, in the eastern and western TP (about 50% of TP), NDVI is interpreted as granger cause of average temperature changes. (4) In the north-eastern and north western parts (about 98% of TP), precipitation is the granger cause of NDVI changes. While for 94% of the plateau, except for a few areas in the southeastern TP, NDVI is supposed to the granger cause of precipitation change at month scale. The precipitation change is considered as granger cause of NDVI in the south-eastern part of the TP (approximate 61% TP) at seasonal scale. In the central and eastern (about 48% of the plateau), NDVI is granger cause of precipitation change. Overall, climate factors have an interactive relationship with vegetation changes. Climatic factors and vegetation greenness can compose a Grainger causality relationship to each other, but climatic factors have stronger Grainger cause effect on vegetation than vegetation's Grainger effect on climatic factors. There is more Granger cause effect region at month scale than seasonal scale over the TP.

Keywords: Tibetan plateau · Climatic changes · Granger causality effect · NDVI

© Springer Nature Singapore Pte Ltd. 2020
Y. Xie et al. (Eds.): GSES 2019/GeoAI 2019, CCIS 1228, pp. 210–220, 2020.
https://doi.org/10.1007/978-981-15-6106-1_16

1 Introduction

Land vegetation is an important component of the earth's surface ecosystem. Its development and succession are mainly affected by climate change and human activities. In the past few decades, climate change has been recognized as a major driver of land vegetation changes [1]. Because of the diversity of vegetation and the complexity of geographical environment, Tibetan Plateau (TP) has become an ideal place for exploring the relationship between vegetation and climate change. Studying the interaction between vegetation change and climatic factors is of great scientific significance to the ecological environment protection and regional ecological scientific development over TP. In recent years, some progress has been made in the study of the correlation between vegetation and climate change based on conventional methods, including multivariate linear analysis and Bayesian inference which can effectively assess the correlation of complex data [2].

Study in northern China show that the response of temperate grassland to precipitation change with the seasonal shifts. Further the change of precipitation would affect the growth of most temperate grassland vegetation in the next year [3]. In high latitudes, vegetation greenness (NDVI) in spring and autumn showed a significant positive correlation with temperature [4]. In most cases, vegetation growth will be affected by temperature and precipitation. Even under the influence of one factor, vegetation growth will change with their time and intensity of the effect [5]. In the study of the response of vegetation NDVI to climate change on the TP, there is uncertainty in estimating the impact of climate change on vegetation greenness based on satellite remote sensing data, so the cause and effect are often controversial. When estimating the correlation between two or more variables, the autocorrelation condition of the time series must be considered, otherwise, it may lead to pseudo-correlation between variables. To address this problem, Granger causality test is suitable for detecting the correlation between variables with eliminating the impact of changes from variables [6, 7]. Actually, it has been a widely used method for attribution analysis of climate change.

With the widespread use of remote sensing data, it is possible to explore the causal relationship between vegetation NDVI and climate at a large scale for TP. This study chooses relevant climatic factors including average temperature and precipitation to analyse their attribution to vegetation status using Granger causality test at monthly and seasonal time scales from 1982 to 2012. We focus on revealing the temporal and spatial patterns of their interaction.

2 Study Area and Data Source

2.1 Study Area

Tibetan Plateau is located in the southwestern China, spanning 3°E–105°E and 26°N–40°N. Its spatial range consists of six provinces, including Tibet, Qinghai, Xinjiang, Sichuan, Yunnan and Gansu, covering an area of 2,500,000 km^2 that is about one fourth of the total area of China. As an alpine mountain area, TP climate is mainly characterised as low temperature, large temperature difference between day and night, scarce precipitation with significant seasonal difference. The annual average temperature gradually decreases from 20 °C in the southeastern part to −6 °C in the northwestern part of the plateau. The annual precipitation decreases gradually from 4000 mm to less than 50 mm. There are a variety of vegetation cover types over the TP, including forest, desert vegetation, grassland (steppe and meadow) which is dominantly covered by grassland (Fig. 1). This study focuses on the vegetation of desert steppe, steppe and meadow and their causal relationship with key climatic factors on the Tibetan Plateau.

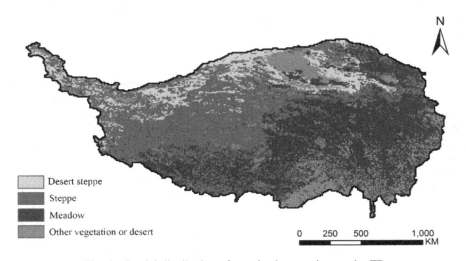

Fig. 1. Spatial distribution of grassland vegetation on the TP

2.2 Data Preprocessing

GIMMS NDVI3g Data. The primary dataset used in this analysis is the third-generation vegetation index dataset (NDVI3g), released by NASA Global Inventory Monitoring and Modeling Systems (GIMMS). This long-time series data spans the period from 1982 to 2012 with a spatial resolution of 0.083° and a temporal interval of 15 days [8, 9]. The GIMMS NDVI3g dataset, has been widely used in vegetation growth and dynamics monitoring, from regional to continental scale [10]. NDVI3g dataset is the longest NDVI time series to date and has been proved to have good ability for the long-term monitoring of vegetation dynamics [11]. In addition, the dataset has been normalized to account for issues such as sensor calibration loss, orbital drift, and atmospheric effects such as volcanic eruptions, so it can provide higher quality data for regions in mid to high latitudes. In this study, based on time series stacks from multitemporal images, we extracted the temporal sequence for each pixel. Savitzky-Golay (S-G) smoothing filtering algorithm was applied to further improve the data consistence on time series.

Meteorological data. In this study, the temperature and precipitation raster data were obtained from the China meteorological forcing dataset produced by merging a variety of data sources [12, 13]. The data sources used to produce the forcing data include CMA (China Meteorological Administration) station record, TRMM satellite precipitation analysis data, Princeton forcing data, and GLDAS data. This dataset has a spatial and temporal resolution of 0.1° and 3-hour, respectively. To combine with GIMMS NDVI3g data, this data source is resampled to 8 km using the nearest method, and then aggregated to monthly data using the mean value for temperature and sum for rainfall, respectively.

3   Methods

3.1   Stationarity Test

The stability of the time series is the premise of Granger causality test. If the stationary test is not performed, the Granger causality test may lead to pseudo-regression [7]. To test whether the time series is stable is to check whether there is a unit root. The commonly used method is the Augmented Dickey-Fuller (ADF) test [14]. For the time series $X = \{xt\}$, the test equation is:

$$\Delta X_t = \alpha + \beta t + \delta x_{t-1} + \sum_{t=1}^{n} \beta \Delta X_{t-1} + \varepsilon_t \qquad (1)$$

where t is a time variable, α is a constant term, βt is a trend term, and εt is a residual term. Assuming the null hypothesis H0: $\delta = 0$, the alternative hypothesis H1: $\delta \neq 0$. The test is divided into three steps. The first step is to test according to formula 1. The second step is to delete the trend item for testing. The third step is to delete the constant item trend item for testing. If the test rejects H0 at any step, it means that there is no unit root in the time series X, that is, X is a stationary time series before the test can be stopped, otherwise the test is continued until the end of the third step. Repeat the same steps and perform the ADF test on the time series $Y = \{yt\}$. If the sequences X and Y are both stationary, they can be tested for Granger causality.

3.2   Granger Causality Test

Whether there is a causal relationship between X and Y can be judged by the Granger causality test [6]. From a statistical point of view, when the correlation coefficients of of X and Y reach a certain level of significance, it can be considered that there is a causal relationship between X and Y. Let two stationary sequences $X = \{xt\}$ and $Y = \{yt\}$, first estimate whether there is causal interpretability between the two time-series data, then introduce the optimal lag order, and judge whether time series X can affect time series Y by Granger causality test. The regression model is as follows:

$$y_t = \alpha + \sum_{t=1}^{n} \beta_t x_{t-1} + \sum_{t=1}^{n} \gamma_t y_{t-1} + \varepsilon_t \tag{2}$$

where n is the maximum lag order of the variables X and Y, α is a constant term, and εt is a residual term. Assume that the null hypothesis H0: X is not the Granger cause of Y (H0: $\forall t \in (1,2,\ldots,n)$, $\gamma t = 0$), and the alternative hypothesis H1: X is the Granger cause of Y (H1: $\exists t \in (1,2,\ldots,n)$, $\gamma t \neq 0$). When X is not considered ($\beta t = 0$), the residual square sum RSr of the regression model is established; when X is considered ($\beta t \neq$), the residual square sum RSu of the regression model are established. Construct F statistic:

$$F = \frac{RS_R - RS_U}{RS_U} \times \frac{N - 2n - 1}{n} \sim F(n, N - 2n - 1) \tag{3}$$

The confidence level is set to 95%. If $F \geq$ F0.05(n, N−2n−1), then there is $t \in (1,2, \ldots,n)$ and $\gamma t \neq 0$, thus the null hypothesis H0 should be rejected. X is the Granger cause of Y, and vice versa, X is not the Granger cause of Y.

4 Results

4.1 Stationary Analysis

The stationarity for the time series of NDVI, TEMP and PREC were tested over the TP. The stationarity analysis between two temporal data series includes stationarity test and co-integration test. The ADF test is suitable for stationarity test and E-G two-step method is applied to co-integration test. At monthly and seasonal scales, the stationarity of NDVI vs TEMP, NDVI vs PREC for each pixel in the TP is analysed. Assuming that the null hypothesis H0 is: both time series are stationary, the alternative hypothesis H1 is: at least one time series exhibits a non-stationarity way. The ADF test is performed on the four comparison tests respectively. If the experimental result rejects H0, the Granger causality test can be performed. Otherwise, it is necessary to judge whether the two sequences are the same order. If it is the same order single sequence, the cointegration test is needed. When the result rejects H0, the Granger causality test can also be performed.

The results at different scales show that the rejection ratio at monthly scales is higher than that at seasonal scales, which is 35.12% and 26.71%, respectively. This shows that the count of stationary pixels in NDVI vs TEMP, NDVI vs PREC at monthly scales is more than that at seasonal scales (Table 1). The rejection ratio of NDVI vs TEMP, NDVI vs PREC is higher than the acceptance ratio at the two scales, especially at the monthly scale. The results indicate that the Granger causality test can be carried out between the NDVI and the average temperature and precipitation of most pixels on the TP. Among them, 99.13% of NDVI vs TEMP, 98.68% of NDVI vs PREC can be tested by Granger causality test at monthly scale, and 64.01% and 71.97% can be tested by Granger causality test at seasonal scale.

Table 1. Statistic of stationarity test between NDVI, temperature and precipitation

H$_0$	Monthly scale		Seasonal scale	
	NDVI vs TEMP	NDVI vs PREC	NDVI vs TEMP	NDVI vs PREC
Reject	99.13%	98.68%	64.01%	71.97%
Accept	0.87%	1.32%	35.99%	28.03%

4.2   Lag Effect of Temperature and Precipitation on NDVI

Granger causality test is substantially sensitive to lag orders. It has been proved that although any lag value can perform the Granger causality test, Granger causality test based on best lag value is more significant [15]. VAR model is applied and select the optimal lag order in NDVI vs TEMP, NDVI vs PREC for each pixel at monthly and seasonal scales, that is, best lag value effect of average temperature on NDVI and precipitation on NDVI, The Akaike information criterion (AIC) and Schwarz Criterion (SC) minimization criteria were used to determine the optimal lag order [16, 18]. The spatial distribution of optimal lag-order for NDVI vs TEMP and NDVI vs PREC at different scales is shown in Fig. 2. At monthly scale, the lag effects of average temperature and precipitation on NDVI in most areas of the TP are concentrated in the 12th to 13rd month. The lag effect of average temperature on NDVI in a few areas in the northern TP was 13rd month and 9–11th month in some southern areas. The lag effect of precipitation on NDVI was 6–7 months in a few areas in the eastern parts and 3–5 months in a few areas in the southern TP. From the seasonal scale, the lag effect of the average temperature on NDVI in the eastern and western regions of the TP is in the third quarter, the lag effect in the central region is in the 5th and 6th quarters, and the southern part of the region is lagged behind in the 6th quarter. In the analysis of the lag

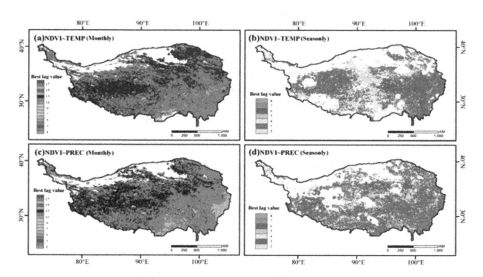

Fig. 2. Spatial distribution of optimal lag order

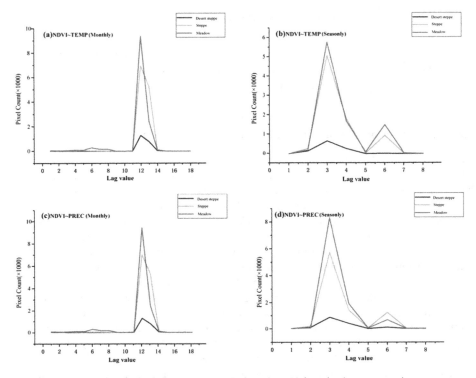

Fig. 3. Statistic of Pixel Count across Various Lag Values in three vegetation types

effect of precipitation on NDVI, the lag effect in the southeastern part of the TP is in the third quarter, and the lag period in the central part is in the fourth and sixth quarters. The lag period that longer than one year, such as 13rd month and 6th quarters, means the long effect of temperature or precipitation on vegetation. Actually, this may be more linked with hydrological year, but not calendar year.

Here, the lag effect of average temperature and precipitation on NDVI in desert steppe, typical steppe and meadow is discussed. The distribution of optimal lag order frequency in the three vegetation types is shown in Fig. 3. It is obvious that the lag effect of average temperature on NDVI and precipitation on NDVI is highly similar to that in Fig. 4, whether on monthly or seasonal scales. The lag effect of average temperature on NDVI and precipitation on NDVI on the monthly scale is mostly concentrated in the 12th-13th month, and the lag effect on the seasonal scale is mainly in the 3-4th and 6th quarters. On the monthly scale, the number of pixels in TEMP $\sim$ NDVI and PREC $\sim$ NDVI in the Tibetan Plateau is highly similar. In the number of pixels with a lag order of 12, meadow > typical grassland > desert grassland; and in the number of pixels with a lag order of 13, Typical grassland > meadow > desert grassland. At the seasonal scale, the heterogeneity of TEMP $\sim$ NDVI and PREC $\sim$ NDVI pixels is larger, and the lags of meadows and typical grasslands are the most in the third quarter, more in the fourth and sixth quarters, the lag order of the desert steppe is mostly concentrated in the third quarter. In addition, the three vegetation types tend to be 0 at the time of lag order 5, which is almost identical with Fig. 4.

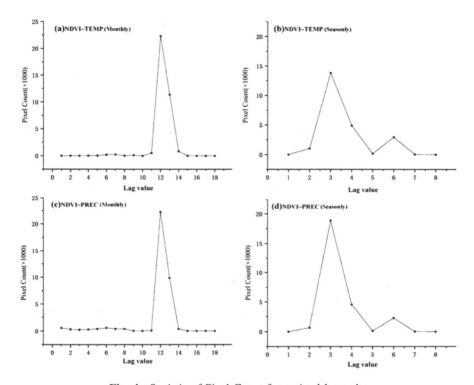

Fig. 4. Statistic of Pixel Count for optimal lag order

4.3   Causality Analysis Between NDVI and Average Temperature

At monthly scale, the average temperature is considered to Granger cause of NDVI change in most areas of the TP (Fig. 5). NDVI is considered to be the Granger cause of NDVI change in most areas, but NDVI is not considered to be the Granger cause of average temperature change in some parts of the southeastern part of the plateau. At the seasonal scale, the average temperature is the Granger cause of NDVI changes in other areas except the central part. In the eastern and Western regions, NDVI is considered to be the Granger cause of the average temperature change, while in other regions NDVI is not considered to Granger cause of the average temperature change.

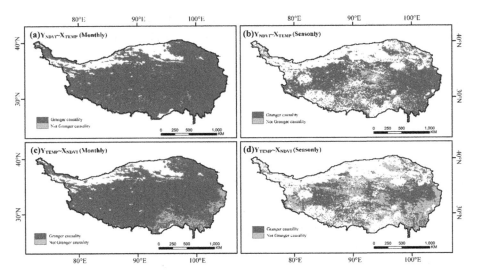

Fig. 5. Effect area of granger cause between NDVI and temperature

4.4 Causality Analysis Between NDVI and Precipitation

Figure 6 is Granger effect areas between NDVI and precipitation. At monthly scale, precipitation is considered to the Granger cause of NDVI changes in the northeastern and northwestern regions of the TP, while precipitation is not considered to Granger cause of NDVI changes in most areas of the southern TP. NDVI is considered to Granger cause of precipitation change in most areas of the plateau, while NDVI is not

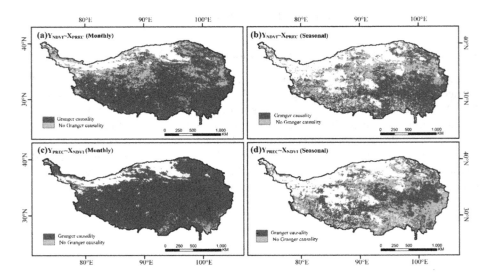

Fig. 6. Granger effect areas between NDVI and precipitation

considered to Granger cause of precipitation change in a few areas in the southeast. At the seasonal scale, precipitation is not Granger cause of NDVI change in the eastern and northwestern regions of the TP, while precipitation is the Granger cause of NDVI change in the southeastern region. NDVI is the Granger cause of precipitation change in the central and eastern regions of the plateau, except in the central and eastern regions of the plateau.

5   Conclusion

Using Granger causality test, the causal relationship between vegetation greenness with temperature and precipitation was analysed at monthly and seasonal scales. The area showing Granger causality between NDVI and climatic factors is more significant at monthly scale than at seasonal scale. The lag period of NDVI to temperature and precipitation mainly concentres in 12–13 months. Temperature presents a more obvious Granger causality than precipitation at both monthly and seasonal scales.

Acknowledgments. This work is supported by National Natural Science Foundation of China (Grant No. 41601478), National Key Research and Development Program (Grant No. 2018YFB0505301, 2016YFC0500103), Key Special Project for Introduced Talents Team of Southern Marine Science and Engineering Guangdong Laboratory (Guangzhou), (Grant No. GML2019ZD0301).

References

1. Nemani, R.R., et al.: Climate-driven increases in global terrestrial net primary production from 1982 to 1999. Science **300**(5625), 1560–1563 (2003)
2. Piao, S., Mohammat, A., Fang, J., Cai, Q., Feng, J.: NDVI-based increase in growth of temperate grasslands and its responses to climate changes in China. Global Environ. Change **16**(4), 340–348 (2006)
3. Sun, Y., et al.: Recent progress in studies of climate change detection and attribution in the globe and China in the past 50 years. Clim. Change Res. **9**, 235–245 (2013)
4. Ichii, K., Kawabata, A., Yamaguchi, Y.: Global correlation analysis for NDVI and climatic variables and NDVI trends: 1982–1990. Int. J. Remote Sens. **23**(18), 3873–3878 (2002)
5. He, B., Chen, A., Jiang, W., Chen, Z.: The response of vegetation growth to shifts in trend of temperature in China. J. Geogr. Sci. **27**(7), 801–816 (2017)
6. Diks, C., Panchenko, V.: A new statistic and practical guidelines for nonparametric Granger causality testing. J. Econ. Dyn. Control **30**(9–10), 1647–1669 (2006)
7. Chen, Y., Rangarajan, G., Feng, J., Ding, M.: Analyzing multiple nonlinear time series with extended Granger causality. Phys. Lett. A **324**(1), 26–35 (2004)
8. Pinzon, J., Tucker, C.: A non-stationary 1981–2012 AVHRR NDVI3g time series. Remote Sens. **6**, 6929–6960 (2014)
9. Peng, D., Zhang, B., Liu, L., Chen, D., Fang, H., Hu, Y.: Seasonal dynamic pattern analysis on global FPAR derived from AVHRR GIMMS NDVI. Int. J. Digit. Earth **5**, 439–455 (2012)
10. De Jong, R., Verbesselt, J., Zeileis, A., Schaepman, M.: Shifts in global vegetation activity trends. Remote Sens. **5**, 1117–1133 (2013)

11. Stow, D., et al.: Variability of the seasonally integrated normalized difference vegetation index across the north slope of Alaska in the 1990s. Int. J. Remote Sens. **24**, 1111–1117 (2003)
12. Yang, K., Wu, H., Qin, J., Lin, C., Tang, W., Chen, Y.: Recent climate changes over the Tibetan Plateau and their impacts on energy and water cycle: a review. Global Planet. Change **112**, 79–91 (2014)
13. Yang, K., et al.: Response of hydrological cycle to recent climate changes in the Tibetan Plateau. Clim. Change **109**, 517–534 (2011)
14. Cheung, Y.W., Lai, K.S.: Lag order and critical values of the augmented Dickey-Fuller test. J. Bus. Econ. Stat. **13**(3), 277–280 (1995)
15. Yi, H.W.: Discussion on how to do granger causality test. J. Postgrad. Zhongnan Univ. Econ. Law **5**, 34–36 (2006)
16. Ferrarini, L.: On the reachability and reversibility problems in a class of Petri nets. IEEE Trans. Syst. Man Cybern. **24**(10), 1474–1482 (2002)
17. Aybar, A., Iftar, A.: Overlapping decompositions and expansions of Petri nets. IEEE Trans. Autom. Control **47**(3), 511–515 (2002)
18. Zhou, J.: The researches on the test power and features on the lagging number selecting criteria about the time series models. System Eng. Theory Pract. **25**, 20–27 (2005)

Exploring the Temporal and Spatial Evolution of the Urbanization of "Belt and Road" Region Based on Nighttime Light Data

Panli Tang[1], Jiejun Huang[1], Xu Zhang[1(✉)], Xining Yang[2], and Xinyu Shi[1]

[1] School of Resource and Environmental Engineering, Wuhan University of Technology, Wuhan 430070, China
x.zhang86@hotmail.com
[2] Department of Geography and Geology and Institute for Geospatial Research and Education, Eastern Michigan University, Ypsilanti, MI 48197, USA

Abstract. The "Belt and Road" area is one of the most economically dynamic and promising regions in the world. Exploring the level of urbanization and the evolution of time and space in the region has important practical significance for the "Belt and Road" construction. Taking the nighttime light detain 2003, 2008 and 2013 as the research data, this study extracts the city lights of the "Belt and Road" countries using the threshold method, and constructs the light intensity and light range index to measure the intensity and range of urbanization level in different countries. Statistical analyses are conducted to reveal the spatial and temporal evolution of the urbanization of the "Belt and Road" area. The results of this study indicates: (1) According to the spatial characteristics of urbanization in different regions, urbanization in the "Belt and Road" area includes three different models: the point model, the linear model, and the planar model. (2) Overall intensity and range of urbanization in the "Belt and Road" area has increased over the past 10 years. However, there are significant variations between different countries and regions. (3) Considering the intensity and range of urbanization, the types of urbanization in the "Belt and Road" area can be into six major categories: comprehensive, extended, lifted, stable, agglomerated, and stagnant. Based on these findings, development proposals are proposed for promoting urbanization in the "Belt and Road" area.

Keywords: Belt and Road · Urbanization · Nighttime light data · Time and space pattern

1 Introduction

In 2013, the "One Belt, One Road" initiative was proposed to strengthen the links between Asian, European and African countries and to form a community of common destiny, which becomes a hot spot of attention in the international community today [1, 2]. The "Belt and Road" area is the most dynamic and promising regions in the world. It is home to 62.3% of the world's population and creates 31.1% of the world's economic output. The average annual economic growth rate has reached 6.28% since the new

© Springer Nature Singapore Pte Ltd. 2020
Y. Xie et al. (Eds.): GSES 2019/GeoAI 2019, CCIS 1228, pp. 221–233, 2020.
https://doi.org/10.1007/978-981-15-6106-1_17

century, which is far above the world average level [3]. At present, most countries along the "Belt and Road" area are experiencing rapid urbanization. Promoting the cooperation in urbanization has a common foundation and internal needs. The characteristics of urbanization development in the "Belt and Road" countries include: (1) The overall level of urbanization development is relatively low, but gross urban population is large. (2) Urbanization is developing at a rapid speed. (3) Huge differences exist between different countries [4]. Promoting the cooperation of China with the "Belt and Road" countries in urbanization is not only a need for China to break through the bottleneck of economic growth, but also a common choice for countries in this area to adapt to economic and social development trends and achieve sustainable economic growth. Therefore, it is of great theoretical and practical significance to explore the characteristics and laws of urbanization development in the "Belt and Road" area.

Due to the incomplete, outdated and inconsistent statistics of different countries and regions, comparative research on the urbanization in different countries and areas has been seriously limited [5–8]. As a new type of data source reflecting the vitality of human social and economic activities, nighttime light data have the characteristics of high data accuracy, wide coverage, long time span, timely update and high-level objectivity [9], which is widely used in urbanization level estimation [10], estimation of socioeconomic parameters [11–13], regional development research [9, 14], urbanization monitoring [15–17], light pollution analysis [18, 19] etc. However, the DMSP/OLS dataset often includes two parts: urban lighting area and rural lighting area. In order to improve the accuracy of study on urbanization level, it is necessary to distinguish between urban and rural lighting areas [20]. At present, the main methods used to extract city lights include the classification comparison method [21], constant area method [22], semi-automatic extraction method of support vector machine (SVM) [23], clustering method [24] and so on. There still lacks unified and effective method for urban nightlight extraction in large and multi-country area.

This study divides "Belt and Road" countries and regions into seven research areas based on geographic location and political boundaries. Using the natural breakpoint method, it divides the DN values of the nighttime light data of each sub-area, extracts the light thresholds of the corresponding area, and identifies the intensity and range of city lights [25–27]. Based on nighttime light data, this study investigates the patterns of urbanization in "Belt and Road" countries from 2003 to 2013, reveals the characteristics of temporal and spatial evolution of urbanization in the "Belt and Road" area over the past ten years, and then offers some recommendations for the urbanization development and the cooperation in urbanization of the "Belt and Road" countries.

2 Study Area

According to the "Development of the Silk Road Economic Belt and the Vision and Action of the 21st Century Maritime Silk Road" jointly issued by the National Development and Reform Commission, the Ministry of Commerce, and the Ministry of Foreign Affairs, this study selects 65 countries in the "Belt and Road" area, including two countries in East Asia, ten countries in the Association of Southeast Asian Nations (ASEAN), 18 countries in West Asia, eight countries in South Asia, five countries in

Central Asia, seven countries in the Commonwealth of Independent States, and 16 countries in Central and Eastern Europe, as the study area (Table 1).

Table 1. Countries in the "Belt and Road" area.

Region	Country
East Asia	China, Mongolia
ASEAN	Singapore, Malaysia, Indonesia, Myanmar, Thailand, Laos, Cambodia, Vietnam, Brunei, Philippines
West Asia	Iran, Iraq, Turkey, Syria, Jordan, Lebanon, Israel, Palestine, Saudi Arabia, Yemen, Oman, United Arab Emirates, Qatar, Kuwait, Bahrain, Greece, Cyprus, Egypt
South Asia	India, Pakistan, Bangladesh, Afghanistan, Sri Lanka, Maldives, Nepal, Bhutan
Central Asia	Kazakhstan, Uzbekistan, Turkmenistan, Tajikistan, Kyrgyzstan
CIS	Russia, Ukraine, Belarus, Georgia, Azerbaijan, Armenia, Moldova
Central and Eastern Europe	Poland, Lithuania, Estonia, Latvia, Czech Republic, Slovakia, Hungary, Slovenia, Croatia, Bosnia and Herzegovina, Montenegro, Serbia, Albania, Romania, Bulgaria, Macedonia

"One Belt and One Road" is an open and inclusive international economic cooperation system. At present, there is no accurate spatial scope for the regional division of each country. This paper combines the seven countries of the Commonwealth of Independent States with other categories according to regional geographic location (Fig. 1), in order to carry out regional spatial analysis.

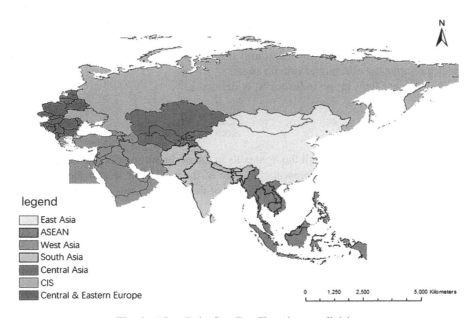

Fig. 1. "One Belt, One Road" study area division

The data of this study are the stable nighttime light data in 2003, 2008 and 2013, which are released by the National Oceanic and Atmospheric Administration (NOAA). These products are divided into 30 arc second grids and acquired by 6 sensors respectively. Therefore, there is a problem of continuity of light intensity in different time periods. The continuity correction of DMSP/OLS stable night light is conducted based on the constant target area method [28, 29].

3 Method

3.1 Threshold Method

The "Natural Breakpoints" category is based on the natural grouping inherent in the data, identifying the classification intervals, grouping the similarities most appropriately, and maximizing the differences between the classes to extract the light thresholds for each region [30]. City lights that are greater than the threshold will be extracted.

3.2 Night Time Light Statistics Method

This study introduces the index of regional average light intensity and standard light range to reflect the intensity and range of regional urbanization.

The regional average light intensity indicator I_j(region j) is defined as:

$$I_j = \sum_{i=1}^{63} DN_i \times \frac{n_i}{N \times 63} \tag{1}$$

DN_i is the gray value of the i-th level in the region ($1 \leq DN_i \leq 63$, 63 is the maximum gray level); n_i is the total number of cells in the i-th gray level in the region; N is the total number of all light cells in the area. I_j represents the proportional relationship with respect to the maximum possible light intensity, which is used to reflect the intensity of regional urbanization.

The regional light range indicator S_j is defined as:

$$S_j = \frac{Area_N}{Area} \tag{2}$$

$Area_N$ is the gross area of all the light cells in the observed region, and Area is gross area of the entire region. S_j represents the ratio of the total light area to the total area, which can be used to reflect the characteristics of the spatial extension of urbanization.

4   Results and Discussion

4.1   Spatial Morphological Characteristics of the Urbanization of the "Belt and Road" Region

According to the spatial characteristics of urbanization in different countries and regions, the urbanization patterns of the "Belt and Road" area can be divided into three categories (Fig. 2): (1) Point urbanization, which is characterized by a relatively discrete distribution of cities, weaker links between adjacent cities, and lower urban agglomeration. The main areas of point urbanization include Central Russia, Western China and Southeast Asia. (2) Linear urbanization, which can be further divided into connected urban belts based on traffic lines, such as the New Eurasia Bridge, river-based watershed type urban belts, such as the Nile and Indus basins, and territorial urban belts formed based on the territorial shape of the country. The common feature of these three types of urbanization is the formation of a relatively long and narrow urban interlocking distribution area. (3) Planar urbanization, which is characterized by the formation of contiguous urbanized areas, often referred to as urban agglomerations in literature. Representative areas of this type include Western Europe, the Persian Gulf, India, and Eastern China.

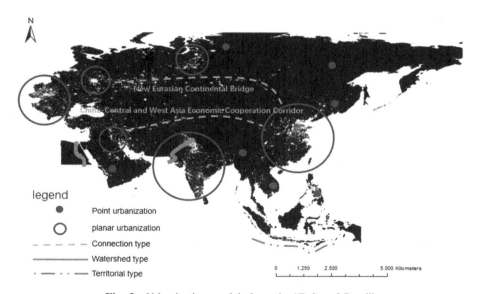

Fig. 2. Urbanization model along the "Belt and Road"

4.2   Spatial Morphological Characteristics of the Urbanization of the "Belt and Road" Region

Figure 3 shows the gross light intensity and the light range of the "Belt and Road" area in 2003 and 2013. During this period, the overall light range of the "Belt and Road" area had increased by 7.70%, and the overall light intensity had increased by 5.66%. It demonstrates that "Belt and Road" the region, in general, is in the stage of rapid urbanization, of which the level of urbanization is constantly improving.

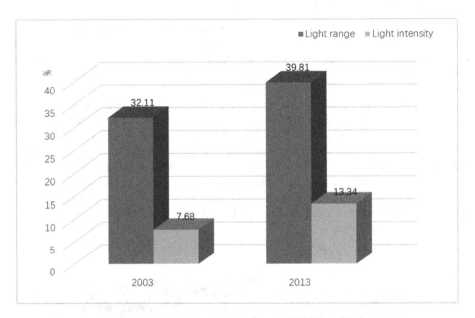

Fig. 3. Light intensity and range in 2003 and 2013

The spatial pattern of light changes in the "Belt and Road" area is shown in Fig. 4. As the result reveals, although the overall light intensity and range of this area had increased, there are obvious variations in terms of the amount of light change across different countries and regions. Some regions, such as East Asia, South Asia had experienced positive changes, while others, such as Central and Eastern Europe, are experiencing negative changes, reflecting the different economic and social conditions and stages of urbanization of different countries.

4.3 Spatial Morphological Characteristics of the Urbanization of the "Belt and Road" Region

Figure 5 shows the top 10 countries with the largest change in the range of urbanization, which is calculated by subtracting the urbanization range in 2013 with the level in 2003. Positive change means the overall urbanization range of the region has increased, indicating an expansion of urbanization. Typical countries of this type include Belarus, Moldova, Thailand, etc. Negative change, in contrast, means the overall urbanization range of the region has reduced, indicating a shrink of urbanization. Typical countries of this type include Hungary, Syria, Czech Republic etc.

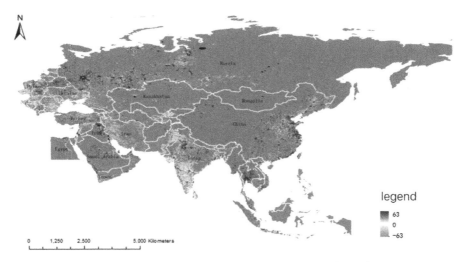

Fig. 4. DN variation of light along the "Belt and Road"

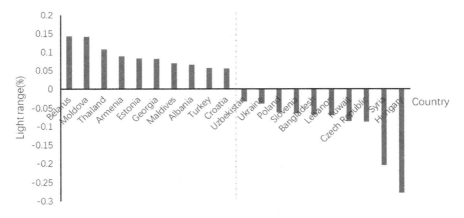

Fig. 5. Top 10 countries with largest positive and negative changes in light range

Figure 6 shows the change of the light range of the "Belt and Road" region between 2003 and 2013. The results show that countries that demonstrate significant increase in light range are mainly distributed in Eastern Europe and Southeast Asia. Countries that only have small changes are mainly concentrated in East Asia, Central Asia and South Asia, such as China, Kazakhstan and India. In comparison, countries have significantly reduced light range are mainly in Eastern Europe and part of the Middle East.

Figure 7 shows the top 10 countries with the largest and least changes in light intensity, which is calculated by subtracting the urbanization intensity in 2013 with that in 2003. The results show that the intensities of urbanization of the "Belt and Road" countries are all positive, which indicates the urbanization intensity of all the "Belt and

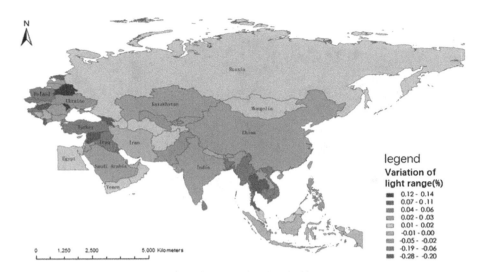

Fig. 6. Change of the light range of the "Belt and Road" region between 2003 and 2013

Road" region has improved. The most remarkable increase can be observed in Qatar, Bahrain, etc., while the smaller changes are in Maldives and Lebanon.

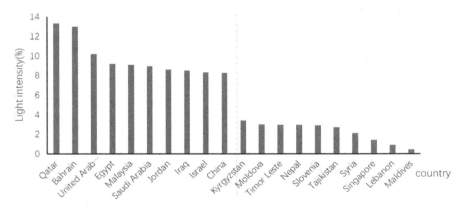

Fig. 7. Top 10 countries with the largest and least changes in light intensity

The change of the light intensity of the "Belt and Road" region between 2003 and 2013 is shown in Fig. 8. The regions with obvious increase in light intensity are East Asia, Southeast Asia and West Asia. The regions with only small changes are Central Asia and some part of the Middle East. There is no country has experienced a significant reduction in light intensity.

Combining the change in light intensity, which includes obvious increase and slight increase, and the change in light range, which includes increase, stability and reduction,

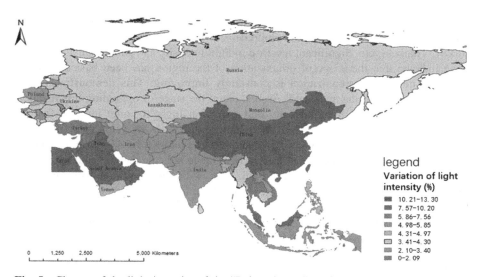

Fig. 8. Change of the light intensity of the "Belt and Road" region between 2003 and 2013

the "Belt and Road" countries can be divided into the six categories according to their urbanization characteristics (Table 2).

Table 2. Typology of National Urbanization

Light intensity	Light range		
	Addition	Stability	Reduction
Obvious addition	Brunei, Iraq, Thailand, Turkey, Vietnam	Bahrain, China, Cyprus, Egypt, Israel, Jordan, Malaysia, Oman, Qatar, Saudi Arabia, United Arab Emirates	Kuwait
Slight addition	Albania, Armenia, Belarus, Cambodia, Croatia, Estonia, Georgia, Moldova, Myanmar, Sri Lanka, Maldives	Singapore, Afghanistan, Azerbaijan, Bhutan, Bosnia & Herzegovina, Bulgaria, India, Indonesia, Iran, Kazakhstan, Kyrgyzstan, Laos, Latvia, Lithuania, Mongolia, Nepal, Pakistan, Philippines, Romania, Russia, Serbia & Montenegro, Slovakia, Tajikistan, Timor Leste, Turkmenistan, Yemen	Bangladesh, Czech Republic, Hungary, Lebanon, Macedonia, Poland, Slovenia, Syria, Ukraine, Uzbekistan

As Fig. 9 shows, significant increase in both the range and the intensity of lighting implies the region is at the stage of comprehensive urbanization. Such countries include Turkey, Thailand, etc. Significant increase in light intensity and smaller change in light range indicate that the scope of urbanization of the region has been basically stable, while the quality of urbanization is constantly improving. This demonstrates these countries, such as China and Saudi Arabia, are at the stage of upgrading urbanization. Significant increase in light intensity and reduction in light range indicate that the country experiencing agglomerated urbanization with growing population migrating from rural areas to cities. Such countries include Iran and Kuwait Slight improvement in light intensity and significant increase in light range indicate that the country, such as Belarus, is experiencing extended urbanization with the continuous expansion of urbanization to rural areas. Little change in both light intensity and light range indicates that the country has reached the stage of stable urbanization, of which the urbanized areas are saturated (e.g., Singapore) or urbanization is proceeding slowly (e.g., Russia). Small change in light intensity and significantly reduction in light range indicate that the country's urban development is limited. These countries, such as Poland, are experiencing stagnated or even declined urbanization.

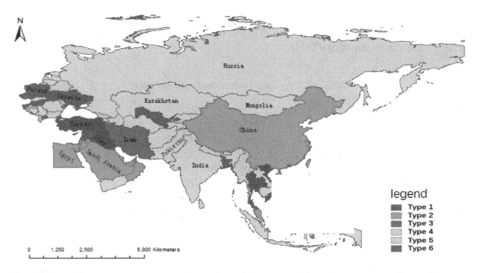

Fig. 9. Typology of the urbanization of the "Belt and Road" countries. Note: Type 1-comprehensive urbanization, Type 2-lifted urbanization, Type 3-agglomerated urbanization, Type 4-extended urbanization, Type 5-stable urbanization, Type 6-stagnant urbanization.

According to the spatial and temporal evolution pattern of the urbanization of "Belt and Road" countries, the following development suggestions are proposed: (1) Taking China's current gradient shift strategy and the development strategy of urban agglomerations and cities as the reference, "Belt and Road" countries can use urban clusters and urban belts as the core, and cross-region urban networks as the basic structure, to comprehensively utilize the combination of point, linear and planar urbanization models to promote sustainable, efficient and healthy development of urbanization. (2) It is necessary to formulate urbanization development strategies based on the local conditions of different countries to promote the gradual transformation of the urbanization of the "Belt and Road" region from scale expansion to quality improvement and realize all-round urbanization. (3) Countries at the stage of comprehensive urbanization and that of stagnant urbanization should strengthen their mutual exchanges. Countries with comprehensive type can further promote their own development, at the same time, stimulate the economic growth of the country with stagnant urbanization through investing in and trading with the latter. For countries of the agglomeration type of urbanization, the expansion of urbanization should be strengthened. For countries of the extended type, the intensity of urbanization should be strengthened. For countries of the stable type, it is necessary to look for new opportunities for economic development through the Belt and Road link. For countries of the lifted type, the intensity and scope of urbanization should be further strengthened.

5 Conclusions

Drawing on the remote sensing data of DMSP/OLS nighttime light data from 2003 to 2013, this paper uses spatial statistical analysis method to spatially analyze the light intensity and light range of the "Belt and Road" countries and regions to reveal the spatial and temporal evolution of urbanization in the "Belt and Road" region. The findings indicate: (1) There are many types of urbanization spatiality in different countries and different regions. They are characterized by point (such as central Russia, western China), linear (including main traffic corridors and major river transportation routes in East Asia and South Asia, such as the Yangtze River, the Yellow River, the Indus River, the Nile River, etc.), and planar urbanization (Western Europe, developed countries in the Middle East). (2) The intensity and range of urbanization in the "Belt and Road" countries and regions have, in general, increased from 2003 to 2013. However, there exist different spatial pattern in terms of the intensity and range of urbanization in different countries and regions, indicating that there are huge gaps in the level of urbanization between these countries and regions. For example, the level of urbanization in coastal areas is significantly higher than that in inland areas. The urbanization level in Western Europe is relatively higher than that in Southeast Asia, reflecting the different in economic and social conditions and stages of urbanization of these countries. (3) According to the changes in light intensity and range, the urbanization of the "Belt and Road" countries can be divided into six types: comprehensive, lifted, agglomerated, extended, stable, and stagnant (declining) urbanization.

However, the study just eliminates the differences of city lighting thresholds between the seven regions, because it is hard to extract city lighting thresholds for each country. The accuracy of the city lighting threshold needs to be improved further. In addition, due to the incomplete, outdated and inconsistent statistics of different countries and regions, quantitative study on the factors of urbanization change in the "Belt and Road" area has been seriously constrained.

Future work should be conducted to obtain and deal with the data after 2013, compare it with this study, and verify the urbanization effect of China's implementation of the Belt and Road development.

Acknowledgements. This work was supported by the National Natural Science Foundation of China (Grant No. 41601163).

References

1. Wang, Y.: China connects the world: what is behind the Belt and Road initiative. Front. Educ. China **14**(2), 335–338 (2019)
2. Li, D.R., Han, R., Li, X.: The spatial-temporal pattern analysis of city development in countries along the belt and road initiative based on nighttime light data. Geomat. Inf. Sci. Wuhan Univ. **42**, 711–720 (2017)
3. Chen, M., Sui, Y., Liu, H., et al.: Urbanization patterns and poverty reduction: a new perspective to explore the countries along the Belt and Road. Habitat Int. **84**, 1–14 (2019)
4. Liu, H., Fang, C., Miao, Y., et al.: Spatio-temporal evolution of population and urbanization in the countries along the Belt and Road 1950–2050. J. Geogr. Sci. **28**(7), 919–936 (2018)
5. Chen, M., Liu, W.: Evolution and assessment on China's urbanization 1960–2010: under-urbanization or over-urbanization. Habitat Int. **178**, 25–33 (2013)
6. Wang, S., Ma, H., Zhao, Y.: Exploring the relationship between urbanization and the eco-environment—A case study of Beijing–Tianjin–Hebei region. Ecol. Ind. **45**, 171–183 (2014)
7. Michaels, G., Rauch, F., Redding, S.J.: Urbanization and structural transformation. Q. J. Econ. **127**(2), 535–586 (2012)
8. Taubenböck, H., Esch, T., Felbier, A., et al.: Monitoring urbanization in mega cities from space. Remote Sens. Environ. **117**, 162–176 (2012)
9. Zhang, Q., Seto, K.C.: Mapping urbanization dynamics at regional and global scales using multi-temporal DMSP/OLS nighttime light data. Remote Sens. Environ. **115**(9), 2320–2329 (2011)
10. Gao, B., Huang, Q., He, C., et al.: Dynamics of urbanization levels in China from 1992 to 2012: perspective from DMSP/OLS nighttime light data. Remote Sens. **7**(2), 1721–1735 (2015)
11. Keola, S., Andersson, M., Hall, O.: Monitoring economic development from space: using nighttime light and land cover data to measure economic growth. World Dev. **66**, 322–334 (2015)
12. Wu, J., Wang, Z., Li, W., et al.: Exploring factors affecting the relationship between light consumption and GDP based on DMSP/OLS nighttime satellite imagery. Remote Sens. Environ. **134**, 111–119 (2013)
13. Ma, T., Zhou, C., Pei, T., et al.: Quantitative estimation of urbanization dynamics using time series of DMSP/OLS nighttime light data: a comparative case study from China's cities. Remote Sens. Environ. **124**, 99–107 (2012)
14. Small, C., Elvidge, C.D.: Night on Earth: mapping decadal changes of anthropogenic night light in Asia. Int. J. Appl. Earth Obs. Geoinf. **22**, 40–52 (2013)
15. Liu, Z.F., He, C.Y., Zhang, Q.F., et al.: Extracting the dynamics of urban expansion in China using DMSP-OLS nighttime light data from 1992 to 2008. Landsc. Urban Plan. **106**(1), 62–72 (2012)

16. Zhao, M., Cheng, W.M., Zhou, C.H., et al.: Assessing spatiotemporal characteristics of urbanization dynamics in Southeast Asia using time series of DMSP/OLS nighttime light data. Remote Sens. **10**(1), 20 (2018)
17. He, C., Li, J., Chen, J., et al.: The urbanization process of Bohai Rim in the 1990s by using DMSP/OLS data. J. Geogr. Sci. **16**, 174–182 (2006)
18. de Miguel, A.S., Zamorano, J., Gómez Castaño, J., et al.: Evolution of the energy consumed by street lighting in Spain estimated with DMSP-OLS data. J. Quant. Spectrosc. Radiat. Transf. **139**, 109–117 (2014)
19. Jiang, W., He, G., Leng, W., et al.: Characterizing light pollution trends across protected areas in China using nighttime light remote sensing data. Int. J. Appl. Earth Obs. Geoinf. **7** (7), 18 (2018)
20. Li, X., Zhou, Y.: Urban mapping using DMSP/OLS stable night-time light: a review. Int. J. Remote Sens. **38**(21), 6030–6046 (2017)
21. Emre, Y., Arzu, E.: Examining urbanization dynamics in Turkey using DMSP–OLS and socio-economic data. J. Indian Soc. Remote Sens. **46**(7), 1159–1169 (2018)
22. Wu, J.S., He, S.B., Peng, G., et al.: Intercalibration of DMSP-OLS night-time light data by the invariant region method. Int. J. Remote Sens. **34**(20), 7356–7368 (2013)
23. Imura, H., Cao, X., Chen, J., et al.: A SVM-based method to extract urban areas from DMSP-OLS and SPOT VGT data. Remote Sens. Environ. **113**(10), 2205–2209 (2009)
24. Zhou, Y., Smith, S.J., Elvidge, C.D., et al.: A cluster-based method to map urban area from DMSP/OLS nightlights. Remote Sens. Environ. **147**, 173–185 (2014)
25. Jang, W., He, G.J., Peng, Y., et al.: Application potentiality and prospects of nighttime light remote sensing in "The Belt and Road" initiative. J. Univ. Chin. Acad. Sci. **3**, 296–303 (2017)
26. Song, J.L.: DBAR initiative: big Earth data for "Belt and Road" development. Bull. Chin. Acad. Sci. **30**(2), 99–105 (2016)
27. Chen, Y.: Study on the economic logic of "The Belt and Road" initiative of China. Chin. Bus. Rev. **12**, 573–584 (2016)
28. Chen, Y.B., Zheng, Z.H., Wu, Zh.F., et al.: Review and prospect of application of nighttime light remote sensing data. Progr. Geogr. **38**, 205–223 (2019)
29. Wu, J., He, S., Peng, J., et al.: Intercalibration of DMSP-OLS night-time light data by the invariant region method. Int. J. Remote Sens. **34**, 7356–7368 (2013)
30. Liu, Y., Delahunty, T., Zhao, N.Z., et al.: These lit areas are undeveloped: delimiting China's urban extents from thresholded nighttime light imagery. Int. J. Appl. Earth Obs. Geoinf. **50**, 39–50 (2016)

Multi-scale Convolutional Neural Network for Remote Sensing Image Change Detection

Xiao Yu, Junfu Fan[(⊠)], Peng Zhang, Liusheng Han, Dafu Zhang, and Guangwei Sun

School of Civil and Architectural Engineering,
Shandong University of Technology, Zibo 255000, China
fanjf@sdut.edu.cn

Abstract. Intelligence method to detect changes in remote sensing images is a difficult but important issue and it is of great significance for natural resources, environmental and social-economy monitoring. In this paper, we presented a novel deep learning model named PSPNet-CONC which combined multi-scale feature deep learning model PSPNet and features extraction module ResNet34 in multi-period remote sensing images. Experiments were designed and conducted systematically for the comparison between the deep learning methods and traditional methods. Our experimental accuracy results show that our model got at least 11% higher in recall index than other state-of-the-art methods. Further more PA also increase by 4.5%, and unchanged accuracy is 1% better than other excellence methods. It demonstrates that with the characteristic of deep learning with multi-scale information, the PSPNet-CONC model could generate higher accuracy and stability detection results than other methods.

Keywords: Multi-scale · Change detection · PSPNet-CONC · Deep neural networks

1 Introduction

Change detection is the process for identifying the changed regions between the given image-pair observing the same scene at different times. Change detection using remote sensing data was widely used in many fields, such as land use classification, illegal building identification, and spatial data updating [1]. In order to monitor land use changes, governments at all levels spend plenty of manpower and material resources on basic survey every year. With the advantages and extensive use of remote sensing technologies, resource survey activities have gradually changed from field operations to internal works. However, the contradiction between the increasingly large amount of remote sensing data and low efficiencies of traditional data processing methods urgently needs to be solved. In addition, image change detection technology is widely used in the fields of ecological environment and disaster monitoring [2], such as assessment of earthquake disaster, glacier melting observation and so on. Therefore, accurate and efficient change detection technology has a high application value.

Change detection methods can be divided into two categories based on the final application purpose: 1) binary change detection methods [3–6]; and 2) multiple change

© Springer Nature Singapore Pte Ltd. 2020
Y. Xie et al. (Eds.): GSES 2019/GeoAI 2019, CCIS 1228, pp. 234–242, 2020.
https://doi.org/10.1007/978-981-15-6106-1_18

detection methods [1, 7–12]. Binary change detection methods consider all kinds of changes as one single change class, thus their aim is to find the changed and unchanged pixels in the considered feature space ignoring the semantic meaning of the possible different kinds of changes [13]. In contrast to binary change detection, the detection of multiple-change classes is a more challenging task. The main tasks of multiple change detection methods are to answer three questions: Whether is the land surface changed? What is it changed for? And what is the change trajectory?

In a previous study to solve the binary change detection, there are three main types methods in change detection research: image algebra method, image transform method and classification-based method. The algebra methods include difference value method, Change Vector Analysis (CVA). CVA is first implemented to separate the change and non-change candidate areas. For the unchanged areas, the image is assigned as the same class as that in the previous map, and the changed areas follow the independent classification [10]. This straightforward approach can provide the opportunity to maintain the consistency of image, and has been demonstrated to be effective in reducing false alarms [14]. However it is not easy to automatically and accurately determine the threshold to identify non-changes in previous works [11, 15], and with the different environment the result of CVA was unstabled.

In order to ignore the correlation between different spectra, then image transform method was introduced. The methods such as GS transform, Principal Components Analysis (PCA) [6, 16], Multivariate Alteration Detection (MAD) [17] are used to eliminate the correlation compression feature between bands. After that, the change and unchanged region are obtained by clustering method. This kind of method only considers the pixel features of the image, which is easy to produce the phenomenon of "salt and pepper noise".

Such methods are mainly based on spectral characteristics, ignoring features such as texture and shape. The methods of object-oriented classification can greatly improve the accuracy of remote sensing image classification for the reason that it contains spectral features texture features and shape features. Therefore, the object-oriented method is used in change detection [8, 18]. In order to solve the "from to" question several methods had proposed, It mainly includes classification before comparison and direct comparison, in which the direct comparison seeks the difference between the results of the two sets of classification, and contains the cumulative error of each classification. The direct comparison method does not have the case of error accumulation, but the analysis scale is single, and the accuracy of different resolution images is not stable. In order to solve the problem of multi-scale characteristics instability, Qi et al. used the scale from fine to coarse image segmentation and obtained the change detection results of each scale according to the change vector analysis [19]. And the methods of fuzzy fusion and decision level fusion are introduced to fuse the binary results of multiple scales. However, the segmentation scale of this method is difficult to determined.

With the development of artificial intelligence technology, deep neural network, which can simultaneously extract spectral, texture and shape information, is widely used in image semantic segmentation. In order to evaluate the applicability of convolutional neural network to change detection, FC-Siam-conc, FC-Siam-conc [4] and siamese neural network [18] based on UNet convolutional neural network model [20] are proposed. Nevertheless, it cannot fuse multi-scale information.

In this paper, in order to solve the aforementioned problems, we present a new deeplearning model with ResNet and PSPNet for multitemporal multispecial remote sensing data. A feature extract module named ResNet34 was used to extract band, shape and texture information. And then instead of determining the class labels of pixel by up sample operation, PSPNet was connected with ResNet34 to fuse multi-scale features. The PSPNet model can fuse multi-scale information from multispecial remote sensing data.

Therefore, the proposed approach consists of two main parts: 1) ResNet34 module to get information from multitemporal remote sensing data; 2) PSPNet module to fuse multi-scale information and generator class labels of pixel.

2   Model and Dataset

2.1   Model Structure

For extracting multi-scale image features, this paper adopts the PSPNet which contains the Pyramid Scene Parsing (PSP) [21] module. And first proposed architecture named PSPNet–CONC. PSPNet–CONC model structure is shown in Fig. 1, which mainly includes the feature extraction module and multi-scale convolution module. Multi-scale convolution module formed by two parts. The first part is multi-scale module which has four scale components. The second part is concat module which responsible for connecting features before POOL and after UPSAMPLE. Multi-period remote sensing images were used as the input of ResNet34 [22] with weight share, then concat features were used as the input of PSPNet. In order to prevent the gradient explosion caused by network too depth, ResNet34 was used as the feature extraction module of the image in this study. The input of the network can then be seen as a single patch of $256 \times 256 \times 3$. In the PSP module, the feature layer was sampled at different scales and the size of features were 1×1, 2×2, 3×3 and 6×6 respectively. After splicing the new features with the features extracted by ResNet34 through the up-sampling operation, a 3×3 convolution operation is conducted to generate a prediction binary image which has the same size as the input image with only one channel. All experiments were performed using PyTorch and with an Nvidia GTX1080 GPU.

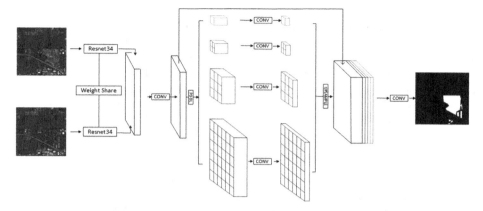

Fig. 1. Architectures of PSPNet-CONC for change detection.

2.2 Dataset

Different methods have different effects on different datasets, for comparing the accuracy of various methods objectively, the OSCD-3 (Onera Satellite Change Detection 3 bands) [5], an open-source dataset was used as training data and precision evaluation data in this study. OSCD dataset comes from Sentinel-2 Satellite at different periods and contains 13 bands between ultraviolet and short-wave infrared, with resolutions between 10 m–60 m. There are 24 groups of 600 × 600 pix images, and each group of images includes two images of different time periods and binary images marked up with changed region and unchanged region. In order to improve the training efficiency, this paper uses python to cut images to 256 × 256 pixels, and the ratio of training data to test data is 8:2. It has been demonstrated in [5] that the three bands of RGB data can still achieve the similar effect of using 13 bands. So, in order to improve the training efficiency, this paper used images with three bands of RGB as the training dataset.

2.3 Compare Method

Hard fusion with CVA is a widely used improved postclassification change detection method for updating LULC maps from multi-temporal remote sensing images [11, 14, 15]. CVA was used to obtain the change intensity, and a threshold is determined to identify the change and non-change candidates. In previous studies, $\mu + a\sigma$ was used as the criterion, and the parameter a was not determined automatically [14]. Therefore, in order to avoid the influence of manual interpretation, we use Kmeans classification algorithm to segment the change map. In addition, PCA as classical method to descending dimension, it also apply to change deletection. After a change intensity map was generator by CVA method, PCA method was used to get a map which the rate of contribution each component more than 75%, finally kmeans method was used to generate binary map. MAD and iteratively reweighted multivariate alteration detection (IRMAD) as excellent image transform methods for change detection. Minimize the correlation index as the condition of MAD method, the result of MAD can be classified by kmeans to generate change and nochange map. Inspired by boosting methods often used in data mining [23] and by expectation maximization [24], the idea in IRMAD is simply in an iterative scheme to put high weights on observations that exhibit little change over time [25]. With the widely used of deeplearning technology, Convolutional Neural Network was used to detect remote sensing change. FC-Siam-conc and FC-siam-diff is the most common method. This paper use both of methods above to compare with the method that we proposed.

2.4 Accuracy Assessment Index

We conducted the accuracy assessment in several ways, to ensure a comprehensive evaluation. For balance change and unchange area accuracy assessment, this paper use three index include Pixel Accuracy (PA), Recall and Unchanged Accuracy. The PA, which has been widely used in pattern recognition, was utilized for comprehensive evaluation through the fusion of truth predicted pixels and false predicted pixels. The Recall is measures of coverage. It counts how much positive pixels divided into positive pixels. Unchange region occupies most of the whole image, so we use unchange index to measure the unchanged region accuracy. The formula of three index are follows:

$$PA = \frac{TP + TN}{TP + TN + FP + FN} \tag{1}$$

$$Recall = \frac{TP}{TP + FP} \tag{2}$$

$$Unchanged = \frac{TN}{TN + FN} \tag{3}$$

Where parameter TP is number of predict changed and truth images also changed pixels, TN is number of predict unchanged and truth images are unchanges pixels, FP is number of predict unchanged but truth images are changed, FN is number of predict changed but truth images are unchanged.

3   Result Analysis

3.1 Loss Compare

To improve the change detection accuracy, image cropping, rotation, scaling and brightness change enhancement methods were performed on the training dataset. 300 times of iterative training calculation were carried out on the training dataset. The losses of train and valid corresponding to PSPNet-CONC and other two deep learning methods were shown in Fig. 2. The loss hold steady after 300 times of iterative. The loss results showed that PSPNet-CONC with multi-scale information could improve the accuracy of change detection.

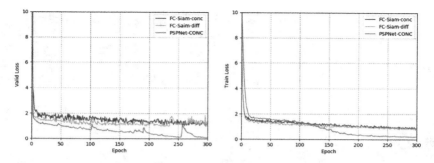

Fig. 2. PSPNet-CONC, FC-Saim-conc and FC-Siam-diff train and valid BCELoss value

3.2 Accuracy Comparison

The performance of convolutional neural network and traditional image change monitoring algorithm on open dataset [5] were analyzed in this paper. Four machine learning methods including MAD [17], IR-MAD [25], PCA [6] and CVA [12], as well as the main convolutional neural network methods including FC-Siam-conc and FC-Siam-diff [4, 5] were implemented and performed on same dataset to systematically compare their accuracies. Experimental results were shown in Table 1.

Table 1. Evaluation metrics for each test cases.

Methods	PA (%)	Recall (%)	Unchanged (%)
IR-MAD	60.85	43.76	61.42
MAD	58.39	44.52	59.13
PCA	58.44	43.32	58.94
CVA	50.14	49.67	50.00
FC-Siam-conc	94.07	47.77	97.46
FC-Siam-diff	94.86	47.94	97.54
PSPNet-CONC (ours)	**98.36**	**59.30**	**98.61**

The results generated using the OSCD dataset show that the method proposed in this paper far outperforms other methods in either recall or unchange accuracy. Especially got at least 11% higher in recall index than other state-of-the-art methods. Further more PA also increase by 4.5%, and unchanged accuracy is 1% better than other excellence methods. Detail illustrations of our results on this dataset can be found in Fig. 3. Deep learning method can solve the "salt and pepper noise" problem. In some mask area, for example cloud cover area, PSPNet-CONC can identify accurately.

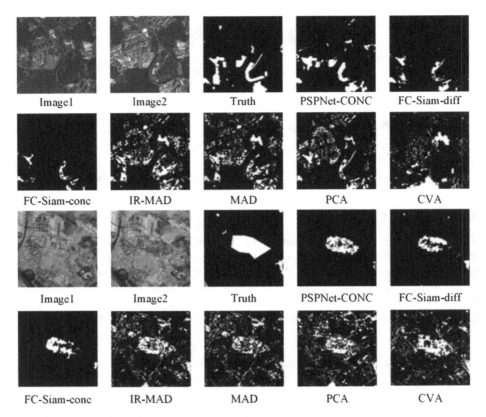

Image1	Image2	Truth	PSPNet-CONC	FC-Siam-diff
FC-Siam-conc	IR-MAD	MAD	PCA	CVA
Image1	Image2	Truth	PSPNet-CONC	FC-Siam-diff
FC-Siam-conc	IR-MAD	MAD	PCA	CVA

Fig. 3. Results of the Montpellier test case of the OSCD dataset using RGB color channels

4 Conclusion and Future Works

In this paper we presented a new end-to-end network architectural called PSPNet-CONC which contains multi-scale module to improve change detection accuracy in remote sensing images. Both of the detection results of changed regions and unchanged regions presented significant improvement on OSCD dataset. Especially on the detection of unchanged regions got much higher improvement of accuracy. The characteristics of the structure of PSPNet-CONC model are that the ResNet34 module was used for feature extraction and the PSPNet module was used for information fusion.

Perspectives for this work include enlarging the dataset both in number of cities and land use types, e.g. enriching it with Sentinel-1 satellites images and add some agricultural land use change dataset. In addition, a natural extension of our work presented in this paper would be to evaluate how these networks perform on detecting semantic changes. The PSPNet-CONC model will be used in evaluating the performance and accuracy on dealing with remote sensing images with more than 3-bands in the future.

References

1. Zhang, P., Gong, M., Su, L., Liu, J., Li, Z.: Change detection based on deep feature representation and mapping transformation for multi-spatial-resolution remote sensing images. ISPRS J. Photogram. **116**, 24–41 (2016)
2. Stramondo, S., Bignami, C., Chini, M., Pierdicca, N., Tertulliani, A.: Satellite radar and optical remote sensing for earthquake damage detection: results from different case studies. Int. J. Remote Sens. **27**(20), 4433–4447 (2006)
3. Celik, T.: Unsupervised change detection in satellite images using principal component analysis and k-means clustering. IEEE Geosci. Remote Sens. Lett. **6**(4), 772–776 (2009)
4. Daudt, R.C., Le Saux, B, Boulch, A.: Fully convolutional siamese networks for change detection. In: 2018 25th IEEE International Conference on Image Processing (ICIP), pp. 4063–4067. IEEE (2018)
5. Daudt, R.C., Le Saux, B, Boulch, A., Gousseau, Y.: Urban change detection for multispectral earth observation using convolutional neural networks. In: 2018 IEEE International Geoscience and Remote Sensing Symposium, IGARSS 2018, pp. 2115–2118. IEEE (2018)
6. Deng, J., Wang, K., Deng, Y., Qi, G.: PCA-based land-use change detection and analysis using multitemporal and multisensor satellite data. Int. J. Remote Sens. **29**(16), 4823–4838 (2008)
7. Desclée, B., Bogaert, P., Defourny, P.: Forest change detection by statistical object-based method. Remote Sens. Environ. **102**(1–2), 1–11 (2006)
8. Huo, C., Zhou, Z., Lu, H., Pan, C., Chen, K.: Fast object-level change detection for VHR images. IEEE Geosci. **7**(1), 118–122 (2009)
9. Wu, C., Du, B., Zhang, L.: Slow feature analysis for change detection in multispectral imagery. IEEE Trans. Geosci. **52**(5), 2858–2874 (2013)
10. Wu, C., Du, B., Cui, X., Zhang, L.: A post-classification change detection method based on iterative slow feature analysis and Bayesian soft fusion. IEEE Trans. Geosci. **199**, 241–255 (2017)
11. Xian, G., Homer, C., Fry, J.: Updating the 2001 National Land Cover Database land cover classification to 2006 by using Landsat imagery change detection methods. Remote Sens. Environ. **113**(6), 1133–1147 (2009)
12. Zhao, M., Zhao, Y.: Object-oriented and multi-feature hierarchical change detection based on CVA for high-resolution remote sensing imagery. Remote Sens. **22**(1), 119–131 (2018)
13. Liu, S., Bruzzone, L., Bovolo, F., Zanetti, M., Du, P.: Sequential spectral change vector analysis for iteratively discovering and detecting multiple changes in hyperspectral images. IEEE Trans. Geosci. Remote Sens. **53**(8), 4363–4378 (2015)
14. Yu, W., Zhou, W., Qian, Y., Yan, J.: A new approach for land cover classification and change analysis: integrating backdating and an object-based method. Remote Sens. Environ. **177**, 37–47 (2016)
15. Xian, G., Homer, C.: Updating the 2001 National Land Cover Database impervious surface products to 2006 using Landsat imagery change detection methods. Remote Sens. Environ. **114**(8), 1676–1686 (2010)
16. Falco, N., Marpu, P.R., Benediktsson, J.A.: A toolbox for unsupervised change detection analysis. Int. J. Remote Sens. **37**(7), 1505–1526 (2016)
17. Nielsen, A.A., Conradsen, K., Simpson, J.J.: Multivariate alteration detection (MAD) and MAF postprocessing in multispectral, bitemporal image data: new approaches to change detection studies. Remote Sens. Environ. **64**(1), 1–19 (1998)

18. Bovolo, F., Bruzzone, L.: A theoretical framework for unsupervised change detection based on change vector analysis in the polar domain. IEEE Trans. Geosci. **45**(1), 218–236 (2006)
19. Qi, Z., Yeh, A.G.-O., Li, X., Zhang, X.: A three-component method for timely detection of land cover changes using polarimetric SAR images. ISPRS J. Photogram. Remote Sens. **107**, 3–21 (2015)
20. Ronneberger, O., Fischer, P., Brox, T.: U-Net: convolutional networks for biomedical image segmentation. In: Navab, N., Hornegger, J., Wells, W., Frangi, A. (eds.) MICCAI 2015. LNCS, vol. 9351, pp. 234–241. Springer, Cham (2015). https://doi.org/10.1007/978-3-319-24574-4_28
21. Zhao, H., Shi, J., Qi, X., Wang, X., Jia, J.: Pyramid scene parsing network. In: Proceedings of the IEEE Conference on Computer Vision and Pattern Recognition, pp. 2881–2890 (2017)
22. He, K., Zhang, X., Ren, S., Sun, J.: Deep residual learning for image recognition. In: Proceedings of the IEEE Conference on Computer Vision and Pattern Recognition, pp. 770–778 (2016)
23. Hastie, T., Tibshirani, R., Friedman, J., Franklin, J.: The elements of statistical learning: data mining, inference and prediction. Math. Intell. **27**(2), 83–85 (2005)
24. Wiemker, R., Speck, A., Kulbach, D., Spitzer, H., Bienlein, J.: Unsupervised robust change detection on multispectral imagery using spectral and spatial features. In: Proceedings of the Third International Airborne Remote Sensing Conference and Exhibition, pp. 640–647 (1997)
25. Nielsen, A.A.: The regularized iteratively reweighted MAD method for change detection in multi-and hyperspectral data. IEEE Trans. Image Process. **16**(2), 463–478 (2007)

Research on Change of Land Use Based on Decision Tree in the Horqin Sandy Land in the Past 25 Years

Shuxiang Wang[1,2], Liusheng Han[1,2(✉)], Ji Yang[2], Yong Li[2],
Congjun Zhu[1,2], Qian Zhao[1,2], Zhenzhen Zhao[1], Li Liu[3],
and Ruiping Zhang[4]

[1] Shandong University of Technology, Zibo 255000, China
hanls@sdut.edu.cn
[2] Guangdong Key Laboratory of Geospatial Information Technology
and Application, Guangzhou Institute of Geography, Guangzhou 510070, China
[3] Ludong University, Yantai 264001, China
[4] Bayannur Electric Power Bureau, Bayannur 015000, China

Abstract. Horqin Sandy Land is a key area for the construction of northeast shelterbelts. Ecological environment of the area is fragile and vulnerable to damage. However, there is a lack of research on the long time series and continuous system in Horqin Sandy Land. Therefore, it is urgent to grasp the impact of policies of the Three-North Shelterbelt Program on Horqin Sandy Land and analysis the differences in land use patterns in the Horqin Sandy Land before and after implementing restoration projects. In this paper, six counties in Horqin Sandy Land are used as the research areas. Six Landsat TM / ETM + / OLI remote sensing images from 1990 to 2015 are collected. Based on the NDVI, NDWI, NDBI, and the principle of spectral hybrid analysis, a decision tree classification method is used to put the satellite image data divided into seven categories: sandy land, cultivated land, waters, forest land, grassland, residential land, and saline-alkali land. The accuracy of the classification results is verified on the basis of high-resolution images and yearbook data. The results show that: (1) Analysis is made from the perspective of land use dynamics. During the period of 1990–2015, the area of cultivated land and residential land continued to increase. The area of salinized land, water body and grassland fluctuated little, and decreased generally. The forest land and desertification area fluctuated greatly and the overall trend was "increase-decrease-increase"; the overall trend of sandy land was "decrease-increase-decrease". The areas of grassland, woodland and sandy land showed a negative correlation. (2) Analysis is made from the perspective of spatial distribution pattern of land use types, each of land types has undergone dramatic changes, among which grassland, cultivated land, forest land, and sandy land are the most important types of change. (3) Twenty-five years after the implementation of the Three-Norths Shelter Program, the ecological environment of Horqin Sandy Land has been initially restored. The ecological environment in this area is extremely fragile and requires long-term and continuous protection at a later stage. This study analyzes the change of land use pattern in Horqin Sandy Land in the past 25 years through an effective decision tree classification method, explores the impact of the Three-North Shelterbelt Program on the evolution of the pattern,

© Springer Nature Singapore Pte Ltd. 2020
Y. Xie et al. (Eds.): GSES 2019/GeoAI 2019, CCIS 1228, pp. 243–258, 2020.
https://doi.org/10.1007/978-981-15-6106-1_19

and provides a recommendation of rationalization for future sustainable land use in Horqin Sandy Land.

Keywords: Decision tree classification · Horqin sandy land · Land use · Three-north Shelterbelt program · Spatial distribution analysis

1 Introduction

Land use change refers to changes in land use and management methods that lead to changes in land cover. Land cover change has always been one of the core areas and hot issues of global change research [1, 2]. Many scholars pay attention to the laws and impacts of land use change in global regions [3, 4]. In 2015, Hegazy used an unsupervised classification method to dynamically monitor urban development and land use change in Egypt from 1985 to 2010. The monitoring information is of great significance for improving future urban sustainable development plans [5]. In 2015, Butt used the maximum likelihood algorithm to monitor the land cover changes of the Simli Basin in Islamabad, Pakistan in 1992 and 2015, and confirmed that land cover changes posed a serious threat to the basin resources [6]. In 2016, Fu classified land cover changes in Landsat image data based on surface reflectance, brightness temperature, and NVI, and analyzed their impact on surface temperature in time series [7]. The above studies have better analyzed the land cover changes in a certain area, but the classification method is simple, the classification accuracy is low, and it cannot reflect the actual situation well [8, 9].

A series of in-depth studies on land use change have been carried out around the world. In 2018, Huang classified remote sensing images based on convolutional neural networks. This classification method avoids over-fitting problems and reduces classification time through fine-tuned and pre-training networks [10]. In 2015, Luus proposed a multi-view, multi-scale deep learning classification method for multi-spectral land use supervision classification, and the prediction accuracy of a single multi-scale view was significantly improved [11]. In 2016, based on cellular automata and GIS, Reine Maria Basse used a decision tree machine learning algorithm to define a new model conversion rule for classifying land use in Luxembourg. The classification results have good accuracy and are universally applicable [12].

The above methods have high classification accuracy, greatly reduce classification time, and is highly practical, but they are only research on the algorithm, and do not have practical application analysis combine with a certain area [13–15].

This area is selected for monitoring, because the Horqin Sandy Land is a key area for the construction of shelterbelts in Northeast China. It belongs to a typical agro-pastoral ecotone, and its ecological environment is fragile and easily damaged. However, since the implementation of the Three-North Shelterbelt Program in 1979, there has not been a long-term series and continuous system study in this area [16].

In conclusion, in order to grasp the influence of Three-North Shelterbelt Program on Horqin Sandy Land, analyze the difference of land use pattern change before and after the policy, and perceive the change of ecological environment in Horqin Sandy Land, this paper uses decision tree classification method to carry out a series of research on the land use change in Horqin Sandy Land in recent 25 years. The spatial pattern changes of long time series and continuous system in Horqin Sandy Land are studied, which is of high accuracy and realization.

2 Materials

2.1 Study Area

The study area in this paper belongs to the Horqin sandy land, including Kezuozhong Banner, Kailu County, Kulun Banner, Naiman Banner, Kezuohou Banner, and Horqin District. The geographical location is 42° 5'N-43° 5'N and 117° 30'E-123° 30'E (Fig. 1). The Horqin sandy area has large terrain fluctuations, and its surface characteristics are high in the north and south, low in the middle, and high in the west and low in the east. Most of the study areas are between 250–650 m above the sea level. The study area is located in the northern hemisphere and belongs to the semi-arid continental monsoon climate in the northern temperate zone. It is dry and windy in spring,

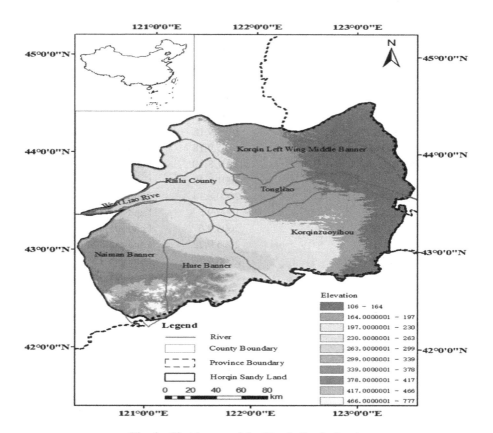

Fig. 1. Sketch map of the Horqin Sandy Land

hot and rainy in summer, cool in short autumn, and dry and severe in winter. The daily temperature difference is large and the precipitation periods are concentrated. The annual average temperature is 5.8–6.4 °C. Because it is closer to the ocean, the average annual precipitation is 350–500 mm. The area has sunshine time of 2900–3100 h throughout the year.

2.2 Data Source

The remote sensing data comes from the "geospatial data cloud" platform. According to the global reference system of the Landsat satellite, there are remote sensing data of 6 scenes including the periods of 1990, 1995, 2000, 2005, 2010, and 2015 covering the study area. The strip number/row number is (120,29), (120,30), (121,29), and (121,30). See Table 1 for details. Socio-economic data comes from the Tongliao Statistical Yearbook and the Inner Mongolia Statistical Yearbook.

Table 1. Remote sensing image data source information by period

Year	Stripe number/Column number	Acquisition time	Sensor type
1990	120.29	1988.10.9	Landsat5 TM
	120.30	1987.8.4	Landsat5 TM
	121.29	1992.8.24	Landsat5 TM
	121.30	1989.7.31	Landsat5 TM
1995	120.29	1995.8.26	Landsat5 TM
	120.30	1995.9.11	Landsat5 TM
	121.29	1995.9.18	Landsat5 TM
	121.30	1995.9.18	Landsat5 TM
2000	120.29	2001.8.10	Landsat5 TM
	120.30	2001.8.10	Landsat5 TM
	121.29	2000.8.14	Landsat5 TM
	121.30	1999.9.5	Landsat7 ETM+
2005	120.29	2005.9.6	Landsat5 TM
	120.30	2005.9.22	Landsat5 TM
	121.29	2006.9.16	Landsat5 TM
	121.30	2006.9.16	Landsat5 TM
2010	120.29	2010.9.12	Landsat7 ETM+
	120.30	2009.9.17	Landsat5 TM
	121.29	2010.8.10	Landsat5 TM
	121,30	2010.8.10	Landsat5 TM
2015	120,29	2014.8.30	Landsat8 OLI
	120,30	2014.8.30	Landsat8 OLI
	121,29	2016.8.26	Landsat8 OLI
	121,30	2015.7.7	Landsat8 OLI

3   Method

3.1   Preprocessing

Download the image data of the corresponding strip number and row number from the geospatial data cloud platform for the six periods (1990, 1995, 2000, 2005,2010, 2015). Due to the function attenuation of satellite sensor, atmospheric effects, and flight attitude [17], the data are pre-processed, and the important processes include radiation calibration, FLAASH atmospheric correction, mosaic, and background value removal. Landsat 5 satellite remote sensing images are synthesized using non-standard pseudo-colors in 4, 5, and 3 bands; Landsat 8 satellite remote sensing images are synthesized using non-standard pseudo-colors in 3, 4, and 5 bands.

3.2   Establishment of Image Interpretation Marks

Due to the differences in the physical structure and chemical composition of various features, they have different degrees of reflection and absorption on the solar spectrum, forming specific spectral features of features [18]. Different land types show different gray levels (brightness) on remote sensing images. Therefore, this study uses different gray levels as the classification and interpretation indicators to establish standards (Table 2).

Table 2. Visual interpretation signs of land use type

Landscape type	Landsat 5 (4.5.3 composite) / Landsat8 (3.4.5 composite) image features	Image
Woodland	Large area, dark green, blurred texture	
Waters	Blue, faceted	
Cultivated field	Bright green, regular geometry	
Residential land	Regular shape, blue (purple) roof and ground	
Sandy land	White, brighter, with corrugated structure	
Saline-alkali land	White, bright, perforated structure	
Grassland	Large area, yellow-green, unclear texture	

3.3   Decision Tree Classification

Combining the processed remote sensing image with DEM data and other high-resolution images, a decision tree classification method is used to classify the remote sensing image. In the classification process, the normalized vegetation index, normalized building index, and normalized water body index are cited. The calculation of the three discriminant functions is as follows [19, 20]:

$$NDVI = \frac{NIR - RED}{NIR + RED} = \frac{TM4 - TM3}{TM4 + TM3} \tag{1}$$

$$NDBI = \frac{TM5 - TM4}{TM5 + TM4} \tag{2}$$

$$NDWI = \frac{TM2 - TM5}{TM2 + TM5} \tag{3}$$

Samples of interest are selected from the seven types of land, and the number of samples is 30. According to NDVI, NDBI, and NDWI, the sample points are assigned values to obtain the box map between each type of feature and the three discriminant functions (Fig. 2). Among them, NDVI can be used to distinguish vegetation information,

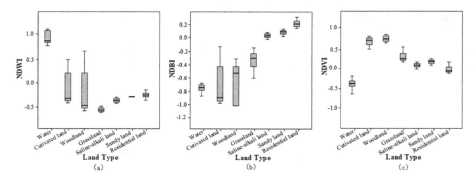

Fig. 2. Threshold relationship between each land category and NDWI, NDBI, NDVI. a. Threshold relationship between each land category and NDWI. b. Threshold relationship between each land category and NDBI. c. Threshold relationship between each land category and NDVI.

NDBI can be used to distinguish residential land, and DEM data and topographic map data can be used to distinguish cultivated land, woodland, and grassland information. NDWI index can be used to distinguish waters. After many experiments and comparisons, the classification threshold classification is finally determined (Fig. 3).

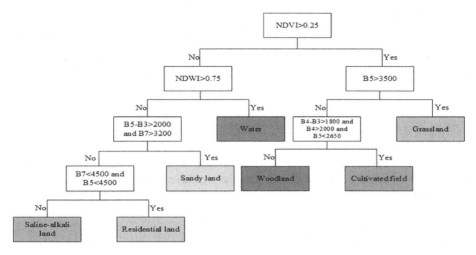

Fig. 3. Threshold division of decision tree classification

3.4   Verification of Accuracy

Using the data of Tongliao Statistical Yearbook and Inner Mongolia Statistical Year-book, the area of cultivated land, waters, woodland, sand land, saline-alkali land, and residential land is extracted from 1990 to 2016, and compared with the result of decision tree classification. The results show that the verification accuracy reaches 94.28%. Due to space limitations, only comparative data are shown in 1990 (Fig. 4).

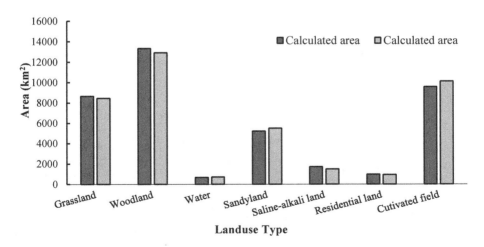

Fig. 4. Comparison of decision tree classification results and statistical data in 1990

3.5   Land Use Dynamics

Land use dynamics is the area change of a certain land use type over a period of time. It reflects the severity of dynamic changes among land use types and can reflect the comprehensive impact of human activities on regional land use type changes [21]. Its calculation formula is:

$$S = \left[\frac{A_{t+\Delta t} - A_t}{A_t}\right] \times \frac{1}{\Delta t} \times 100\% \qquad (4)$$

Where $A_{t+\Delta t}$ is the area of a certain land use type in period $t + \Delta t$, A_t is the area of a certain land use type in period t, and S is the dynamic degree of land use within a certain period of time for a certain land use type.

4   Results and Analysis

4.1   Dynamic Changes in Landscape Pattern

According to the image classification system and the feature characteristics of the study area, NDVI, NDBI, and NDWI are introduced, and the decision tree classification method is supplemented with visual interpretation. The pixel classification method and the knowledge method are integrated to divide the remote sensing image into residential land, cultivated land, woodland, waters, sandy land, grassland, and saline-alkali land, and finally obtained the land use map of Horqin area from 1990 to 2015 (Fig. 5).

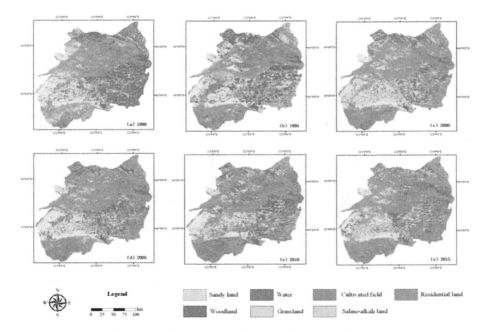

Fig. 5. Land use map of Horqin area from 1990 to 2015

Based on the six-stage land use classification map, we can count the number of pixels in each land category, calculate the area and proportion of each land category based on the remote sensing image resolution, and study the dynamic changes of each landscape pattern type in Horqin Sandy Land (Table 3, Table 4, Fig. 6).

Table 3. Areas and proportions of each class from 1990 to 2000 in the Horqin Sandy Land (Area: km^2, Proportion: % of Horqin Sandy Land)

Land Type	1990		1995		2000	
	Area	Proportion	Area	Proportion	Area	Proportion
Grassland	8653.08	21.54	8928.32	22.08	9615.15	22.89
Woodland	13341.4	33.22	13700.11	33.89	14069.39	33.49
Water	674.85	1.68	723.65	1.79	613.3	1.46
Sandy land	5220.7	13	3989.94	9.87	3795.12	9.03
Saline-alkali land	1717.99	4.28	1363.32	3.37	1478.89	3.52
Residential land	979.95	2.44	1022.14	2.53	1065.17	2.54
Cultivated field	9575.59	23.84	10700.11	26.47	13371.33	27.07

Table 4. Areas and proportions of each class from 2005 to 2015 in the Horqin Sandy Land (Area: km^2, Proportion: % of Horqin Sandy Land)

Land Type	2005		2010		2015	
	Area	Proportion	Area	Proportion	Area	Proportion
Grassland	9769.31	22.83	8385.6	20.19	8397.83	19.69
Woodland	15308.1	35.78	15027.4	36.19	15591.2	36.55
Water	413.94	0.97	238.76	0.57	270.54	0.63
Sandy land	2949.1	6.89	3207.9	7.73	2636.2	6.18
Saline-alkali land	1088.66	2.54	890.54	2.14	1230.94	2.89
Residential land	1113.43	2.60	1146.65	2.76	1212.7	2.84
Cultivated field	12147.4	28.39	12626.9	30.41	13316.2	31.22

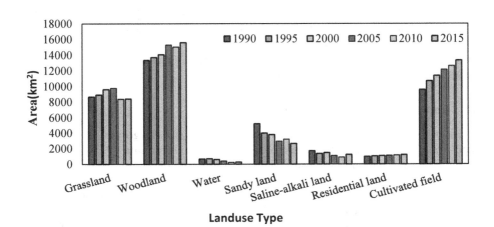

Fig. 6. Statistics of areas of land use types in the Horqin Sandy Land from 1990 to 2015

As can be seen from Table 3 and Table 4, and Fig. 5, grassland, woodland, sandy land, and cultivated land are the most important land types in Horqin Sandy Land. These four land types account for more than 90% of the area, of which the area of woodland has always been the largest. However, in the 25-year change of land type, the area of cultivated land has continued to increase. By 2015, its area proportion is almost the same as that of forest land. The area change of grassland is a decreasing trend in general, and its area ratio has decreased from 21.54% to 19.69% and has decreased by 2.95%; the area change of woodland has an increasing trend in general, and its area ratio has increased from 33.2% to 36.55%, which has increased by 16.86%. The proportion of the area of sandy land is shrinking. Its area proportion dropped from 13% to 6.18% and has decreased sharply by 49.5%. The area change of cultivated land is generally increasing, and its area proportion has increased from 23.84% to 31.22%, which has increased sharply by 39.06%; the area of waters has greatly reduced, and its area proportion has plummeted from 1.68% to 0.63%, which has decreased by 59.91%. The area of saline-alkali land has decreased, and its area proportion has dropped from 4.28% to 2.89% and has decreased by 28.35%; the area of residential land has increased steadily, and its area ratio has increased from 2.44% to 2.84% and has increased by 23.76%.

It can be seen from Fig. 6 that from 1990 to 2015, the area of residential land and cultivated land continued to increase. The cultivated land area increased 3740.64 km^2, a sharp increase of 39.06% over 1990; the area of residential land has increased 232.75 km^2, a steady increase of 23.76% over 1990; the area of waters and saline-alkali land area showed a trend of decreasing fluctuations. The area of waters increased from 1990 to 1995, decreased sharply from 1995 to 2010, and increased from 2010 to 2015. In general, it's a decreasing trend. Compared with 1990, the area decreased by 59.91% in 2015; the area of saline-alkali land decreased from 1990 to 1995, increased from 1995 to 2000, decreased from 2000 to 2005, increased from 2005 to 2010, and decreased from 2010 to 2015. Generally, it decreased by 28.35% over 1990; the area of woodland and grassland area have a positive correlation. From 1990 to 2005, its area increased, and from 2005 to 2010, its area decreased and its area increased from 2010 to 2015. On the whole, there is a trend of "increasing-decreasing-increasing"; the area of sandy land has a negative correlation with the area of woodland and grassland. From 1990 to 2005, its area decreased sharply. From 2005 to 2010, the area changes reversed. From 2010 to 2015, its area decreased again, generally showing a development trend of "decrease-increase-decrease".

4.2 Land Use Dynamics Analysis

Based on the remote sensing images, the area of each land category in Horqin Sandy Land from 1990 to 2015 was calculated. Combining with formula (4), we can obtain the land area change and land use dynamics of the area in the six periods (Table 5, Table 6).

It can be seen from Table 5 that from 1990 to 1995, the area of grassland, woodland, waters, residential land and cultivated land in Horqin Sandy Land increased, with dynamic degrees of 0.64%, 0.54%, 1.45%, 0.86% and 2.35%. The area of saline-alkali land decreased, and the dynamic degree was −4.71% and −4.13%. Among them,

Table 5. Statistics of area changes and land use dynamics in Horqin Sandy Land in each period from 1990 to 2005 (Change area: km²; Dynamic degree: %)

Land type	1990–1995		1995–2000		2000–2005	
	Change area	Dynamic degree	Change area	Dynamic degree	Change area	Dynamic degree
Grassland	275.24	0.64	686.83	1.54	154.16	0.32
Woodland	358.66	0.54	369.28	0.54	1238.79	1.76
Water	48.8	1.45	−110.35	−3.05	−199.36	−6.5
Sandy land	−1230.76	−4.71	−194.82	−0.98	−86.02	−4.46
Saline-alkali land	−354.67	−4.13	115.57	1.70	−30.23	−5.28
Residential land	42.19	0.86	43.03	0.84	48.26	0.91
Cultivated field	1124.52	2.35	671.22	1.25	776.05	1.36

Table 6. Statistics of area changes and land use dynamics in Horqin Sandy Land in each period from 2005 to 2015 (Change area: km²; Dynamic degree: %)

Land type	2005–2010		2010–2015		1990–2015	
	Change area	Dynamic degree	Change area	Dynamic degree	Change area	Dynamic degree
Grassland	−1383.71	−2.83	12.23	0.03	−255.25	−0.12
Woodland	−280.78	−0.37	563.81	0.75	2249.76	0.67
Water	−175.18	−8.46	31.78	2.66	−404.31	−2.4
Sandy land	258.8	1.76	−571.7	−3.56	−2584.5	−1.98
Saline-alkali land	−198.12	−3.64	340.4	7.64	−487.05	−1.13
Residential land	33.22	0.6	66.05	1.15	232.75	0.95
Cultivated field	479.47	0.79	689.38	1.09	3740.64	1.56

the woodland area changed the least, and the sandy land area changed the most. From 1995 to 2000, the area of grassland, woodland, saline-alkali land, residential land and cultivated land increased in the study area, with dynamic degrees of 1.54%, 0.54%, 1.7%, 0.84% and 1.25%, while the area of waters and sandy land decreased with dynamic degrees of −3.05% and −0.98%, of which the change in woodland area is the smallest and the change in waters area is the largest. From 2000 to 2005, the area of grassland, woodland, residential land and cultivated land increased, with dynamic degrees of 0.32%, 1.76%, 0.91% and 1.36%, while the area of waters, sandy land and saline-alkali land decreases, with dynamic degrees of −6.5%, −4.46%, and −5.28%, among which the annual change rate of grassland area is the smallest, and waters area changes the most.

It can be seen from Table 6 that from 2005 to 2010, the area of sandy land, residential land and cultivated land increased, with a dynamic degree of 1.76%, 0.6%, and 0.79%, and the area of grassland, woodland, waters, and saline-alkali land decreased with a dynamic degree of −2.83%, −0.37%, −8.46%, −3.64%, of which the change in woodland area is the smallest and the change in waters area is the largest. From 2010 to 2015, the area of grassland, forest land, waters, saline-alkali land, residential land and cultivated land increased, with a dynamic range of 0.03%, 0.75%, 2.66%, 7.64%, 1.15%, and 1.09%, while the area of sandy land decreased. The dynamic degree was −3.56%, of which the grassland area changed the least, and the saline-alkali land area changed the most.

4.3   Analysis of Spatial Transfer Characteristics

Using the land use classification map of Horqin Sandy Land from 1990 to 2015, and overlay analysis in Arcgis, we can get the spatial transfer distribution map of various land types from 1990 to 2015. The figure shows the spatial transfer characteristics among land types, including amount of conversion, source of conversion, and destination of conversion (Fig. 7).

From 1990 to 2015, Horqin Sandy Land has experienced drastic changes in land use types during 25 years. The most important types of spatial transfer are woodland to cultivated land, grassland to cultivated land, woodland to grassland, sandy land to grassland, sandy land to woodland, grassland to sandy land, cultivated land to woodland, saline-alkali land to cultivated land. Among them, the expansion of cultivated land and the reduction of sandy land are the most drastic. The expansion of cultivated land is embodied by the conversion of woodland into cultivated land, grassland into cultivated land and saline-alkali land into cultivated land. Woodland change to cultivated land mainly occurs in the northeast, southeast, and central river basins of Horqin Sandy Land. Grassland change to cultivated land mainly occurs in the northwestern and central regions of the Horqin Sandy Land; the saline-alkaline land change to cultivated land mainly occur in the southern regions. The reduction of sandy land is mainly manifested in two main transfer directions: sandy land to woodland and sandy land to grassland. Sandy land change to woodland mainly occurs in the central and western regions of Horqin Sandy Land; sandy land change to grassland mainly occurs in the eastern part of Horqin Sandy Land. In addition, the shifts of cultivated land to woodland, woodland to grassland, and grassland to sandy land are more prominent. Among them, cultivated land change to woodland mainly occurs in the central and northern regions and southern mountainous areas of Horqin Sandy Land; woodland change to grassland occurs mainly in eastern and northern regions of Horqin Sandy Land. The grassland change to sandy land mainly occurs in the northwestern and western areas of the Horqin Sandy Land.

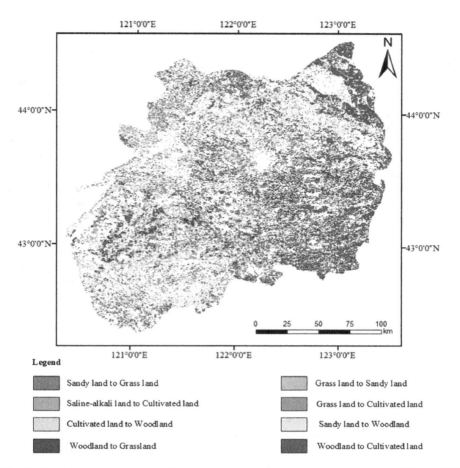

Fig. 7. Spatial transfer characteristics of land types in Horqin Sandy Land from 1990 to 2015

5 Discussion

Horqin Sandy Land is a key area for the construction of shelterbelts in Northeast China. It belongs to a typical agriculture and husbandry interlace zone. The ecological environment is fragile and vulnerable to damage. It is extremely sensitive to human activities and climate change.

In the past 25 years, the most prominent manifestation of land use change in the Horqin Sandy Land is the sharp decline in sandy land and the increase in the area of woodland. The main reason is that China has implemented a large-scale green project of the Three-North Shelterbelts Program since 1979, organizing afforestation. Secondly, the Chinese government has also implemented related policies such as prohibiting overgrazing, which has played a certain role in inhibiting grassland degradation. The sharp decrease in waters and increase in cultivated land area is another prominent manifestation of land use change in the region. The main reasons for the performance in two aspects are due to human activities, excessive extraction of water resources, waste of water resources, resulting in a sharp decrease in the area of waters, and a series of climatic factors such as rainfall, temperature and evaporation also have a certain impact on its sharp decline; the increase in cultivated land is mainly caused by excessive human reclamation. As the population continues to increase, in order to meet the food needs of more people, a large amount of wasteland, forest land and lightly saline-alkali land need to be cultivated, which is one of the reasons for the decrease in the area of saline-alkali land. The land use change in Horqin Sandy Land is another manifestation of a decrease in the area of saline-alkali land and a significant increase in residential land. Among them, the saline-alkali land is unstable. The main reason for the decrease in saline-alkali land area is human reclamation. Secondly, it has a very positive correlation with rainfall, air temperature, evaporation, and annual sunshine hours. However, the area of saline-alkali land shows a decreasing development trend. The impact of human activities is greater than that of natural climatic factors.

It can be seen from the spatial transfer characteristics map that the changes in cultivated land are the most prominent, and the change of woodland to cultivated land is the most obvious. This shows that at the same time as the implementation of the Three-North Shelterbelt Program, deforestation also exists. This is the reason that sandy land is located in the central and western parts of the Horqin Sandy Land and the construction of the Three-North Shelterbelt Program also occurs here, the flat terrain in the eastern region is more suitable for reclamation of wasteland and human living.

Land changes in waters and sandy land have been the most dramatic in the past 25 years. Although 25 years after the implementation of the Three-North Shelterbelt Program, land desertification in the region has reversed, it can be seen that environmental changes in the region are extremely unstable through the path of spatial transfer. In particular, the phenomenon of grassland turning into sandy land should be taken seriously to prevent the expansion of desertified land, which requires long-term and continuous protection at a later stage. At the same time, attention should be paid to strengthening the education of human environmental protection awareness so that people cherish and love water resources, use water resources reasonably, and reduce water waste.

6 Conclusion

Based on the data of Landsat TM / ETM + / OLI in 1990, 1995, 2000, 2005, 2010 and 2015, and the statistical yearbooks of Tongliao City and Inner Mongolia, this paper takes six counties (Kezuozhong Banner, Kailu County, Kulun Banner, Naiman Banner, Kezuohou Banner, and Horqin District) in Horqin Sandy Land as the research area, Based on NDVI, NDWI, NDBI, and the principles of spectral hybrid analysis, a decision tree classification method is used to put the satellite image data divided into seven categories: sandy land, cultivated land, waters, woodland, grassland, residential land, and saline-alkali land. The classification results are evaluated and verified based on high-resolution imagery and yearbook data. This paper further explores the dynamic changes of land use in the area in terms of comprehensive land use dynamics and spatial patterns of land use types, and analyzes the impact of shelter policies on the evolution of landscape patterns. The research results show that: (1) From the analysis of land use dynamics, the area of cultivated land and residential land continued to increase from 1990 to 2015. The area of Saline-alkali land, waters, and grassland fluctuated relatively little, and generally showed a decreasing trend. The woodland and sandy land area fluctuated greatly, and the woodland area generally showed a trend of "increasing-decreasing-increasing"; the sandy land area generally showed a trend of "decreasing-increasing-decreasing". The areas of grassland, woodland and sandy land showed a negative correlation. (2) From the analysis of land use space transfer, each land type has undergone dramatic changes, among which grassland, cultivated land, woodland, and sandy land are the most important types of change. (3) Twenty-five years after the implementation of the Three-Norths Shelter Program, the ecological environment of Horqin Sandy Land has been initially restored. The ecological environment in this area is extremely fragile and requires long-term and continuous protection at a later stage. This study analyzes the change of land use pattern in Horqin Sandy Land in the past 25 years through an effective decision tree classification method, explores the impact of the Three-Norths Shelter Program on the evolution of pattern, and provides suggestions for sustainable land use in Horqin Sandy Land in the future.

Acknowledgments. This study was jointly supported by China Southern Power Grid Guangzhou Power Supply Bureau Co., Ltd. Key Technology Project (0877002018030101SRJS0 0002); Guangdong Provincial Science and Technology Program (2017B010117008); Guangzhou Science and Technology Program (201806010106, 201902010033); the National Natural Science Foundation of China (41976189,41976190); the Guangdong Innovative and Entrepreneurial Research Team Program (2016ZT06D336); the Southern Marine Science and Engineering Guangdong Laboratory (Guangzhou) (GML2019ZD0301); the GDAS's Project of Science and Technology Development (2016GDASRC-0211,2018GDASCX-0403,2019GDASYL-0301001, 2017GDASCX-0101,2018GDAS CX-0101,2019GDASYL-0103003).

References

1. Lambin, E.F., Meyfroidt, P.: Global land use change, economic globalization, and the looming land scarcity. Proc. Natl. Acad. Sci. **108**(9), 3465 (2011)
2. Prestele, R., et al.: Hotspots of uncertainty in land-use and land-cover change projections: a global-scale model comparison. Glob. Change Biol. **22**(12), 3967–3983 (2016)
3. Kehoe, L., et al.: Global patterns of agricultural land-use intensity and vertebrate diversity. Divers. Distrib. **21**(11), 1308–1318 (2015)
4. Lindquist, J.E., Annunzio, R.D.: Assessing global forest land-use change by object-based image analysis. In: Remote Sensing (2016)
5. Hegazy, I.R., Kaloop, M.R.: Monitoring urban growth and land use change detection with GIS and remote sensing techniques in Daqahlia governorate Egypt. Int. J. Sustain. Built Environ. **4**(1), 117–124 (2015)
6. Butt, A., et al.: Land use change mapping and analysis using remote sensing and GIS: a case study of Simly watershed, Islamabad, Pakistan. Egypt. J. Remote Sens. Space Sci. **18**(2), 251–259 (2015)

7. Fu, P., Weng, Q.: A time series analysis of urbanization induced land use and land cover change and its impact on land surface temperature with Landsat imagery. Remote Sens. Environ. **175**, 205–214 (2016)
8. Khatami, R., Mountrakis, G., Stehman, S.V.: A meta-analysis of remote sensing research on supervised pixel-based land-cover image classification processes: general guidelines for practitioners and future research. Remote Sens. Environ. **177**, 89–100 (2016)
9. Xu, X., et al.: Multiple morphological component analysis based decomposition for remote sensing image classification. IEEE Trans. Geosci. Remote Sens. **54**(5), 3083–3102 (2016)
10. Huang, B., Zhao, B., Song, Y.: Urban land-use mapping using a deep convolutional neural network with high spatial resolution multispectral remote sensing imagery. Remote Sens. Environ. **214**, 73–86 (2018)
11. Luus, F.P.S., et al.: Multiview deep learning for land-use classification. IEEE Geosci. Remote Sens. Lett. **12**(12), 2448–2452 (2015)
12. Basse, R.M., Charif, O., Bódis, K.: Spatial and temporal dimensions of land use change in cross border region of Luxembourg. Development of a hybrid approach integrating GIS, cellular automata and decision learning tree models. Appl. Geogr. **67**, 94–108 (2016)
13. Rawat, J.S., Kumar, M.: Monitoring land use/cover change using remote sensing and GIS techniques: a case study of Hawalbagh block, district Almora, Uttarakhand, India. Egypt. J. Remote Sens. Space Sci. **18**(1), 77–84 (2015)
14. Tolessa, T., Senbeta, F., Kidane, M.: The impact of land use/land cover change on ecosystem services in the central highlands of Ethiopia. Ecosyst. Serv. **23**, 47–54 (2017)
15. Pervez, M.S., Henebry, G.M.: Assessing the impacts of climate and land use and land cover change on the freshwater availability in the Brahmaputra River basin. J. Hydrol.: Reg. Stud. **3**, 285–311 (2015)
16. Bai, X., et al.: A detailed and high-resolution land use and land cover change analysis over the past 16 years in the Horqin Sandy Land, Inner Mongolia. Math. Prob. Eng. **2017**, 13 (2017)
17. Zhou, J., et al.: Land use model research in agro-pastoral ecotone in northern China: a case study of Horqin Left Back Banner. J. Environ. Manag. **237**, 139–146 (2019)
18. Chang, Y., et al.: Anisotropic spectral-spatial total variation model for multispectral remote sensing image destriping. IEEE Trans. Image Process. **24**(6), 1852–1866 (2015)
19. Upadhyay, A., et al.: Land use and land cover classification of LISS-III satellite image using KNN and decision tree. In: 2016 3rd International Conference on Computing for Sustainable Global Development (INDIACom) (2016)
20. Gu, H., et al.: An object-based semantic classification method for high resolution remote sensing imagery using ontology. In: Remote Sensing (2017)
21. Minta, M., et al.: Land use and land cover dynamics in Dendi-Jeldu hilly-mountainous areas in the central Ethiopian highlands. Geoderma **314**, 27–36 (2018)

Extracting River Illegal Buildings from UAV Image Based on Deeplabv3+

Zhiyong Liu[1], Wenxiang Liu[2(✉)], Hongchang Qi[1], Yanfei Li[1],
Gengbin Zhang[1], and Tao Zhang[1]

[1] China Southern Power Grid Guangzhou Power Supply Bureau Co., Ltd.
Power Transmission Management Station II, Guangzhou 510000, China
[2] Nanchang Hangkong University, Nanchang 330063, China
2308341311@qq.com

Abstract. At present, the area extraction and contour identification of illegal buildings in rivers is generally a combination of manual identification and professional software. This method identifies illegal houses with low efficiency, large workload, huge human resource consumption and high requirements for the overall quality of staff. Aiming at the above problems, this paper proposes a method for extracting and identifying the area of illegal buildings in rivers based on deep learning. Identify the Pixel Accuracy (PA) and the Mean Intersection over Union (MIoU) of illegal house method reaching 94.71% and 89.09%. After the deeplabv3+network learns the illegal building features, it automatically detects and identifies the building and generates the building outline shp file. The shp file and the auxiliary arcgis software can be used to extract the illegal area and contour of the building. Based on the method of this paper, the contour marking of river house is shortened from 0.5–1 h to 2.5–5 min compared with the manual identification time. Compared with the manual method for extracting the illegal building area, the area extraction rate is basically above 90%. The results of this method are reliable and in line with actual needs.

Keywords: Building area extraction · Contour recognition · Deep learning · Deeplabv3+ · Arcgis

1 Introduction

At present, as the society and the public pay more and more attention to environmental pollution, the government departments are increasingly strengthening the comprehensive treatment of environmental pollution, especially the investigation of water pollution control and water safety hazards. The investigation of illegal houses in rivers is regarded as the key work of potential safety hazards. At this stage, the general method for extracting the area and contour of illegally built houses in rivers is to obtain orthophotos of rivers based on low-altitude photogrammetry of UAV, and to combine artificial and professional software to identify illegal buildings. The specific process of the method is to import the orthophoto into professional software such as Arcgis, then mark the contour of houses near rivers by manual and vectorize them into SHP files,

© Springer Nature Singapore Pte Ltd. 2020
Y. Xie et al. (Eds.): GSES 2019/GeoAI 2019, CCIS 1228, pp. 259–272, 2020.
https://doi.org/10.1007/978-981-15-6106-1_20

and finally extract the overlapping area by vectoring the river buffer. At the same time, the outline of the house in the overlapping area is marked as a illegal building. Due to the large length of the river and the dense buildings on both sides of the river, and the process of manually marking the outline and contour of the house is time consuming and manpower. According to statistics, manual identification and vectorization of house contour consume 40% of the time and 60% of the human resources in the whole process. Besides, manual identification and vectorization of house contour are the most important part of the whole process. Therefore, the use of manual identification of house is heavy and error-prone, which cannot meet the actual needs.

In recent years, with the improvement of computer hardware, deep learning algorithms have been rapidly developed in the field of machine vision, and the practical results in related fields are endless and quite rich. The semantic segmentation algorithm based on the full convolutional neural network has also made great progress and improvement. Convolutional neural network (CNN) [1] has significant ability in feature extraction, and the image classification [2] and target recognition algorithm [3] based on convolution neural network have made great progress. With the actual production needs, image classification and target recognition algorithms can not meet the actual needs, people want to accurately identify the target object contour and specific range (i.e., semantic segmentation), which requires the accuracy to pixels. Therefore, researchers put forward an image semantic segmentation algorithm model, which can be roughly divided into two structures. The first is the encoder-decoder structure, which uses CNN as feature extraction layer and uses pooling layer to gradually reduce the spatial dimension of input data, this process is called encoding; and then uses network layer such as deconvolution to gradually recover the details of the target and the corresponding spatial dimension, which is called decoding. The semantic segmentation models of this structure include: FCN-8 s [4], U-Net [5], SegNet [6], RefineNet [7], etc. The second is dilated convolution structure, which abandons the filling operation in the encoder in order to obtain a dense feature map, and then follows the dilated convolution operation to ensure that the network receptive field is unchanged, and finally uses the difference directly in the feature map. Upsampling discards the deconvolution upsampling in the decoder. The dilated convolution structure is represented by DeepLab [8], DeepLabv2 [9], DeepLabv3 [10], and DeepLabv3+ [11].

This article will use INRIA aerial image dataset as the initial analysis dataset of the algorithm. This dataset has remote sensing images of different cities, and with only two kinds of labels are the house and the background. Based on the INRIA aerial image dataset dataset, this paper compares the network models such as FCN-8 s, SegNet, DeepLabv3, and DeepLabv3+, and selects the superior network from the above network as the building identification network model in this paper. Finally, on the basis of this network, the data production, processing and model training adjustment and verification process are elaborated, and the error analysis of the method of identifying the house and extracting t the area of illegal building is also carried out.

2 Research Area and Data Processing

2.1 Research Area

The research area of this paper is Guangzhou, which is located in 112°57′−114°3′E and 22°26′−23°56′N. It is located in the south of China, the south-central part of Guangdong Province, and the north-north edge of the Pearl River Delta. It is the confluence of Xijiang, Beijiang and Dongjiang rivers. Therefore, there are many large and small rivers in the city, water pollution and water safety hazards are also prominent, Guangzhou's geographical location is shown in Fig. 1.

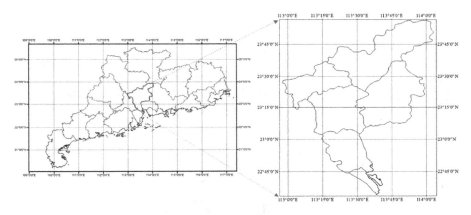

Fig. 1. Location map of Guangzhou

2.2 Data Collection and Processing

In this paper, DJI Phantom 4 RTK was used to collect orthophoto data of 54 rivers with serious water pollution and water safety hazards in Guangzhou. The total length of the river is 35.4 km. Using intelligent aerial triangulation processing, the orthophoto map of the research area is obtained through automatic engineering establishment, image preprocessing, automatic matching of connection points, free network adjustment, control point measurement and regional network adjustment, as shown in Fig. 2.

In this paper, 54 rivers are divided into training sets and test sets, the number is 50 and 4 respectively. Because the river is longer, it has a larger image resolution (generally greater than 20000*15000), which is not conducive to network training. Therefore, 50 complete river images are sequentially cut into 1000*1000 resolution images, and 9600 images are obtained. It can be seen from Fig. 2 that due to the low overlap of the most edge images on both sides of the river and the large error between the images, the image is distorted at the edge of the house. Therefore, the image of the deformed and distorted house must be removed to obtain 4400 images. The resulting image is marked with a labelme and the corresponding 4400 labels are generated. Finally, the 4400 images and labels were divided into training sets and verification sets of 4000 and 400 respectively.

Data enhancement technology can be used to expand the data set on the basis of limited data, improve the generalization ability of the model to prevent over-fitting, and at the same time increase the noise of the data to improve the robustness of the model to some extent. In order not to increase the workload of data annotation and the memory consumption of the machine during training, this paper will use soft data enhancement technology for the data set, which is different from the hard data enhancement method directly to the number of data sets themselves, but in the data training process, the data is enhanced after each batch of data is trained, and the data is randomly cut into 513*513 resolution images, randomly turned left and right, considering the characteristics of image color, shape and texture. Add noise, color jitter, and normalize the image R, G, and B bands by subtracting 123.68, 116.78, and 103.9, respectively [2].

Fig. 2. Part of the river orthophoto image

3   Research Methods

3.1   Metric

There are many indicators used to measure the quality of models in deep learning. Different metrics are needed for different tasks. In this paper, PA (pixel accuracy), mIoU (mean intersection over union) and speed are used as the criteria to measure the quality of the identification model of illegal houses.

Assuming that the image k+1 class (including the background class), P_{ij} represents the number of pixels that belonging to class i but are divided into class j. P_{ii} represents the true number of cases, P_{ij} represents the false positive number, and P_{ji} represents the false negative number, then PA is expressed as:

$$PA = \frac{\sum_{i=0}^{k} P_{ii}}{\sum_{i-0}^{k} \sum_{j=0}^{k} P_{ij}} \tag{1}$$

mIoU is expressed as:

$$mIoU = \frac{1}{K+1}\sum_{i=0}^{k}\frac{P_{ii}}{\sum_{j=0}^{k}P_{ij} + \sum_{j=0}^{k}P_{ji} - P_{ii}} \tag{2}$$

Through the calculation of PA can judge the prediction accuracy of the model for the whole image, and calculation of mIoU can judge the accuracy of the model for predicting the class contour (i.e., the predicted class coincides with the labeled class as much as possible). The calculation of the speed can judge the speed of the model. The above metrics can be a more balanced measure of the model.

3.2   Important Hyperparameters

Learning Rate. Learning rate is an important hyperparameter in supervised learning and deep learning, which determines whether the objective function converges or not. When a large learning rate is chosen, the objective function can converge quickly, but if the learning rate is too large, which is easy to cause the loss value to explode and the objective function to oscillate around the local minimum or minimum value. When choosing a smaller learning rate, the convergence speed of the objective function is slow, which can cause the objective function to converge to a local minimum value, and also make the network over-fitting. The different learning rate functions converge as shown in Fig. 3.

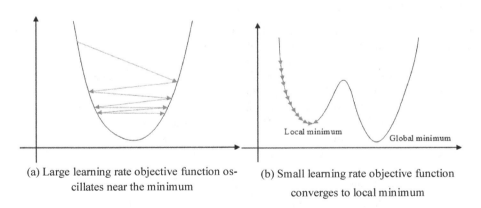

(a) Large learning rate objective function oscillates near the minimum

(b) Small learning rate objective function converges to local minimum

Fig. 3. The effect of learning rate on the objective function

Learning Rate Decay Step. Different learning rates have a great influence on the convergence of the objective function. Therefore, two learning rate setting methods are proposed: fixed learning rate (always maintain a learning rate during training) and learning rate decay (learning rate decreases with the increase of training rounds in the training process). In the initial stage of network training, you need to use the large learning rate to accelerate the training. In the late stage of network training, the network can fit most of the data, and you should use the small learning rate training. In this paper, the polynomial attenuation strategy is used to train the network. The number of attenuation steps determines the speed of attenuation of the learning rate. The polynomial attenuation formula is as follows:

$$lr = (lr - end\_lr) \times \left(1 - \frac{global\_step}{decay\_step}\right)^{power} + end\_lr \qquad (3)$$

Batch Normalization Decay. As the convolutional neural network deepens, the overall distribution of the eigenvalues gradually approaches the upper and lower limits of the interval of the nonlinear function during training (i.e., the overall distribution of the values gradually shifts or changes), which results in the disappearance of the gradient of the lower layer neural network and the slow convergence of the deep layer neural network in the back propagation. Batch normalization normalizes the input value distribution of any neuron in each layer to the standard normal distribution with a mean of 0 and variance of 1, which makes the training gradient larger and avoids the problem of gradient disappearance, and greatly accelerate the training speed. In order to further improve the batch normalization, let the last mean and variance have the effect of attenuating the current value, so that after each value is connected, the moving average will be relatively smooth, and the moving average can effectively prevent a certain number of abnormal data training. Moving average formula:

$$V_t = batch\_norm\_decay * V_{t-1} + (1 - batch\_norm\_decay) * V_t \qquad (4)$$

3.3   Comparison and Selection of Model Results

This paper uses the French public dataset INRIA aerial image dataset to train the semantic segmentation networks such as FCN-8s, SegNet, DeepLabv3 and DeepLabv3 +. The dataset is labeled as house and background. It contains 360 remote sensing images with 5000*5000 pixels. Before the network training, 360 images and corresponding labels should be divided into 1000*1000 small-resolution images for network batch training. This article uses the machine environment for Ubuntu16.04LTS, CPU parameters: Core i7-6800k, 3.4 GHZ × 12, memory 64G, GPU parameters: GeForce GTX 1080 Ti, memory 11G. The training results of the above model on the INRIA aerial image dataset data set are shown in Table 1.

Table 1. Training results of each model on the INRIA aerial image dataset

Method	mIoU/%	Builing/%	PA/%	Time/(FPS)
FCN-8s	78.21	66.95	92.31	7.69
SegNet	79.54	67.65	92.74	9.72
deeplabv3(16)	81.82	71.53	93.41	5.29
deeplabv3(8)	80.69	69.57	93.10	3.82
deeplabv3+(16)	**83.33**	**73.84**	**94.04**	4.86
deeplabv3+(8)	82.04	71.97	93.44	3.45

Note: (.) represents 1/(.) times the feature map relative to the original image.

It can be seen from Table 1 that the deeplabv3+ model has a pixel accuracy (PA) of 94.04% when the house is identified for the entire image, which is 0.63%–1.73% higher than other models. At the same time, the mIoU reaches 83.33%. Compared with the other four models, it is also the best in predicting the outline of the house. The predicted outline of the house is shown in Fig. 4. When the house is identified for the orthophoto image, each ortho-spliced image is cut into 1000*1000 resolution images and 600–1200 images are obtained. The use of deeplabv3+(16) to predict the image of a house takes 2.5–5 min for each river. The prediction speed is acceptable. Therefore, this paper uses deeplabv3+ as the model for illegal building identification.

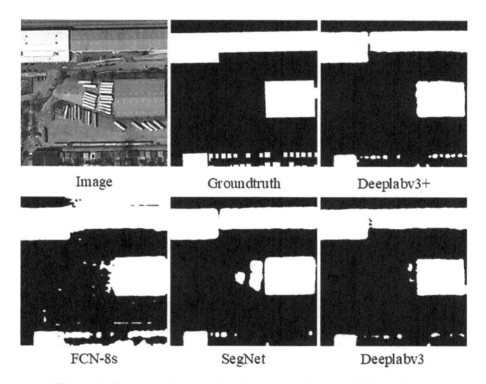

Fig. 4. Prediction result image of each model on INRIA aerial image dataset

3.4 Deeplabv3+ Principle

Deeplabv3+ is the latest semantic segmentation model proposed by Google in 2018. It uses the encoding-decoding structure based on the previous deeplab series. The model structure is shown in Fig. 5. The model principle is as follows:

- **Step 1.** The original image input network extracts features through the deep convolution network to obtain the feature map FM1;
  Deep convolutional network uses ResNet [12] as the feature extraction network of the model. ResNet network is proposed to solve the problem of deep network

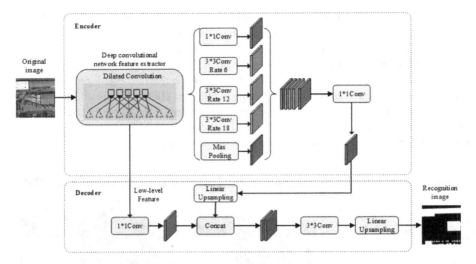

Fig. 5. Deeplabv3 + network principle structure

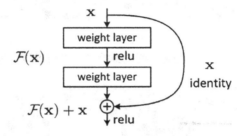

Fig. 6. Schematic diagram of the ResNet network jump connection

training gradient disappearance and network degradation. It is divided into two parts: Identity mapping and residual mapping, ResNet network jump connection as shown in Fig. 6, the training formula is: $y = x + F(x)$. If the network has reached the optimal level and continues to deepen the network layer, the static mapping will be regarded as 0 ($F(x) = 0$). At this time, only identity mapping ($y = x$) is left, so that the network is still optimal and the network performance will not decrease as depth increases.

- **Step 2.** Send FM1 into atrous spatial pyramid pool (ASPP) to get the feature maps FM21, FM22, FM23, FM24, FM25 under different convolutions.

  The atrous spatial pyramid pool convolutes FM1 into feature maps FM21, 3 * 3 hole convolution (expansion rate 6) to feature maps FM22, 3 * 3 hole convolution (expansion rate 12) to feature maps FM23, 3 * 3 hole convolution (expansion rate 18) to feature maps FM24, and the maximum pooling (global average pooling of FM1) to feature maps FM25. The convolution kernels with different expansion rates are shown in Fig. 7.

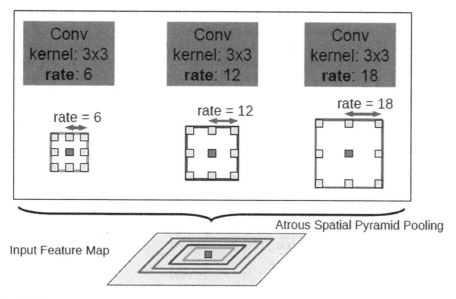

Fig. 7. Schematic diagrams of hole convolution kernels with expansion rates of 6, 12 and 18, respectively

- **Step 3.** After splicing operation of FM21, FM22, FM23, FM24, FM25, the feature map FM21-5 is obtained, and then perform 1*1 convolution on FM21-5 to compress the channel number to 256-dimensional to obtain the feature map FM3;
- **Step 4.** Extract the low-level feature map LFM in the deep convolution ResNet network and perform 1*1 convolution to obtain the channel number 48-dimensional feature map LFM-48, and then perform 4 times bilinear interpolation of FM3 to get the same size as the LFM-48. The score map SM1, and finally the SM1 and LFM-48 are spliced to obtain the score map SM2 (the original size is 1/4);
- **Step 5.** After SM3 is convolved twice by 3*3, 4 times bilinear interpolation is performed to obtain the prediction graph PM which is consistent with the original image size.

4   Results Analysis

4.1   Model Training

Considering the balance of machine performance and accuracy and the ease of migration learning training when training the network, the model uses ResNet-101 as the feature extraction network. In this paper, three hyperparametric objects (learning rate, learning rate decay step, batch normalized decay) which are more important to the model are selected and six schemes are designed for experiment. The training results are shown in Table 2. The experimental results in Table 2 show that the experimental method 4 uses the appropriate learning rate of 0.005 and the larger learning rate attenuation step of

30000, and the batch normalized attenuation index of 0.95 makes the network model get the best results on the data validation set (mIoU = 89.09%, PA = 94.71%).

Method 4 is used to train the network with different sample sizes, and the demand analysis of the network training data volume is carried out. The model effect is shown in Fig. 8. It is obvious from Fig. 8 that as the amount of training data increases, the accuracy of the network on the verification set continues to improve, but the amount of model accuracy increases gradually, and the demand for data volume is constantly saturated. However, due to the total sample size, the model has not yet reached the best accuracy value, and there is still room for improvement.

Table 2. Comparison of results of different training methods

No.	Learning rate	Learning rate decay step	Batch normalization decay	mIoU (%)	PA (%)
Training method 1	0.05	30000	0.9997	59.45	76.76
Training method 2	0.01	30000	0.9997	83.83	91.81
Training method 3	0.01	30000	0.95	87.81	94.32
Training method 4	0.005	30000	0.95	**89.09**	**94.71**
Training method 5	0.001	30000	0.95	87.36	93.80
Training method 6	0.005	20000	0.95	88.25	94.30

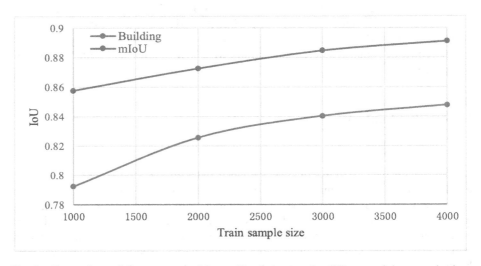

Fig. 8. Comparison of the accuracy of the verification set under different training sample sizes

4.2   Contour Extraction and Area Calculation of Illegal Buildings

In order to facilitate the model prediction, four river orthophoto images in the test set are divided into 1000*1000 resolution images and input into deeplabv3+ pre-training network for prediction respectively. Then, the predicted images are spliced to obtain a complete orthophoto prediction map. The river 3 housing prediction is shown in the Fig. 9, then the contour image is extracted and vectorized into a shp file, and the original image and shp file are imported into the arcgis software for the calculation of the illegal building area and the outline of the illegal building. The error analysis is shown in Table 3.

Fig. 9. River 3 housing forecast image

Table 3. Comparison of the area and error of illegal construction of rivers in the test set

River number	The area of illegal buildings extracted by different methods (m²)		Error (%)
	Manual label extraction method	DeepLabv3 + deep learning algorithm	
River 1	358.86	326.45	9.03
River 2	1094.52	646.26	40.95
River 3	1564.71	1487.44	4.93
River 4	1911.14	1742.42	8.82

4.3   Error Analysis

From Table 3, it can be seen that the error of extraction of illegally constructed house area based on deep learning algorithm is less than 10% compared with manual label extraction method, but it is 40.95% on river 2. The outline of illegally constructed house in river 2 is shown in Fig. 10. As can be seen from the figure, the error mainly comes from the frame of houses in the graph, and the total area is 441.87 m^2. The main reason is that trees cover the house. Artificial marking can effectively infer the outline of the whole house, and the machine can only recognize the uncovered part.

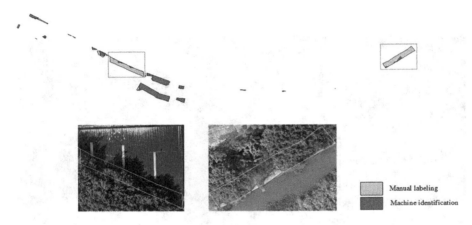

Fig. 10. River 2 illegal building outline and some houses example

In general, the river error mainly comes from the following three points: Firstly, the upper part of the tree is partially occluded; secondly, it can be seen from Fig. 8 that the model is not optimal due to the sample size limitation, which will lead to greater errors in the identification of houses; thirdly, because of the influence of wind speed on the UAV orthophoto photograph, there is a certain error in image overlap, mosaic image. For example, there are some pixel deformations, which lead to unclear houses and their contours, while some houses collapse, which will inevitably lead to errors in machine identification. An example of the error source is shown in Fig. 11.

Vegetation occlusion   Building deformation and collapse

Fig. 11. Example of machine identification and manual labeling for extracting illegal building outline and area error

5 Conclusion and Summary

It takes a lot of manpower and time to extract the area and contour of illegally constructed houses by manual labeling, and mainly concentrated in the stage of building outline labeling. According to statistics, this stage consumes 40% and 60% of the time and manpower. At the same time, due to the large amount of labeling work and the long time, there are also many missing and mislabeled manual outlines of the house, which require a certain amount of manpower and time for re-inspection work. The accuracy and MIOU of the method based on in-depth learning are up to 94.71% and 89.09% respectively. If the number of training samples is enough, the model accuracy and MIoU can be even higher. Compared with the manual method, this method of extracting the area and contour of illegal buildings is more than 90%. The method in this paper can greatly reduce the time and manpower of building contour labeling and greatly simplify the workload.

This paper proposes a method based on the deep learning network model deeplabv3 + to automatically identify the house outline and extract the area and contour of the illegal building. Through the actual test and analysis of this method and manual labeling method to extract the outline and area of the building illegally constructed, it is proved that this method can identify the outline of the building accurately and quickly, and can meet the actual demand. However, this method has defects. It is unable to accurately identify the outline of the house in the case of house occlusion and deformation collapse. In the later stage, we can further analyze and try to eliminate the defective image by using traditional image processing technology.

Acknowledgements. This study was jointly supported by China Southern Power Grid Guangzhou Power Supply Bureau Co., Ltd. Key Technology Project (080000KK52190001); Guangdong Provincial Science and Technology Program (2017B010117008); Guangzhou Science and

Technology Program (201806010106, 201902010033); the National Natural Science Foundation of China (41976189, 41976190); the Guangdong Innovative and Entrepreneurial Research Team Program (2016ZT06D336); the Southern Marine Science and Engineering Guangdong Laboratory (Guangzhou) (GML2019ZD0301); the GDAS's Project of Science and Technology Development (2016GDASRC-0211, 2018GDASCX-0403, 2019GDASYL-0301001, 2017GDASCX-0101, 2018GDAS CX-0101).

References

1. Krizhevsky, A., Sutskever, I., Hinton, G.E.: ImageNet classification with deep convolutional neural networks. In: Advances in Neural Information Processing Systems, pp. 1097–1105 (2012)
2. Simonyan, K., Zisserman, A.: Very deep convolutional networks for large-scale image recognition. arXiv preprint arXiv:1409.1556 (2014)
3. Redmon, J., Divvala, S., Girshick, R., et al.: You only look once: unified, real-time object detection, pp. 779–788 (2015)
4. Long, J., Shelhamer, E., Darrell, T.: Fully convolutional networks for semantic segmentation. IEEE Trans. Pattern Anal. Mach. Intell. **PP**(99), 1 (2014)
5. Ronneberger, O., Fischer, P., Brox, T.: U-Net: convolutional networks for biomedical image segmentation. In: Navab, N., Hornegger, J., Wells, W.M., Frangi, A.F. (eds.) MICCAI 2015. LNCS, vol. 9351, pp. 234–241. Springer, Cham (2015). https://doi.org/10.1007/978-3-319-24574-4_28
6. Badrinarayanan, V., Kendall, A., Cipolla, R.: SegNet: a deep convolutional encoder-decoder architecture for image segmentation. arXiv preprint arXiv:1511.00561 (2015)
7. Lin, G., Milan, A., Shen, C., et al.: RefineNet: multi-path refinement networks for high-resolution semantic segmentation arXiv preprint arXiv:1611.06612 (2016)
8. Chen, L.C., Papandreou, G., Kokkinos, I., et al.: Semantic image segmentation with deep convolutional nets and fully connected CRFs. Comput. Sci. **4**, 357–361 (2014)
9. Chen, L.C., Papandreou, G., Kokkinos, I., et al.: DeepLab: semantic image segmentation with deep convolutional nets, atrous convolution, and fully connected CRFs. IEEE Trans. Pattern Anal. Mach. Intell. **40**(4), 834–848 (2016)
10. Chen, L.C., Papandreou, G., Schroff, F., et al.: Rethinking atrous convolution for semantic image segmentation. arXiv preprint arXiv:1706.05587 (2017)
11. Chen, L.C., Zhu, Y., Papandreou, G., et al.: Encoder-decoder with atrous separable convolution for semantic image segmentation. arXiv preprint arXiv:1802.02611 (2018)
12. He, K., Zhang, X., Ren, S., et al.: Deep residual learning for image recognition. arXiv preprint arXiv:1512.03385 (2015)

Research on Spatiotemporal Evolution Law of Black and Odorous Water Body in Guangzhou City

Qian Zhao[1,2], Liusheng Han[1,2(✉)], Yong Li[2], Ji Yang[2],
Shuxiang Wang[1,2], Congjun Zhu[1,2], and Yinguo Li[3]

[1] Shandong University of Technology, Zibo 255000, China
hanls@sdut.edu.cn
[2] Guangdong Key Laboratory of Geospatial Information Technology
and Application, Guangzhou Institute of Geography, Guangzhou 510070, China
[3] State Grid Chongqing Yongchuan Power Supply Company,
Yongchuan 402106, Chongqing, China

Abstract. The urban black and odorous water bodies affect the daily life of urban residents, destroy the aquatic ecosystem and damage the image of the city. This research used the evaluation method of black and odorous water body proposed by the Ministry of Housing and Urban-Rural Development of the People's Republic of China (MOHURD) to evaluate the water quality of 50 key creeks (53 sections) in Guangzhou in 2016 and 2017 and analyzed the spatial distribution of black and odorous water bodies and their changes over time. The results showed that the water quality of Conghua District, Panyu District, Nansha District, and Yuexiu District was relatively good, while the water quality of Baiyun District, Haizhu District, Huadu District, Huangpu District, Liwan District, Tianhe District, and Zengcheng District was relatively poor. Compared with 2016, the water quality of creeks in Huangpu District and Baiyun District deteriorated in 2017, while the water quality of creeks in Tianhe District, Liwan District, and Huadu District improved. The water quality of creeks in other districts did not change significantly. In 2017, ammonia nitrogen (NH_3-N), total phosphorus (TP), and chemical oxygen demand (COD) content decreased, and dissolved oxygen (DO) content increased compared to 2016. In 2017, the number of black-odor water bodies was greatly reduced, and the water quality of creeks was generally improved. The results can provide a reference for the treatment of the black and odorous creeks in Guangzhou.

Keywords: Black and odorous water body · Spatiotemporal evolution law · Guangzhou City

1 Introduction

The urban black and odorous water body is a collective name for the unpleasant color and/or the unpleasant smell of water in the built-up area of the city [1]. The black and odorous water bodies affect the daily life of urban residents, destroy the aquatic ecosystem and damage the image of the city. The Action Plan for Prevention and Control of Water Pollution promulgated by the State Council in 2015 proposed a

© Springer Nature Singapore Pte Ltd. 2020
Y. Xie et al. (Eds.): GSES 2019/GeoAI 2019, CCIS 1228, pp. 273–289, 2020.
https://doi.org/10.1007/978-981-15-6106-1_21

control objective that "by 2020, the black-odor water bodies in the built-up areas of prefecture-level and above cities will be controlled within 10% [2]. By 2030, the black and odorous water bodies in the urban built-up areas will be eliminated." The renovation of black-odor water body has become a key link in the improvement of the urban environment. However, because the black color and odor of urban water body are affected by many factors, the task of remediation is very difficult.

As early as the 1960s, domestic and foreign scholars began to study the urban black and odorous water bodies. Previously, domestic and foreign research on black and odorous water bodies mainly focused on the following four aspects: 1) The concept of black and odorous water bodies [3]; 2) Study on the black-odor causes in different types of water bodies [4–6] and how to cause black odor [7, 8]; 3) Evaluation models of black and odorous water bodies [9, 10]; 4) Study on the treatment of black and odorous water bodies [11]. Few studies have been carried out from the perspective of the temporal and spatial evolution of the black and odorous water body. With the rapid population growth and rapid economic and social development, the creeks in Guangzhou have been seriously polluted. Many creeks have become black and odorous water bodies, and environmental problems have become increasingly prominent [12]. This study applied the evaluation method of black and odorous water body in urban built-up areas proposed by the Ministry of Housing and Urban-Rural Development of the People's Republic of China to evaluate the water quality of 50 creeks (53 sections) in Guangzhou, and analyzed pollution status and temporal and spatial evolution, to provide advice on the improvement of black and odorous creeks.

2 Data Sources

The water quality data of creeks used in this study are from the "Water Quality Monitoring Information of Key Creeks" (http://www.gzepb.gov.cn/gzepb/comm/dbszjc.html) on the website of Guangzhou Municipal Ecological Environment Bureau. The time span of the data is in 2016–2017. In July 2015, Guangzhou Municipal Ecological Environment Bureau selected 50 key creeks. Since then, water quality monitoring has been carried out once a month, and the monitoring results have been published in the "Water Quality Monitoring Information of Key Creeks" on the website of Guangzhou Municipal Ecological Environment Bureau. Monitoring indexes for creeks include transparency, dissolved oxygen, ammonia nitrogen, total phosphorus, and chemical oxygen demand. Among them, in April 2016, Subway A creek did not have monitoring conditions due to dredging works. In November 2017, Subway A creek was not monitored due to dry up. In December 2017, Subway A creek and Daling River were not monitored due to construction impact. Therefore, this study does not evaluate the water quality of Subway A creek and Daling River in the above-mentioned months. In addition, because the Liuxi River (Lixiba) is located at the cross section of Baiyun District and Huadu District, this study classifies the Liuxi River (Lixiba) into Huadu District.

Pollutant emission data of industrial pollution sources come from the Communique of the First National Pollution Source Survey of Guangdong Province (http://www.gzepb.gov.cn/gzepb/hjzs/201006/e265490354d0436b8a1b91d17bfccf37.shtml). By looking at the industrial production value of various districts in Guangzhou in recent years, the top industries in terms of production value were found. The industrial information of Huadu District comes from the Statistical Bulletin on National Economic and Social Development of Huadu District, Guangzhou in 2017 (https://www.huadu.gov.cn/xxgk/tjxx/tjnb/201806/t20180613_549615.html). The industrial information of Zengcheng District comes from the Statistical Bulletin on National Economic and Social Development of Zengcheng District in 2018 (http://www.zc.gov.cn/gk/sj/cyjj/sjtj/201907/t20190711_224196.html). The industrial information of Huangpu District comes from the special statistics "A Brief Analysis of Industrial Production above Designated Size in Huangpu District, Guangzhou Development Zone in 2018" (http://www.gdd.goven/hp00/0702/201901/fa1e07c16b6d42d690ab4f560d017c97.shtml). The industrial information of Liwan District comes from the Statistical Bulletin on National Economic and Social Development of Liwan District in 2018 (http://www.lw.gov.cn/lwqtjj/008/201908/cbca55e29d714f8c91e46261790b807f/files/75d18ab7c5cf4b32bdde611e37ebff84.pdf). The industrial information of Yuexiu District comes from the Statistical Bulletin on National Economic and Social Development of Yuexiu District in 2017 (http://www.yuexiu.gov.cn/yxxxw/pc/zwgk/sjfb/tjgb_sj/20190708/detail-83635.shtml). The industrial information of Tianhe District comes from Tianhe Yearbook (2018) (http://xxgk.thnet.gov.cn/xxgkk/07/201904/d756fd804817447589d49d551954ab1e.shtml). The industrial information of Baiyun District comes from Baiyun District's Annual Statistical Yearbook in 2017. The industrial information of Haizhu District comes from the Statistical Bulletin on National Economic and Social Development of Haizhu District, Guangzhou in 2015 (http://www.haizhu.gov.cn/site/hzqzf/zwgk/sjfb/tjgb_22940/201811/P020181112392082074962.pdf). The industrial information of Panyu District comes from the Statistical Bulletin on National Economic and Social Development of Panyu District, Guangzhou in 2017 (http://www.panyu.gov.cn/gzpy/zwgk_sjkf_tjgb/201812/706cfdb25e694941b3b96b2b969e2886.shtml). The industrial information of Nansha District comes from the Statistical Yearbook on National Economic and Social Development of Nansha District, Guangzhou in 2017 (http://www.gzns.gov.cn/nssj/jjjs/gmjjzyzb/201903/t20190315_383876.html). The industrial information of Conghua District comes from the Statistical Bulletin on National Economic and Social Development of Conghua District, Guangzhou in 2016 (http://www.conghua.gov.cn/zgch/tjgb/201711/8d5415a255194aea929ab9a733316dd7.shtml). Industrial statistics are mainly based on 2017. However, since the information for 2017 is incomplete, it is replaced with information close to the year.

3   Overview of Study Area

Guangzhou is located at 112° 57′ E to 114° 3 'E and 22° 26′N to 23° 56'N. It is located in southern China and south-central Guangdong. Guangzhou's climate is warm and rainy. The river system in the territory is developed. The Pearl River and its many tributaries run through Guangzhou. The water area of the city is 74,400 hm^2, accounting for 10.05% of the city's land area [13]. There are 231 creeks in the central urban area of Guangzhou. The total length of creeks is 918 km, and the width of creeks is 5–100 m. In recent years, Guangzhou City has vigorously rectified polluted creeks and has selected 50 key creeks. Figure 1 shows the locations of 50 key creeks (53 sections).

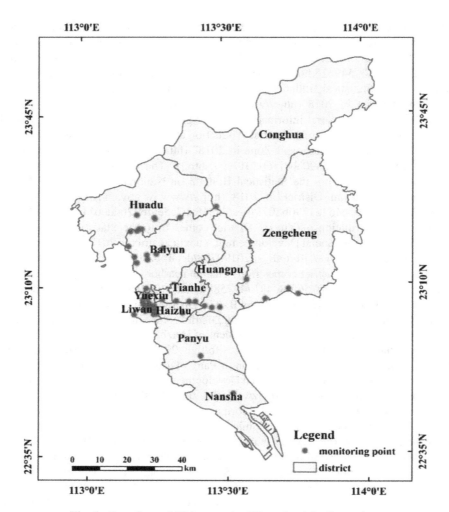

Fig. 1. Locations of 50 key creeks (53 sections) in Guangzhou

4   Research Methods

4.1   Water Quality Index Analysis Methods Used by Guangzhou Municipal Ecological Environment Bureau

Guangzhou Municipal Ecological Environment Bureau monitors 50 creeks (53 sections) once a month and uses the methods in the following table (Table 1) to measure and analyze the water quality indexes.

Table 1. Water quality index analysis methods

Water quality index	Analysis method	Method source
Transparency	Plug's plate method	"Water and Wastewater Monitoring and Analysis Methods (Fourth Edition, Supplementary Edition)"
Dissolved oxygen	Iodometric method	GB 7489-87
Ammonia nitrogen	Nadler's reagent colorimetry	GB 7479-87
Total phosphorus	Ammonium molybdate spectrophotometry method	GB 11893-89
Chemical oxygen demand	Potassium dichromate method	GB 11914-89

4.2 MOHURD's Evaluation Method of Black and Odorous Water Body

MOHURD has stipulated the evaluation indexes and grading standards for pollution levels of urban black and odorous water bodies in the "Guide for the Remediation of Urban Black and Odorous Water Body" (referred to as the Guide). The evaluation indexes include transparency, dissolved oxygen (DO), oxidation-reduction potential (ORP) and ammonia nitrogen (NH3-N). The black and odorous water bodies can be subdivided into "slightly black and odorous" and "heavily black and odorous". The grading standards are shown in Table 2.

Table 2. Grading standards for pollution levels of urban black and odorous water body

Characteristic index	Slightly black and odorous	Heavily black and odorous
Transparency (cm)	25–10	<10
Dissolved oxygen (mg/L)	0.2–2.0	<0.2
Oxidation-reduction potential (mV)	−200–50	<−200
Ammonia nitrogen (mg/L)	8.0–15	>15

When the water depth is less than 25 cm, the transparency is valued at 40% of the water depth.

The Guide specifies 4 indexes. Due to the lack of oxidation-reduction potential in the data of Guangzhou Municipal Ecological Environment Bureau, this study uses only transparency, dissolved oxygen, and ammonia nitrogen for evaluation. This method divides the creeks into three types: not black and odorous, slightly black and odorous, and heavily black and odorous. Of the three indexes of the creek, as long as one of them reaches the standards of heavily black and odorous creek, the creek is judged to be heavily black and odorous. If none of the three indexes reaches the standards of heavily black and odorous creek, as long as one of them reached the standards of slightly black and odorous creek, the creek is judged to be slightly black and odorous. If none of them reached the standards of heavily or slightly black and odorous creek, the creek is judged to be not black and odorous. This method was used to evaluate the monthly water quality categories of the 50 key creeks (53 sections) in Guangzhou in 2016 and 2017.

4.3   Spatial and Time Series Analysis Methods

In order to analyze the overall distribution of black-odor creeks, we count the annual average values of the total number of black-odor creeks in each district throughout the year and separately count them according to the three types of not black and odorous, slightly black and odorous, and heavily black and odorous. In order to analyze the spatial pattern of black-odor creeks, we use ArcMap software to draw the spatial distribution map of black-odor creeks and combine the industrial information of various districts in Guangzhou to study the relationship between black-odor creeks and industrial types in space.

In order to explore the regularity of creek water quality over time, we calculate the average values of ammonia nitrogen, total phosphorus, chemical oxygen demand, and dissolved oxygen concentration of all creeks each month. According to the standard limits of the above-mentioned four indexes in the Surface Water Environmental Quality Standard (GB3838-2002) (Table 3), we analyze the relationship between the concentration of creek water quality indexes and the time. The period from April to September in Guangzhou is divided into the wet season, and the remaining seasons are classified as the dry season. We count the numbers and proportions of black-odor creeks during the wet season and the dry season each year and separately count them according to the three types of not black and odorous, slightly black and odorous, and heavily black and odorous. Then we analyze the changes in creek water quality during the wet seasons and the dry seasons in 2016 and 2017.

Table 3. Classification standards for pollution levels of urban black and odorous water body

Water quality index	Class I	Class II	Class III	Class IV	Class V
Ammonia nitrogen (mg/L)	0.15	0.5	1.0	1.5	2.0
Total phosphorus (mg/L)	0.02	0.1	0.2	0.3	0.4
Chemical oxygen demand (mg/L)	15	15	20	30	40
Dissolved oxygen (mg/L)	7.5	6	5	3	2

5   Results and Analysis

5.1   Spatial Analysis of Black and Odorous Water Body

Table 4 shows the annual average number of black-odor creeks in various districts of Guangzhou in 2016. In Baiyun District, there were 3 sections of creeks that were slightly black and odorous and 3 sections of creeks that were not black and odorous every month. In Haizhu District, there was a section of creek that was slightly black and odorous and a section of creek that was not black and odorous every month. In Huadu District, there were 5 sections of slightly black and odorous creeks, a section of heavily black and odorous creek, and 5 sections of creeks that were not black and odorous. In Huangpu District, there were 3 sections of slightly black and odorous creeks and a section of creek which was not black and odorous. In Liwan District, there were 9 sections of slightly black and odorous creeks, a section of heavily black and odorous creek, and 10 sections of creeks that were not black and odorous. In Tianhe District, there were 2 sections of slightly black and odorous creeks and a section of heavily black and odorous creek. In Zengcheng District, there was a section of creek which was slightly black and odorous and 2 sections of creeks that were not black and odorous. There was basically no black and odorous creek in Conghua District, Panyu District, Nansha District and Yuexiu District and the water quality was relatively good.

Table 4. Statistics of the annual average number of black-odor creeks in various districts of Guangzhou in 2016

District	Not black and odorous (section)	Slightly black and odorous (section)	Heavily black and odorous (section)	Total (section)
Baiyun District	3	3	0	6
Conghua District	1	0	0	1
Panyu District	1	0	0	1
Haizhu District	1	1	0	2
Huadu District	5	5	1	11
Huangpu District	1	3	0	4
Liwan District	10	9	1	20
Nansha District	1	0	0	1
Tianhe District	0	2	1	3
Yuexiu District	1	0	0	1
Zengcheng District	2	1	0	3
Total	26	24	3	53

Table 5 shows the annual average number of black-odor creeks in various districts of Guangzhou in 2017. In Baiyun District, there were 2 sections of creeks that were slightly black and odorous, a section of creek which was heavily black and odorous, and 3 sections of creeks that were not black and odorous every month. In Haizhu District, there was a section of creek that was slightly black and odorous and a section of creek that was not black and odorous every month. In Huadu District, there were 4 sections of slightly black and odorous creeks, a section of heavily black and odorous creek, and 6 sections of creeks that were not black and odorous. In Huangpu District, there were 2 sections of slightly black and odorous creeks, a section of heavily black and odorous creek, and a section of creek which was not black and odorous. In Liwan District, there were 7 sections of slightly black and odorous creeks and 13 sections of creeks which were not black and odorous. In Tianhe District, there were 2 sections of slightly black and odorous creeks and a section of creek which was not black and odorous. In Zengcheng District, there was a section of creek which was slightly black and odorous and 2 sections of creeks that were not black and odorous. There were basically no black and odorous creeks in Conghua District, Panyu District, Nansha District and Yuexiu District and the water quality was relatively good. Through the annual average value evaluation of the overall situation of black and odorous water bodies, we found that the creek water quality of Huangpu District and Baiyun District in 2017 was worse than the previous year. In 2017, the creek water quality of Tianhe District, Liwan District, and Huadu District improved compared with the previous year. There was no significant change in creek water quality in other districts.

Table 5. Statistics of the annual average number of black-odor creeks in various districts of Guangzhou in 2017

District	Not black and odorous (section)	Slightly black and odorous (section)	Heavily black and odorous (section)	Total (section)
Baiyun District	3	2	1	6
Conghua District	1	0	0	1
Panyu District	1	0	0	1
Haizhu District	1	1	0	2
Huadu District	6	4	1	11
Huangpu District	1	2	1	4
Liwan District	13	7	0	20
Nansha District	1	0	0	1
Tianhe District	1	2	0	3
Yuexiu District	1	0	0	1
Zengcheng District	2	1	0	3
Total	31	19	3	53

Figure 2 and Fig. 3 show the spatial distribution of black and odorous creeks in Guangzhou in 2016 and the spatial distribution of black and odorous creeks in Guangzhou in 2017. According to the Communique of the First National Pollution Source Survey in Guangdong Province, among the main water pollutants of industrial pollution sources, the industries with the highest COD emissions are as follows: The emissions from the textile industry are 101,900 tons; the emissions from the paper-making and paper products industry are 72,100 tons; the emissions from the farm and sideline food processing industry are 46,200 tons; the emissions from the metal products industry are 16,100 tons; the emissions from beverage manufacturing are 15,700 tons; the emissions from leather, fur, feather (velvet) and their products industry are 14,500 tons; the emissions from chemical raw materials and chemical products industry are 13,600 tons; the emissions from communications equipment, computers, and other electronic equipment industry are 12 thousand tons. The total COD emissions from the above 8 industries account for 79.1% of the COD emissions from the discharge outlets of industrial wastewater plants. The industries with the highest NH3-N emissions are as follows: The emissions from the textile industry are 2597.95 tons; the emissions from the farm and sideline food processing industry are 1817.00 tons; the emissions from the papermaking and paper products industry are 1618.98 tons; the emissions from chemical raw materials and chemical products industry are 1423.55 tons; the emissions from leather, fur, feather (velvet) and their products industry are 1184.19 tons; the emissions from communications equipment, computers, and other electronic equipment industry are 639.94 tons; the emissions from the non-ferrous metal smelting and rolling processing industry are 581.52 tons; the emissions from beverage manufacturing are 575.61 tons; the emissions from petroleum processing, coking, and nuclear fuel processing industry are 568.48 tons. The total NH3-N emissions from the above 9 industries account for 81.3% of the NH3-N emissions from the discharge outlets of industrial wastewater plants. By searching the industrial information of each district, we got the main industries with higher production values in each district. The major industries in Huadu District are automobile manufacturing; computer, communications, and other electronic equipment manufacturing; and leather, fur, feather (velvet), and their products industry. The major industries in Zengcheng are automobiles and their parts industry, motorcycle industry and textile industry. The major industries of Huangpu District are electronics and communication equipment manufacturing, automobile manufacturing, and power and heat production and supply industry. The major industries of Liwan District are tobacco products industry, ferrous metal smelting and rolling processing industry, and pharmaceutical manufacturing industry. The major industries in Yuexiu District are water production and supply industry, printing and recording media reproduction industry, and furniture manufacturing. The major industries in Tianhe District are the power industry, gas and water production and supply industry, and computer communications and other electronic equipment manufacturing. The major industries of Baiyun District are furniture manufacturing, electrical machinery and equipment manufacturing, and chemical raw materials and chemical products manufacturing. The major industries in Haizhu District are beverage manufacturing, pharmaceutical manufacturing, and communications equipment computers and other electronic equipment manufacturing. The major industries of Panyu District are automobile manufacturing, general equipment industry, and electrical machinery. The major industries of Nansha District are automobile manufacturing, chemical raw materials and chemical products manufacturing, and electrical machinery and equipment manufacturing. The major industries of Conghua District are the biomedicine, cosmetics, and daily necessities industry; automobile manufacturing and motorcycle parts industry; and new materials and new energy industry. According to the analysis, computers, communications, and other electronic equipment manufacturing industries, as well as leather, fur, feather (velvet), and their products industry had high levels of COD and NH3-N emissions in Huadu District. COD and NH3-N emissions of the textile industry were relatively high in Zengcheng District. Huangpu District and Tianhe District had higher COD and NH3-N emissions in communications equipment, computers, and other electronic equipment manufacturing industry. COD emissions of the pharmaceutical manufacturing industry in Liwan District were relatively high. COD and NH3-N emissions of chemical raw materials and chemical products manufacturing in Baiyun District were relatively high. COD and NH3-N emissions of beverage manufacturing and communications equipment computer and other electronic equipment manufacturing were high, and COD emissions of pharmaceutical manufacturing were high in Haizhu District. COD and NH3-N emissions of the main industries were relatively low in Panyu District and Yuexiu District. COD and NH3-N emissions of chemical raw materials and chemical products manufacturing were relatively high in Nansha District. The COD emissions of the biomedicine and cosmetics and daily necessities industries were relatively high in Conghua District. Huadu District, Zengcheng District, Huangpu District, Liwan District, Tianhe District, Baiyun District, and Haizhu District had large COD and NH3-N emissions, and more black and odorous water bodies. The emissions of major water pollutants in Yuexiu District and Panyu District were low, so there was almost no black and odorous water body. The creek pollution in Conghua District was less, which may be due to the lower discharge of sewage. The water quality was relatively good in Nansha District.

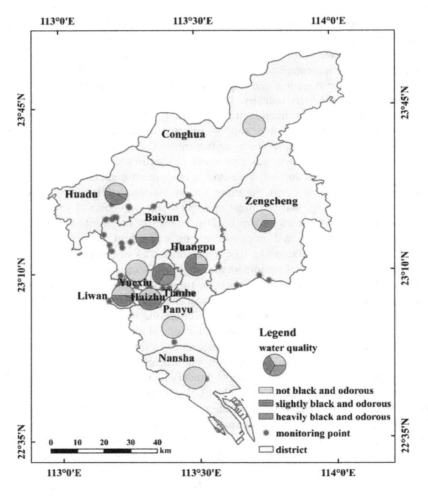

Fig. 2. Spatial distribution of black and odorous creeks in Guangzhou in 2016

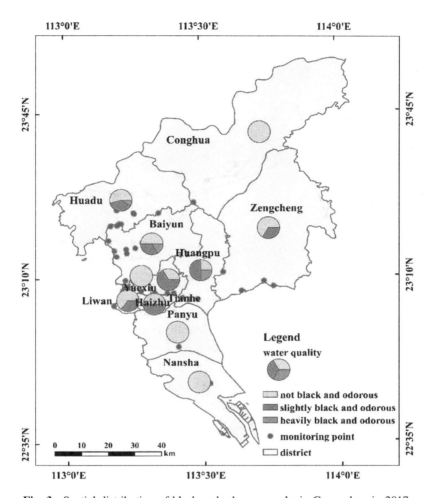

Fig. 3. Spatial distribution of black and odorous creeks in Guangzhou in 2017

5.2   Time Series Analysis of Water Quality Indexes

The variation of the concentration of NH3-N, TP, COD, and DO with time in the creeks is shown in Fig. 4. In 2016, from the monthly average, NH3-N and TP were inferior to Class V every month. The COD was inferior Class V in March, and the rest of the months were higher than the inferior Class V standard. The concentration of DO was higher than the inferior Class V standard every month. The content of NH3-N showed a trend of decreasing first and then increasing. During the dry season, the content of NH3-N was relatively high, but it decreased sharply in February and November. The content of NH3-N in the wet season was relatively low overall. The content of TP decreased first and then increased. The content of TP in the dry season was relatively high, but there was a significant decrease in November. The content of

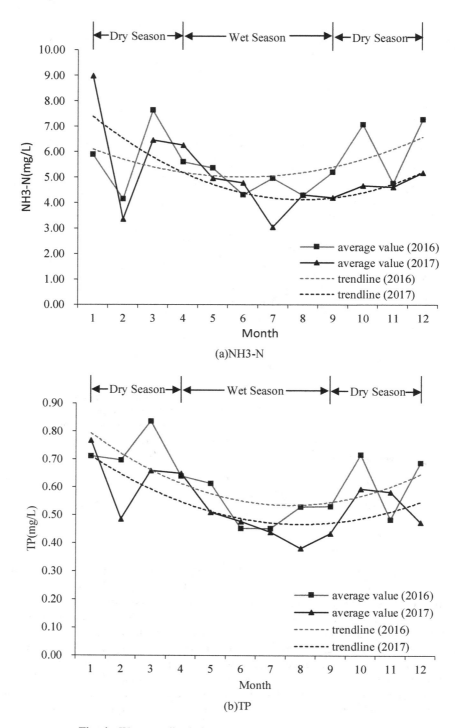

Fig. 4. Water quality index concentration changes over time

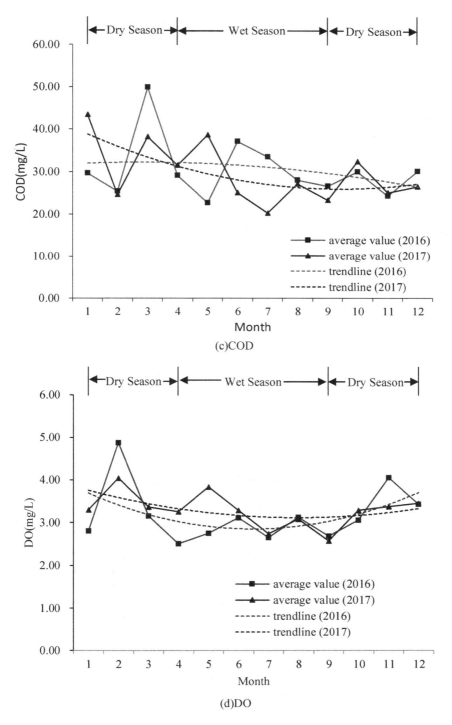

(c)COD

(d)DO

Fig. 4. (*continued*)

Table 6. Comparison and analysis of creek water quality during the wet season and the dry season in 2016

Category	The wet season		The dry season	
	Number of creeks (section)	Proportion (%)	Number of creeks (section)	Proportion (%)
Not black and odorous	149	47	166	52.2
Slightly black and odorous	156	49.2	126	39.6
Heavily black and odorous	12	3.8	26	8.2
Total	317	100	318	100

Table 7. Comparison and analysis of creek water quality during the wet season and the dry season in 2017

Category	The wet season		The dry season	
	Number of creeks (section)	Proportion (%)	Number of creeks (section)	Proportion (%)
Not black and odorous	187	58.8	192	61
Slightly black and odorous	120	37.7	101	32
Heavily black and odorous	11	3.5	22	7
Total	318	100	315	100

TP in the wet season was relatively low, reaching a minimum in June and July. The overall content of COD showed a trend of rising first and then falling, reaching the maximum and minimum values in March and May, respectively. The content of DO appeared to decrease first and then increase, reaching the maximum and minimum values in February and April, respectively.

In 2017, from the monthly average, the content of NH3-N was inferior to Class V every month and reached the slightly black and odorous standard in January. The content of TP reached the Class V water standard in August, and the rest of the months were inferior to Class V. COD reached the inferior Class V standard in January, and the rest of the months were above the inferior Class V standard. The content of DO was higher than the inferior Class V standard every month. The content of NH3-N decreased first and then increased, and it was relatively high during the dry season. The content of NH3-N was relatively low in the wet season, and it fell sharply in February and July. The content of TP decreased first and then increased. The content of TP in the dry season was relatively high, but it fell sharply in February. The content of TP in the wet season was relatively low, reaching a minimum in August. The content of COD appeared to decrease first and then increase, reaching the maximum and minimum values in January and July, respectively. The content of DO showed a decrease first and then a slow rise, reaching the minimum value in September, and there was no significant difference between the wet and dry seasons. On the whole, the content of NH3-N, TP, and COD decreased in 2017 compared to 2016, and the content of DO increased compared to 2016. The major exceeding factors of creeks were NH3-N and TP.

Based on MOHURD's evaluation method of black and odorous water body, we obtained the comparative analysis of the creek water quality in the dry season and the wet season in 2016 and 2017, respectively, as shown in Table 6 and Table 7. It can be seen that the number of slightly black and odorous creeks during the wet season increased by 9.6% and 5.7% respectively compared with the dry season in 2016 and 2017. The number of heavily black and odorous creeks during the wet season decreased by 4.4% and 3.5% respectively compared with the dry season in 2016 and 2017. During the wet season, the number of heavily black and odorous creeks was reduced, but the number of slightly black and odorous creeks was increased and the total number of black and odorous creeks was increased. This situation may be due to the enterprises secretly discharging sewage during the flood season. In 2016, the number of black and odorous creeks in Guangzhou was 320. In 2017, the number of black and odorous creeks in Guangzhou was 254. The number of black-odor water bodies was greatly reduced, and the water quality of the creeks was generally improved.

6   Conclusions and Recommendations

(1) In this paper, the black and odorous water body evaluation method of the Ministry of Housing and Urban-Rural Development of the People's Republic of China was used to evaluate the water quality of 50 key creeks (53 sections) in Guangzhou in 2016 and 2017. We analyzed the spatial distribution of black and odorous water bodies and their changes with time. The results showed that the water quality of Conghua District, Panyu District, Nansha District, and Yuexiu District was relatively good, while the water quality of Baiyun District, Haizhu District, Huadu District, Huangpu District, Liwan District, Tianhe District, and Zengcheng District was relatively poor. Through the annual average value evaluation of the overall situation of black and odorous water bodies, we found that the creek water quality of Huangpu District and Baiyun District in 2017 was worse than the previous year. In 2017, the creek water quality of Tianhe District, Liwan District, and Huadu District improved compared with the previous year. There was no significant change in creek water quality in other districts. Mainly affected by industrial wastewater, Huadu District, Zengcheng District, Huangpu District, Liwan District, Tianhe District, Baiyun District, and Haizhu District had large COD and NH3-N emissions, and they have more black and odorous water bodies. The discharge of major water pollutants in Yuexiu District and Panyu District was small, so there is almost no black-odor water body. The creek pollution in Conghua District was less, which may be due to the lower discharge of sewage. The water quality was relatively good in Nansha District.

(2) On the whole, the content of NH3-N, TP, and COD decreased in 2017 compared to 2016. The content of DO increased compared to 2016. The major exceeding factors

of creeks are NH3-N and TP. During the wet season, the number of heavily black and odorous creeks was reduced, but the number of slightly black-odor creeks was increased and the total number of black-odor creeks was increased. This situation may be due to the enterprises secretly discharging sewage during the flood season. In 2016, the number of black and odorous creeks in Guangzhou was 320. In 2017, the number of black and odorous creeks in Guangzhou was 254. The number of black-odor water bodies was greatly reduced, and the water quality of the creeks was generally improved.

(3) Based on the study of the spatiotemporal evolution law of the black and odorous water bodies, we make suggestions for the treatment of the black-odor creeks in Guangzhou. The government should vigorously remediate industrial enterprises that are involved in the "scattering pollution" of water, strictly purify and treat industrial wastewater, and require that they are discharged after meeting the standards. Other pollution sources must also be strictly controlled, and appropriate treatment methods should be selected for treatment.

Acknowledgments. This study was jointly supported by China Southern Power Grid Guangzhou Power Supply Bureau Co., Ltd. Key Technology Project (0877002018030101SRJS00002); Guangdong Provincial Science and Technology Program (2017B010117008); Guangzhou Science and Technology Program (201806010106, 201902010033); the National Natural Science Foundation of China (41976189, 41976190); the Guangdong Innovative and Entrepreneurial Research Team Program (2016ZT06D336); the Southern Marine Science and Engineering Guangdong Laboratory (Guangzhou) (GML2019ZD0301); the GDAS's Project of Science and Technology Development (2016GDASRC-0211, 2018GDASCX-0403, 2019GDASYL-0301001, 2017GDASCX-0101, 2018GDASCX-0101, 2019GDASYL-0103003).

References

1. Wang, X., Wang, Y.G., Sun, C.H., et al.: Formation mechanism and assessment method for urban black and odorous water body: a review. Chinese J. Appl. Ecol. **27**(04), 1331–1340 (2016)
2. Zhou, H.C.: Water pollution control action plan. Green China **08**, 50–53 (2017)
3. Hao, Y.Q., Zhang, L., Sun, C., et al.: Study on evaluation standard of urban black-odor rivers in Jiangsu Province. Environ. Sci. Technol. **26**(06), 46–50 (2013)
4. Wood, S., Williams, S.T., White, W.R., et al.: Factors influencing geosmin production by a streptomycete and their relevance to the occurrence of earthy taints in reservoirs. Water Sci. Technol. **15**, 191–198 (1983)
5. Hishida, Y., Ashitani, K., Fujiwara, K.: Occurrence of musty odor in the Yodo River. Water Sci. Technol. **8**, 193–196 (1988)
6. Watts, S.F.: The mass budgets of carbonyl sulfide, dimethyl sulfide, carbon disulfide and hydrogen sulfide. Atmosph. Environ. **34**, 761–779 (2000)
7. Chen, J., Xie, P., Ma, Z.M., et al.: A systematic study on spatial and seasonal patterns of eight taste and odor compounds with relation to various biotic and abiotic parameters in Gonghu Bay of Lake Taihu. China. Sci. Total Environ. **409**, 314–325 (2010)

8. Gao, J.H., Jia, J.J., Kettner, A.J., et al.: Changes in water and sediment exchange between the Changjiang River and Poyang Lake under natural and anthropogenic conditions. China. Sci. Total Environ. **481**, 542–553 (2014)
9. Sugiura, N., Utsumi, M., Wei, B., et al.: Assessment for the complicated occurrence of nuisance odours from phytoplankton and environmental factors in a eutrophic lake. Lakes Reservoirs: Res. Manag. **9**, 195–201 (2004)
10. Ruan, R.L., Huang, C.Y.: Study on black-odor assessment and stand of water quality of Suzhou River. Shanghai Water **18**(3), 32–36 (2002)
11. Xu, M., Yao, R.H., Song, L.L., et al.: Primary exploration of general plan of the urban black-odor river treatment in China. Chinese J. Environ. Manag. **7**(2), 74–78 (2015)
12. Chen, H.Z., Wang, Y.J., Song, H.Y.: Examples and reflections on the release of monitoring information for creek water quality in Guangzhou City. Guangzhou Chem. Ind. **42**(20), 152–153 + 248 (2014)
13. Zhang, W.H., An, G.F., Zhou, L.: Status of water environment and treatment countermeasures in Guangzhou City. Municipal Technol. **34**(03), 138–140 + 143 (2016)

Spatial-Temporal Analysis of Water Supply Services at Different Scales in the Wuhua River Basin

Zhengdong Zhang[1(✉)], Yang Yang[1], Yuchan Chen[2], Tengfei Kuang[1], Jun Cao[1], Songjia Chen[1], and Qingpu Li[3]

[1] College of Geography, South China Normal University, Guangzhou 510631, China
zhangzdedu@163.com
[2] Guangzhou Institute of Geography, Guangzhou 510070, China
[3] Guangdong Institute of Ecological Environment and Soil, Guangzhou 510650, China

Abstract. Studying the change of water supply services and its impact mechanism is of great significance for assessing the quality of regional ecological environment. Based on the water production module of the lnVEST model, this paper analyzes the spatial and temporal evolution characteristics of water supply services at different scales (watersheds, sub-watersheds, and hydrological response units) in the Wuhua River Basin, and discusses the reasons for the spatial and temporal changes of water supply services in the study area. The results show that: First, from 1976 to 2016, the precipitation in the Wuhua River Basin first increased, then decreased, and then increased; the average temperature and the average annual evapotranspiration from 1980 to 2015 were $0.0188 \, °C \cdot a^{-1}$ and $0.5094 \, mm \cdot a^{-1}$ rate increasing, respectively. Second, the main land-use type in the Wuhua River Basin were woodland and cultivated land, which together account for more than 98% of the basin area. From 1980 to 2000, it was mainly the conversion between cultivated land, woodland and grassland. From 2000 to 2015, the land use type was frequently changed, mainly converted to construction land. Third, the water supply services at different scales in the Wuhua River Basin from 1980 to 2015 basically changed. Consistently, They all decrease first and then increase. The spatial distribution of water supply services in the basins in/of different years was significantly different. Finally, from 1980 to 2015, the changes in precipitation and water supply services at different scales in the river basin were almost the same, indicating that precipitation is the main factor affecting water supply services. The land-use type with the strongest water supply capacity is construction land, followed by cultivated land, grassland, woodland, and waters. The research aims to provide a scientific basis for establishing a water resource utilization and economic development model, so as to achieve sustainable social and economic development.

Keywords: Water supply service · Invest model · Hydrological response unit · Spatial-temporal change

© Springer Nature Singapore Pte Ltd. 2020
Y. Xie et al. (Eds.): GSES 2019/GeoAI 2019, CCIS 1228, pp. 290–306, 2020.
https://doi.org/10.1007/978-981-15-6106-1_22

1 Introduction

With the rapid development of the economy, the problem of water resources has become increasingly serious, and water supply services have received more and more attention. How to systematically recognize the changing laws of freshwater ecosystems, comprehensively understand the response relationship of key processes to changes in ecosystem services, how to assess the impact of changes in water resources on natural ecosystems and human social systems, and how to achieve the sustainable development of natural-social-economic complex ecosystems, these issues have become one of the hot topics of current researches [1–5]. At present, domestic and foreign scholars have conducted extensive researches on water supply services in river basins, but most of the studies have only analyzed the spatial and temporal characteristics of water supply services from the basin and sub-watershed scales [6–10], and few studies have considered spatial heterogeneity. On the basis of nature [11], the impact of each land use type on water supply services is quantified from the scale of the hydrological response unit, which is the focus of this paper and the key problem to be solved.

In recent years, some progress has been made in the simulation of ecosystem service models. The main models are InVEST, SoIVES, GUMBO, ARIES, SAORES [12–15], etc. Currently, the most widely used is the Integrate Valuation of Ecosystem Services and Tradeoffs Tool (InVEST) jointly developed by Stanford University, World Wildlife Fund, and the Nature Conservation Association. It has many outputs and can perform multi-level analysis [16, 17]. Its water production module comprehensively considers factors such as land-use types, climate change, vegetation cover, and soil characteristics. It has been widely used in many regions at home and abroad, and has achieved good simulation results [11, 18, 19].

The Wuhua River Basin is located in a humid subtropical region, with abundant precipitation, large terrain fluctuations, and small vegetation coverages. It is an area with severe soil erosion in the Han River Basin [20]. The impact of changes in the ecological environment in the basin on water supply services is unclear. Therefore, in this paper, based on the ArcGIS software and the InVEST model of the water production module, the temporal and spatial dynamic characteristics of water supply services at different scales in the river basin are analyzed, and the impact of each land use type on the water supply services is quantified at the hydrological response unit scale. The causes of the spatiotemporal heterogeneity of supply services are expected to provide scientific basis for ecological protection and construction in the subtropical hilly areas of southern China.

2 Materials and Methods

2.1 Study Area

The Wuhua River Basin is the primary tributary of the Mei River in the upper reaches of the Han River. It rises at Yajizhai in the northeast of Huilong Town, Longchuan County, Guangdong Province. With a catchment area of 1031 km^2, Hezikou Hydrological station is the most important control station in the river basin. The terrain in the basin is high in the southwest and low in the north. The watershed belongs to the red soil distribution area in the south subtropics. The soil is mainly red soil, and the soil types such as yellow soil, purple soil, and paddy soil are staggered. The land-use type in the river basin is main woodland, the cultivated land is distributed along the river, and the construction land is mainly distributed in the south of the river basin. It belongs to a typical humid subtropical monsoon climate zone with abundant precipitation. The average annual temperature is 21.1 °C and the average annual precipitation is about 1518 mm, and the precipitation during the year is mainly concentrated in March–September, accounting for approximately 82.5% of the annual precipitation. Precipitation is the main source of water supply in the basin, and plays a key role in the supply of water. Figure 1 shows the location of Wuhua River Basin.

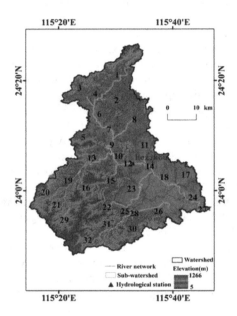

Fig. 1. The location of Wuhua River Basin (The serial number in the figure is the sub-watershed number)

2.2   Data Sources

The research data mainly include meteorological data, land-use data, soil data, digital elevation model (DEM) data and hydrological data. (1) Meteorological data are from China Meteorological Science Data Sharing Service Network (http://cdc.cma.gov.cn), including precipitation and temperature from five meteorological stations in Heyuan, Shantou, Huilai, Wuhua County, and Meixian County from 1976 to 2016; (2) Land use data for the Wuhua River Basin was derived from the supervised classification of Landsat images of 1980, 1990, 2000, 2005, 2010, and 2015 with 30 m spatial resolution from remote sensing interpretation data of the Chinese Academy of Sciences, its scale is 1: 100,000; (3) soil data is of Guangdong Province in 2010 with a scale of 1: 100,000, which is from the Chinese Academy of Sciences, And the soil attribute data mainly comes from a literature named "Record of Guangdong Soil Species", which is recorded the result of the second soil census in Guangdong Province [21]; (4) DEM data: GDEM V2 digital elevation data with a resolution of 30 m, derived from computer network information of the Chinese Academy of Sciences Geospatial data cloud platform (http://www.gscloud.cn); (5) Hydrological data are from the Meizhou Hydrological Bureau, including the runoff and runoff depth in the Hezikou hydrological station (Fig. 1).

2.3   Research Methods

Division of Sub-watershed and Hydrological Response Unit. The ArcSWAT model is used to divide the sub-watershed and hydrological response units in the Wuhua River Basin. First, watershed definition: River network definition based on DEM data, combined with the actual situation of the river network in Wuhua River Basin, finally set the minimum sub-catchment area to 3660hm^2 and divide 32 sub-watersheds (Fig. 1) [22]; Second, editing the river network points to determine the total watershed Water outlets and calculation of sub-watershed parameters; (3) Finally, dividing hydrologic response units (HRU) of the basin. HRU refers to the land surface complex with the same vegetation type and soil conditions. It is the smallest plot unit divided on the basis of sub-watersheds [23]. The land-use type map and soil type map are imported and superimposed to obtain the watershed hydrological response unit feature class layer [24]. This study divides the watershed into 247, 242, 242, 242, 247, and 245 hydrological response units based on land-use cover conditions in 1980, 1990, 2000, 2005, 2010, and 2015.

InVEST Model Water Production Module

Model Algorithm. In this paper, the InVEST model 3.2.0 version of the water production module is used to calculate the water supply service. The water production module is evaluated based on the annual average precipitation data and the Budyko hydrothermal coupling equilibrium assumption [25]. It also considers the key factors of spatial heterogeneity of the confluence, such as soil type, precipitation, and vegetation cover. The main algorithm of the water production module is as follows:

$$Y_{xj} = \left(1 - AET_{xj}/P_x\right) * P_x \tag{1}$$

$$\frac{AETxj}{Px} = \frac{1 + \omega x Rxj}{1 + \omega x Rxj + \frac{1}{Rxaj}} \tag{2}$$

$$\omega x = Z \frac{AWCx}{Px} \tag{3}$$

$$R_{xj} = \frac{Kxj * ET0}{Px} \tag{4}$$

$$AWC_x = \min(MSD_x, RD_x) * PAWC_x \tag{5}$$

In Eqs. (1)–(5), Yxj is the annual water production of the land cover type j in the grid unit x; AETxj is the actual evapotranspiration of the land cover type j in the grid unit x; Px is the grid unit x Precipitation; ωx is the non-physical parameter of natural climate-soil properties; Rxj is the Bydyko drying index; Z is the Zhang coefficient; AWCx is the soil effective water content of the grid element x, determined by soil depth and physical and chemical properties; Kxj is the grid Vegetation evapotranspiration coefficient of land cover type j in unit x; ET0 is the reference crop evapotranspiration; MSDx is the maximum soil depth; RDx is the root depth; PAWCx is the vegetation available water content, which is obtained indirectly using a weighted average of soil texture and soil porosity.

Model Inputs and Parameter Determination. The input parameters of the InVEST model water production module include precipitation, potential evapotranspiration, land-use type, soil depth, available soil water content, and biophysical parameter table.

(1) Average annual precipitation (P).

Taking the 1976–2016 precipitation data of five meteorological stations in and around the Wuhua River Basin as data, in order to avoid the low representativeness of the single-year data, this paper selects six time periods (1976–1985, 1986–1995, 1996–2005, 2000–2009, 2006–2015, 2010–2016) corresponding to the time of land cover (1980, 1990, 2000, 2005, 2010, 2015), respectively, and averaged for many years before using ANUSPLIN4.4. The model was spatially interpolated to obtain the 6-phase spatial grid layer of precipitation in Wuhua River Basin [26, 27] (Fig. 2a).

(2) Average annual potential evapotranspiration (ET0).

This paper uses the modified Thornthwaite formula [28, 29] The formula is as follows:

$$ET0 = k\left(\frac{10T}{I}\right)a * u * N/360 \tag{6}$$

$$I = \sum_{1}^{12} i_j \tag{7}$$

$$i_j = 0.098 * T \tag{8}$$

$$a = 0.016 * I + 0.5 \tag{9}$$

In Eqs. (6)–(9), k = 16, T is the monthly average temperature; I is the temperature efficiency index; a is the heat index function; u is the number of days per month; N is the adjustment coefficient related to the day length and latitude.

This paper selects six periods corresponding to the annual precipitation, and calculates the potential evapotranspiration of 5 stations in Heyuan county, Shantou county, Huilai county, Wuhua county, and Meixian. Finally, the ordinary Kerry method in Arcgis software is used for spatial interpolation (Fig. 2b).

(3) Land use/land cover

The land use is classified according to the China Land Use Remote Sensing Classification System proposed by Liu Jiyuan [30], et al., and then converted into raster data (Fig. 2c).

(4) Root restricting layer depth and plant available water content

According to the detailed information of the soil layer in "Record of Guangdong soil species" [21], the Root restricting layer depth is the total depth of the soil layer without root distribution. The plant available water content is calculated using the Soil Water Characteristics in the SPAW Hydrology model [31], and its properties are converted into raster plots (Fig. 2d and Fig. 2e).

(5) Biophysical parameter table.

It mainly includes land-use code, root depth, vegetation evapotranspiration coefficient, etc. The data on the root depth mainly refer to previous research results. The vegetation evapotranspiration coefficient mainly references the value of the FAO evapotranspiration coefficient (crop coefficient) and the material of the lnVEST model [26].

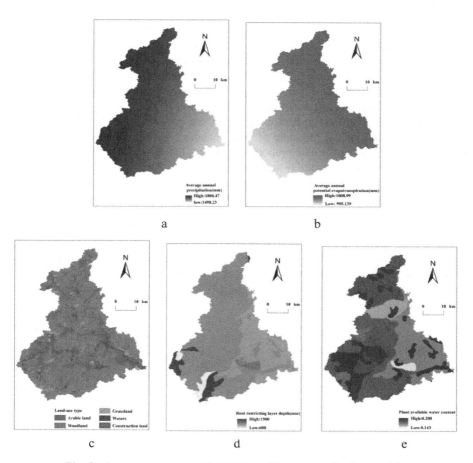

Fig. 2. Space parameters of InVEST model water production module

(6) Z parameter

Z is an empirical constant that captures the local precipitation pattern and hydrogeological characteristics, with typical values ranging from 1 to 30 [17]. The multi-year average runoff data of the Wuhua River Basin can be obtained based on the data of the basin's area, the catchment's area controlled by the Hezikou Hydrological Station, and the average runoff data for many years at this site, and then the model's Z coefficient was repeatedly checked. Finally the simulation effect of the lnVEST water production module for many years has reached more than 86%, indicating the results of this research are reliable.

3 Results

3.1 Analysis on the Change Trend of Climatic Factors

The annual average precipitation in the Wuhua River Basin increased first, then decreased and then increased, and the lowest point occurred in 1991, the highest point appeared in 2006, and the period of large interannual changes in precipitation has shortened in the past 15 years (Fig. 3a); the average temperature rise rate is 0.0188 °C·a^{-1}, and climate warming has an impact on ecosystem functions in the basin (Fig. 3b); the annual average potential evapotranspiration was 1019.3 mm from 1980 to 2015 in the Wuhua River Basin, and it increased at a rate of 0.5094 mm·a^{-1} (Fig. 3c). According to the principle of water balance, water supply is the difference between precipitation and actual evapotranspiration [17]. There is a close relationship between actual evapotranspiration and potential evapotranspiration, which characterizes the evapotranspiration capacity of the surface ecosystem. As the main process of the water cycle, precipitation is the main source of water supply in the Wuhua River Basin. Potential evapotranspiration is an important part of water output, and temperature is a key factor affecting potential evapotranspiration. Therefore, precipitation, potential evapotranspiration, and temperature all have an important impact on the change of water supply services in the basin, and their changing trends are also closely related to the change of water supply services [34, 35].

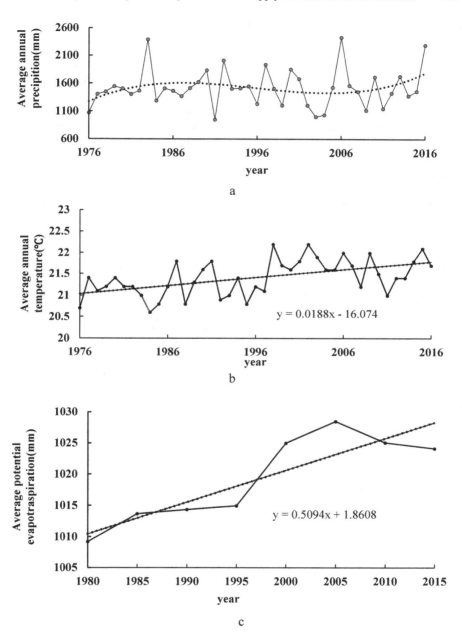

Fig. 3. Annual variations of major climatic factors in the Wuhua River Basin

3.2   Analysis on the Change of Land Use

The land use data from 1980 to 2015 of the research area was imported into ArcGIS for spatial calculation. It was found that the woodland was the largest area in the Wuhua River Basin, accounting for nearly 77% of the total area; followed by cultivated land, accounting for the total area nearly 18%, and the construction land area is less than 1% (Table 1). Over the past 35 years, the land-use types have changed differently in the Wuhua River Basin. Among them, the area of grassland has changed from 2.67% to 1.91% and the area of construction land was from 0.67% to 1.28%. The area of cultivated land has gradually decreased. The changing trend of woodland area is increasing first and then decreasing, while the changing state of water area is relatively stable, which was at about 0.82% (Table 1). Woodland area increased significantly due to grassland conversion to woodland from 1980 to 2000 in the watershed. Different land-use types are frequently converted, and the main manifestations are that grassland, woodland and cultivated land are converted to construction land to varying degrees (Table 2) from 2000 to 2015, which is consistent with the change trend of area of each land-use type in Table 1. Comprehensive analysis shows that large-scale soil erosion control work was carried out in the study area from 1980 to 2000, and a large increase in the area of woodland, which shows that the forest restoration policy in the area had achieved significant results. From 2000 to 2015, with the rapid economic development, human activities gradually strengthened, and the frequency of conversion between different land-use types in the region also increased. However, after 2005, it even appeared at the cost of reducing the area of woodland and cultivated land in order to meet the demand for construction land. In the past 35 years, the land use patterns in the Wuhua River Basin have undergone significant changes, and this change will also have an important impact on water supply services in the basin.

Table 1. The land use type area and proportion of Wuhua River Basin from 1980 to 2015

Land-use type	1980		1990		2000		2005		2010		2015	
	Area/km^2	Rate/%	Area/km^2	Rate/%	Area/km^2	Rate/%	Area/km^2	Rate/%	Area/km^2	Rate/%	Area/km^2	Rate/%
Cultivated land	345.43	18.75	345.13	18.73	345.08	18.73	344.49	18.70	342.92	18.61	343.11	18.62
Woodland	1419.50	77.04	1426.91	77.44	1427.12	77.45	1428.50	77.53	1427.98	77.50	1425.48	77.36
Grassland	49.25	2.67	42.89	2.33	42.68	2.32	40.35	2.19	37.16	2.02	35.21	1.91
Watershed	15.96	0.87	15.20	0.82	15.20	0.82	15.20	0.82	15.13	0.82	15.13	0.82
Construction land	12.44	0.67	12.46	0.68	12.50	0.68	14.04	0.76	19.38	1.05	23.66	1.28

Table 2. The land use transfer matrix of Wuhua River Basin from 1980 to 2015 (km^2)

Period	Land-use type	Cultivated land	Woodland	Grassland	Watershed	Construction land
1980–1990	Cultivated land	345.3048	0.0072	0.1197	0	0
	Woodland	0	1419.4942	0.0018	0	0
	Grassland	0	6.3747	42.87967	0	0
	Watershed	0	0.4167	0	15.54828	0
	Construction land	0	0	0	0	12.43723
1990–2000	Cultivated land	345.0895	0	0	0	0.0378
	Woodland	0	1426.912	0	0	0
	Grassland	0	0.0981	42.78992	0	0
	Watershed	0	0	0	15.19778	0
	Construction land	0	0	0	0	12.49614
2000–2005	Cultivated land	344.405	0.3033	0	0	0.3753
	Woodland	0.0828	1426.0126	0.0063	0	1.0143
	Grassland	0.0009	2.1942	40.37935	0	0.108
	Watershed	0	0	0	15.19778	0
	Construction land	0	0	0	0	12.50208
2005–2010	Cultivated land	342.9643	0.8568	0.0009	0	0.6714
	Woodland	0.0045	1424.6681	0.0018	0.0009	3.8277
	Grassland	0	0	37.22988	0	0.0162
	Watershed	0	0.0405	0	15.13298	0.0243
	Construction land	0	0	0	0	14.03686
2010–2015	Cultivated land	341.3326	0.0009	0	0	1.5894
	Woodland	1.7046	1424.4216	0	0	1.8558
	Grassland	0	0	35.39353	0	1.7694
	Watershed	0	0	0	15.13423	0
	construction land	0.063	1.0215	0	0	18.29774

3.3 Spatial and Temporal Variations of Water Supply in Sub-watershed Scale

After processing data such as precipitation, potential evapotranspiration, plant effective water content, land use, and root restricting layer depth with the help of ArcGIS, the

InVEST water production module was used to obtain six periods of water supply at watershed and sub-watershed scales in Wuhua River Basin. With reference to the relevant water supply grading standards [7, 8], the water supply in the sub-watershed of the study area is divided into five levels, such as >700 mm, 700–800 mm, 800–900 mm, 900–1000 mm, 1000–1100 mm (Fig. 4). From the perspective of geographical distribution, the water supply of sub-watershed in each year is generally inconsistent, but it is partially consistent. The water supply in the sub-watershed of the whole study area has an obvious lamellar spatial distribution trend, which is similar to the spatial distribution of precipitation. Among them, the water supply in the sub-watershed of the entire study area has an obvious lamellar spatial distribution trend, which is similar to the spatial distribution of precipitation. The water supply in the sub-watersheds on the 14th, 19th, 25th, and 27th is more than that of other sub-watersheds in the same year, but the water supply in the sub-watershed on the17th, 26th, 28th, and 30th is less, and its spatial distribution is related to the land-use in the sub-watershed [36].

On the time scale, the average water supply in the Wuhua River Basin in 1980, 1990, 2000, 2005, 2010, and 2015 is respectively 887.82 mm, 817.98 mm, 740.67 mm, 754.08 mm, 793.44 mm, and 858.87 mm (Fig. 5). It shows a decreasing trend and then an increasing trend. It can be seen from Fig. 4 that the water supply in each sub-watershed is higher than 700 mm, and 1/3 of the water supply in the sub-watershed is higher than 900 mm in 1980; the water supply in the sub-watershed is basically between 800–900 mm in 1990; The supply of water sources in each sub-watershed is basically 700–800 mm, but the water supply on 17th, 26th and 28th sub-watersheds are less than 700 mm; In 2010, the number of sub-watersheds with water supply of 800–900 mm increased significantly compared with 2005; In 2015, the water supply in almost all sub-watersheds is more than 800 mm, and even the water supply on 14th sub-watershed is higher than 1000 mm.

3.4   Spatial and Temporal Variations of Water Supply in Hydrological Response Unit Scale

At present, the InVEST water production model only guarantees the accuracy of the estimated water supply services at the watershed scale and the sub-watershed scale [17]. However, it finds that the effects of climate and land use on water supply services based on the hydrological response unit scale estimates are consistent with the findings of Xu Jie [8], so this estimation method is considered. The InVEST water production module was imported into the hydrological response unit from 1980 to 2015, and got the water supply in 1980, 1990, 2000, 2005, 2010, and 2015 at hydrological response unit scale (Fig. 6).

According to the actual situation of water supply in the hydrological response unit scale of the study area, it is divided into five levels(>600 mm, 600–900 mm, 900–1000 mm, 1000–1100 mm, 1100–1500 mm) (Fig. 6). From the perspective of geospatial distribution, it is found that the water supply of the hydrological response unit in the entire study area is similar to the spatial distribution of land use. The areas that are more water supply at the scale of hydrological response unit correspond to the areas where cultivated land, construction land and grassland are distributed, and the areas with less water supply correspond to the areas with woodland distribution. On the

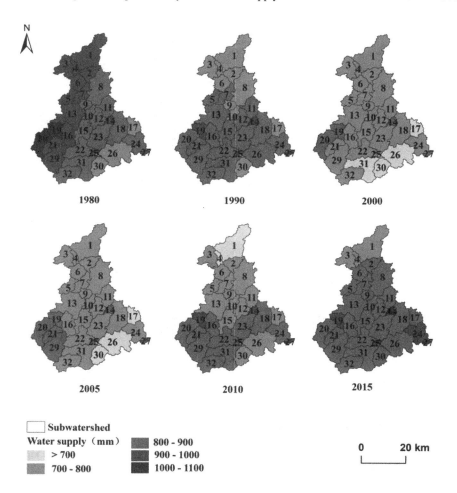

Fig. 4. The spatial distribution of water supply in Wuhua sub-watershed from 1980 to 2015

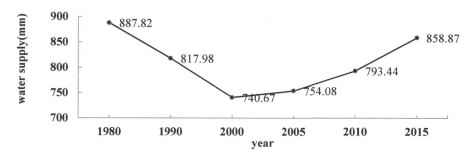

Fig. 5. Water supply quantity of Wuhua River Basin from 1980 to 2015

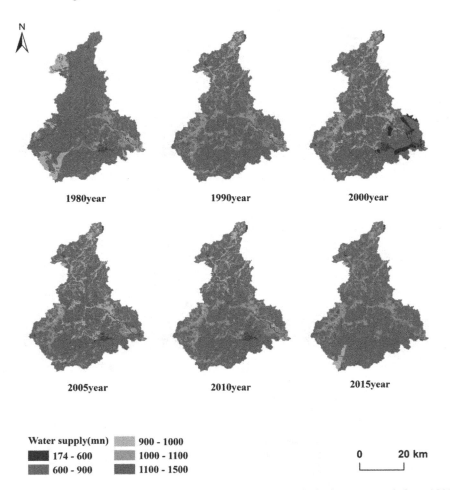

Fig. 6. Spatial distribution of water supply in the scale of hydrological response unit from 1980 to 2015

time scale, the total water supply decreased first and then increased at the hydrological response unit scale from 1980 to 2015. In 1980, the water supply of most hydrological response units was 600–900 mm. In 1990, the number of hydrological response units with water supply of 1100–1500 mm decreased significantly. The number of hydrological response units with water supply of 172–600 mm and 900–1000 mm increased significantly in 2000. In 2005, the most number of hydrological response units on water supply were 600–900 mm and 900–1000 mm. In 2010, the water supply for the most number of hydrological response units was 600–900 mm. The number of hydrological response units with water supply of 1100–1500 mm increased significantly in 2015.

3.5   The Reason for Spatial-Temporal Changes of Water Supply Services in River Basin

From the perspective of water balance, precipitation and actual evapotranspiration are the key links in determining the water supply of the ecosystem. Climate factors affect water supply mainly by changing precipitation and potential evapotranspiration. Precipitation is the main source of ecosystem water cycle. Potential evapotranspiration represents the water consumption capacity of the regional ecosystem, the greater the potential evapotranspiration, the more water it consumes. From 1980 to 2016, the rainfall in the Wuhua River Basin first decreased and then increased, which is basically consistent with the overall time trend of water supply at different scales, indicating that rainfall can directly affect the amount of water supply.

Based on the hydrological response unit scale, the water supply of each land-use type in the river basin is calculated. According to Table 3, the water supply of different land-use type in the study area from 1980 to 2015 in order from most to least is construction land, arable land, grassland, woodland and watershed; From 1980 to 2015, the water supply of each land-use type shows a trend of first decline and then increase, which was consistent with the trend of rainfall.

Table 3. Sources of water supply for various land use/cover types of Wuhua River Basin based on hydrological response unit scale from 1980 to 2015 (mm)

Land-use type	1980	1990	2000	2005	2010	2015
Cultivated land	1110.27	1042.33	961.04	977.30	1019.48	1085.63
Woodland	823.73	758.81	675.91	692.43	734.93	795.37
Grassland	1053.09	1006.96	926.08	943.37	986.85	1047.50
Watershed	617.90	601.98	483.44	509.98	573.11	638.45
Construction land	1351.73	1296.94	1210.92	1234.17	1281.21	1347.23

4   Discussion

The climatic factors have the most significant impact on water supply services in the watershed, and the impacts of different land-use types on water supply services are also significantly different, which is almost consistent with the results of Wang Yuchun [37]. In this study, using InVEST models, we found that the precipitation trend of the watershed from 1980 to 2015 was almost the same as the change trend of water supply services at different scales, indicating that climate factors have the greatest impact on the water supply services Precipitation affects water supply service capacity. At the same time, the impacts of various land-use types on water supply services in the Wuhua River Basin were quantified from the hydrological response unit scale. Among them, construction land has the greatest impact on water supply services. Land use directly affects actual evapotranspiration by changing the underlying surface conditions, and then affects water sources. In the future, it is possible to accurately quantify the changes in water supply services caused by different land use changes based on scenario simulation methods, which will have important practical significance for solving water and soil loss problems in the watershed.

Ecosystem water supply service is a complex comprehensive adjustment process, but the water production module in the InVEST model used in this paper does not consider the impact of plant leaf types and groundwater exchange on water supply service, simplifying the hydrological cycle process and reducing the results. In the future, we can make full use of the methods and achievements of ecology and hydrology. Through the combination of observation data and system simulation, we can comprehensively study the formation, changes and influencing factors of water supply services, and provide ecological System service research provides reliable data foundation and theoretical support.

This study explores the spatial and temporal evolution characteristics of water supply services at different scales in the Wuhua River Basin, and quantifies the impact of various land-use types on water supply services at the hydrological response unit scale. This study will be help to resolve the contradiction between water supply and demand in the research area.

5   Conclusions

Taking the Wuhua River Basin as a research area, this paper analyzes the spatial and temporal evolution characteristics of water supply services at different scales of the Wuhua River Basin, and quantifies the impact of land-use types on water supply services from the scale of the hydrological response unit. The research results have important effects on the stability of the ecosystem and the prevention of soil erosion policies in the subtropical hilly areas of southern China. The main conclusions are as follows:

(1) From 1976 to 2016, the precipitation in the Wuhua River Basin first increased, then decreased, and then increased; the average temperature and the average annual evapotranspiration were 0.0188 $°C·a^{-1}$ and 0.5094 $mm·a^{-1}$ rate increasing from 1980 to 2015, respectively.

(2) The main land use type in the Wuhua River Basin is woodland, which accounts for almost 77% of the total area; From 1980 to 2000, a large amount of grassland was converted into woodland, which caused a significant change in the area of woodland; From 2000 to 2015, the area of water area hardly changed, but other land use types changed frequently, and the area of construction land increased significantly.

(3) From 1980 to 2015, the change trend of water supply services was the same at different scales in the Wuhua River Basin, showing a trend of first decline and then increase. The spatial distribution of water supply services is obvious in different years in Wuhua River Basin.

(4) From 1980 to 2015, the changes in precipitation and water supply services at different scales in the river basin were almost the same, indicating that precipitation is the main factor affecting water supply services. Among different land-use types, construction land and woodland have a greater impact on water supply services, while waters have a smaller impact on water supply services.

Acknowledgments. This study was jointly supported by the National Natural Foundation of China (41471147), and the Graduate Innovation Program of South China Normal University (2018LKXM038).

References

1. Millennium Ecosystem Assessment: Ecosystems and Human Well-Being: A Framework for Assessment. Island Press, Washington, DC (2003)
2. Daily, G.C.: Nature's Services Societal Dependence on Natural Ecosystems. Island Press, Washington, DC (1997)
3. Yihui, W., Guihuan, L., Rui, W.: Eco-Compensati on in Guanting reservoir watershed based on spatiotemporal variations of water yield and purification services. J. Resour. Ecol. **9**(4), 416–425 (2018)
4. Marquès, M., Bangash, R.F., Kumar, V., Sharp, R., Schuhmacher, M.: The impact of climate change on water provision under a low flow regime: a case study of the ecosystems services in the Francoli river basin. J. Hazard Mater. **263**, 224–232 (2013)
5. Li, S., Liu, J., Zhang, C., et al.: Research trends of ecosystem services and geographical research paradigm. Acta Geogr. Sin. **66**(12), 1618–1630 (2011)
6. Pan, T., Wu, S., Dai, E., et al.: Spatial-temporal change of water supply and service of ecosystem in sanjiangyuan district based on InVEST model. Chin. J. Appl. Ecol. **24**(01), 183–189 (2013)
7. Bao, Y., Li, T., Liu, H., et al.: Spatial-temporal change of water conservation function in loess plateau of northern shanxi based on InVEST model. Geogr. Res. **35**(04), 664–676 (2016)
8. Jie, X., Xiao, Yu., Gaodi, X., et al.: Spatial and temporal pattern analysis of water supply and service in dong jiang river basin. Acta Ecol. Sin. **15**(36), 4892–4906 (2016)
9. Zhang, L., Cheng, L., Chiew, F., Fu, B.: Understanding the impacts of climate and landuse change on water yiled. Environ. Sustain. **33**, 167–174 (2018)
10. Yin, Y., Wu, S., Zhao, D., Dai, E.: Impacts of climate change on water conservation in the source area of the Yellow River over the past 30 years. Geogr. Res. **35**(01), 49–57 (2016)
11. He, S., Ye, L., Zhu, W., et al.: Study on soil erosion and water supply change in the shanqi river basin of Taihang from 2000 to 2015. Geogr. Res. **37**(09), 1775–1788 (2018)
12. Zhao, W., Liu, Y., Feng, Q., et al.: Ecosystem services under the coupling framework of human and earth systems. Progr. Geogr. **37**(01), 139–151 (2018)
13. Wu, B., Wang, J., Qi, S., et al.: A review of quantitative methods for ecosystem service tradeoffs and future model development. J. Resour. Ecol. **10**(02), 225–233 (2019)
14. Sherrouse, B.C., Clement, J.M., Semmens, D.J.: A GIS application for assessing, mapping, and quantifying the social values of ecosystem services. Appl. Geogr. **31**(2), 748–760 (2011)
15. Li, T., Lu, Y.: Research progress of ecosystem service modeling technology. Acta Ecologica Sinica **38**(15), 5287–5296 (2018)
16. Liang, Ma., Taotao, J., Yihui, W., et al.: InVEST model research progress. Ecol. Econ. **10** (31), 126–131 (2015)
17. Sharp, R., Tallis, H.T., Ricketts, T., et al.: InVEST3.2.0 user's guide. The Natural Capital Project, Stanford (2015)
18. Yuchu, X., Jie, G., Suxin, Z., et al.: Spatial-temporal pattern of biodiversity in Bailong river basin landscape based on remote sensing and InVEST models. Geogr. Sci. **38**(06), 979–986 (2018)

19. Hou, H., Dai, E., Zhang, M.: Research progress of InVEST model application. J. Cap. Norm. Univ. Nat. Sci. Ed. **04**(39), 62–67 (2018)
20. Chen, Y., Zhang, Z., Wan, L., et al.: Identification of non-point source pollution risk areas and risk paths in Wuhua river basin. Acta Geogr. Sin. **73**(09), 1765–1777 (2018)
21. Liu, A.: Record of Guangdong Soil Species. Science Press, Beijing (1996)
22. Li, M., Han, H., Liu, X., et al.: Study on optimal sub-watershed division scheme for SWAT model-a case study of Erhai river basin in Yunnan province. China Agrometeorol. **33**(02), 185–189 (2012)
23. Li, J., Zhou, Z.: Landscape pattern and ecological hydrological process analysis of Yanhe river basin. Acta Geographica Sinica **69**(07), 933–944 (2014)
24. Wang, J., Xie, D., Ni, J., et al.: Identification of soil erosion risk pattern in watershed based on Yuanhui landscape unit. Acta Ecol. Sin. **37**(24), 8216–8226 (2017)
25. Budyko, M.I.: Climate and Life. Academic Press, New York (1974)
26. Liu, Z., Yu, X., Wang, S., et al.: Comparison of precipitation interpolation accuracy of three covariate methods in thin disk smooth spline interpolation. Progr. Geogr. **31**(01), 56–62 (2012)
27. Hutchinson, M.F., Xu, T.B.: Anusplin Version 4.4 User Guide. The Australian National University, Canberra (2013)
28. Chen, Z., Zhang, Z.: Drought index and drought monitoring and control system. China Desert **15**(01), 10–18 (1995)
29. Thornthwaite, C.W.: An approach toward a rational classification of climate. Geogr. Rev. **01**(38), 55–94 (1948)
30. Liu, J., Ning, J., Kuang, W., et al.: Spatial and temporal patterns and new characteristics of land use change in China from 2010 to 2015. Acta Geogr. Sin. **73**(05), 789–802 (2018)
31. Zhou, W.: Study on soil available moisture content of main soil types in China based on GIS. Nan Jing Agricultural University (2003)
32. Canadell, J., Jackson, R.B., Ehleringer, J.B., Mooney, H.A., Sala, O.E., Schulze, E.D.: Maximum rooting depth of vegetation types at the global scale. Oecologia **108**(4), 583–595 (1996). https://doi.org/10.1007/BF00329030
33. Allen, R.G., Pereira, L.S., Raes, D., Smith, M.: Crop evapotranspiration-guidelines for computing crop water requirements-FAO Irrigation and drainage paper 56. FAO Rome **300**(9), 1–15 (1998)
34. Zhou, J., Gao, J., Gao, Z., Yang, W.: Analysis of water conservation services in forest ecosystems. Acta Ecologica Sinica **38**(05), 1679–1686 (2018)
35. Liu, H., Wu, J., Chen, X.: Spatio-temporal changes and trade-offs of ecosystem services in the Danjiangkou water source area. Acta Ecologica Sinica **38**(13), 4609–4624 (2018)
36. Ye, L.: Spatial heterogeneity analysis of water supply and service in Qi river basin. HeNan University (2018)
37. Wang, Y., Zhao, J., Fu, J., Wei, W.: Quantitative assessment and spatial difference of water conservation function in Shiyang River Basin. Acta Ecologica Sinica **38**(13), 1–12 (2018)

Intelligent Perceptions and Services of Spatial Information

Spatialization Method of Atmospheric Quality Public Opinion Based on Natural Language Processing

Yong Sun[1], Pengfei Song[1(✉)], Min Ji[1], Yan Zheng[1], and Liguo Zhang[2]

[1] Shandong University of Science and Technology, Qingdao 266590, China
1015609473@qq.com
[2] Shandong Institute of Land Surveying and Mapping, Jinan 250102, China

Abstract. With the increasing environmental awareness of the country and the people and the implementation of strong environmental protection measures, China's atmospheric quality was improved obviously, but local atmospheric pollution events still occurred frequently. In order to make up for the missing detection of local atmospheric pollution events caused by sparse fixed State-controlled monitoring stations, the paper proposed a spatialization method for atmospheric quality public opinion information based on natural language processing. Using Chinese word segmentation, part of speech tagging and other methods, the paper extracted addresses from public atmospheric pollution complaints data. Through an effective combination of those addresses, the paper realized address matching of those complaint points, and spatialized those key complaint areas in Shandong Province in the form of heat map. Through comparison and analyzing with the atmospheric quality monitoring data of national control stations, it showed that the key areas of public complaints were highly consistent with the key pollution areas which were monitored by national control stations. The research result showed that the public could perceive the atmospheric quality directly, and reflect the local atmospheric pollution at a smaller space-time scale effectively, which was a robust supplementation to the monitoring data of the national control station.

Keywords: Local atmospheric pollution · Public perception · Spatialization · Natural language processing · Address matching

1 Introduction

With the rapid development of China's economy, local atmospheric pollution events [1] had aroused the attention of the government and the public. At present, atmospheric quality monitoring mainly relies on state-controlled stations, most of these are located in cities and distributed. Through the long time series monitoring of these stations can provide the basis for macro-regional atmospheric quality change analysis, but lack of the monitor of local atmospheric pollution events. The public can perceive atmospheric quality directly [2], their evaluation can serve as the basis for government departments to control atmospheric pollution [3]. Since entering the Internet era, people often talk

© Springer Nature Singapore Pte Ltd. 2020
Y. Xie et al. (Eds.): GSES 2019/GeoAI 2019, CCIS 1228, pp. 309–320, 2020.
https://doi.org/10.1007/978-981-15-6106-1_23

and evaluate atmospheric quality on the Internet, by collecting public opinion and spatializing them can effectively reflect the local atmospheric pollution distribution. It can not only improve people's political participation [4], but also provide decision basis for government departments to regulate and control local atmospheric pollution incidents.

Scholars had carried out many researches on spatial methods. Wang used the night lighting data to spatialize the township population data of Shandong province [5]; Yin utilized FME software to spatialize the traffic data of floating vehicles [6]. Atmospheric quality complaint data contained a large amount of address information, the process of public opinion spatialization was based on address information for address matching and spatial location. Due to public complaint data is natural language, the natural language processing (NLP) based on artificial intelligence (AI) had also been developed rapidly, especially in Chinese word segmentation, part of Speech tagging, parsing, spatial information natural language query interface, query grammar rules and other aspects [7–9], therefore different from the traditional address matching method for establishing a standard geographical address library, Dilek integrated NLP into geocoding, and implemented the address standardization through address analysis, error correction and address reorganization based on NLP [10]; Xu realized the massive Chinese address fast matching based on Bayesian algorithm in the big data environment[11]; Cheng proposed a fuzzy Chinese address segmentation matching method based on rules[12]; Li used the k-tree method to research fuzzy address matching [13]. At present, there is a lack of literature on address matching and spatial location of public atmospheric quality complaint information. For researching the address matching and spatial visualization of emphasis complaint areas, we utilized Chinese word segmentation and part of speech tagging based on atmospheric quality public opinion data from Shandong provincial environmental protection department, which can provide some references for atmospheric quality monitoring in Shandong province from macro to smaller scale.

2   Opinion Data Source and Processing Method

2.1   Data Source

Atmospheric pollution public opinion data were obtained from the public complaint's platform of Shandong provincial environmental protection department, which contains type of complaints, time, contents, the handling status, and response status of the complaints. Types of atmospheric pollution include: flying dust, odor, soot, motor vehicle/mobile source, industrial waste gas, smoke and other atmospheric pollution; the complaint content exists in natural language and needs to be analyzed and processed to extract Chinese address information and atmospheric pollution topical opinion information.

In the process of public opinion data processing, it is necessary to control the data quality, and exclude the data that do not contain address information and the data of non-air pollution. Calculated the similarity of each complaint data containing address information with a representative statement of each type of air pollution type. If the similarity with one of the air pollution types reaches 70% or more, the data is useable. At the same time, it is need to remove some common words and special symbols, prevent them from interfering with address extraction, improving efficiency and accuracy. This paper selected 5,500 complaint data from February 2017 to October 2018, and selected 5,028 valid information.

2.2 Natural Language Public Opinion Information Processing Method

At present, NLP mainly includes Chinese word segmentation, part of speech tagging, and named entity recognition, commonly used Chinese word segmentation methods include dictionary-based word segmentation and statistical-based word segmentation [14, 15], the statistical-based word segmentation method has greatly improved the recognition of unknown words than the dictionary-based method [16], and its main algorithms include Hidden Markov Model (HMM), Maximum Entropy Model, and Conditional Random Field Model (CRF) [17]. In view of the performance advantages of the CRF model, this paper chooses it to segment word, and used the HMM model to mark unknown words.

Conditional Random Field Model (CRF). When CRF model segmenting word, it can not only statistical word frequency, but also take context into consideration, it has good segmentation effect on some unknown words in the data. For example, the public complaint data contained many factory names and enterprise names, CRF model can accurately classify these names in the process of Chinese word segmentation to improve the accuracy of address extraction. The most commonly used is the linear condition random field. The conditional probability formula of the model is as follow:

$$p(y|x) = \frac{1}{Z(x)} \exp[\sum_{i,k} \lambda_k t_k(y_{i-1}, y_i, x, i) + \sum_{i,l} \mu_l s_l(y_i, x, i)] \tag{1}$$

the normalization factor Z(x) is as expressed in Eq. (2):

$$Z(x) = \sum_{y} [\sum_{i,k} \lambda_k t_k(y_{i-1}, y_i, x, i) + \sum_{i,l} \mu_l s_l(y_i, x, i)] \tag{2}$$

The formula is a conditional probability that the random variable Y takes the value y under the condition that the random variable X takes the value of x, where λ_k and μ_l are corresponding weights, t_k and s_l are characteristic functions, and i represents all possible y Values, k and l represent the number of the characteristic function, and the first three parameters in $t_k(y_{i-1}, y_i, x, i)$ are the largest group of linear nodes connected by three nodes, $s_l(y_i, x, i)$ indicates the probability that different x is labeled as y_i [18].

Hidden Markov Model (HMM). HMM contains a visible state sequence and a hidden state sequence, which are related by probability. In the process of part-of-speech tagging, HMM has a good effect on the tagging of unregistered words, and can mark the factory name in the complaint data. HMM consists of five parameters [19], can be expressed by a five-tuple {N, M, η, A, B}. Where N represents the number of hidden states, the probability of each state may have a certain value, and may also be determined by analysis; M represents the number of visible states, obtained by the training set; $\eta = \{\eta_i\}$ represents the probability of occurrence at the initial moment of each hidden state; $A = \{a_{ij}\}$ represents the transition matrix of the hidden state, that is, the probability of occurrence of the event from the hidden state 1 to the hidden state 2; $B = \{b_{ij}\}$ represents the confusion matrix, that is, the probability of the visible state occurs under the condition of a hidden state.

3 Atmospheric Pollution Public Opinion Data Processing Flow Design

3.1 Data Processing Flow

The spatial analysis and evaluation of atmospheric pollution requires extracting address in the public opinion data, and realizes the spatial position conversion through address matching method. The data processing process includes data preprocessing, natural language processing, part of speech extraction, address matching, etc. Among them, data preprocessing includes two steps: content extraction and remove stop-word, NLP includes Chinese word segmentation and part of speech tagging. The specific data processing flow is shown in Fig. 1.

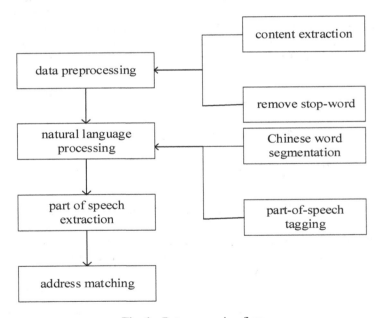

Fig. 1. Data processing flow

3.2   Data Preprocessing

Data preprocessing is extracting the content of the public complaints and remove the stop-word. Firstly, designed the corresponding database structure in order to facilitate management, data reading and writing convenience. Then, transferred public complaint data to the database, and the content data containing location information were stored in separate fields to improve the reading and writing efficiency in the data analysis process, cleaned the data by removing some special symbols and stop words at the same time.

3.3   Natural Language Processing (NLP)

After obtaining the content of the public complaint data, further processing of the data is required. NLP includes two processes of Chinese word segmentation and part-of-speech tagging. CRF is used to segment the content of public complaints, then, the word segmentation results are matched in the Baidu lexicon to find the part of speech. The words that cannot be recognized are unknown words, and then the HMM is used to mark unknown words. Because of the same word has a variety of part of speech in different statements, and there are many unknown words in different application fields, resulting in part of speech tagging has many difficulties. For this reason, people divide the part of speech into ordinary nouns, ordinary verbs, and adjectives and so on 24 kinds, as shown in Table 1. In order to further determine the exclusive name, people also divide the person names, place names, organization names, time, four types of proper noun speech, concrete as shown in Table 2, these classifications provide strong support for part of speech tagging.

Table 1. Table part of speech tag.

Part of Speech	Implication	Part of Speech	Implication	Part of Speech	Implication
n	common noun	v	regular verbs	r	pronoun
nr	personal name	ad	deputy form word	xc	other function words
nz	other proper noun	q	quantifiers	t	time noun
a	adjectives	u	auxiliary word	nw	title
m	numeral	s	premises noun	vn	verbnoun
c	conjunctions	nt	institutional group name	d	adverb
f	bearing the noun	vd	adverb	p	preposition
ns	toponymy	an	noun adjectives	w	punctuation

Table 2. Noun description table.

Abbreviations	Implication
PER	personal name
LOC	address
ORG	institution name
TIME	time

3.4   Part of Speech Extraction and Address Matching

After part of speech tagging analysis, the LOC and ORG words have been extracted, due to the contents of public complaints include pollution emission of factories, the extracted factory name was marked as the organization name. After the part of speech extraction was completed, the spatial positioning of complaint information shall be realized according to the extracted location name and organization name, that is the address matching.

Address matching mainly include the following combinations according to the extracted place name and organization name:

(1) If the parsed part of speech has only the place name and no organization name, only the place name is matched with the address;
(2) If the parsed part of speech has only the name of the institution and no place name, only the name of the institution is matched;
(3) If there are both place names and institution names, the institution name and institution name shall be judged: if the institution name contains place names, the institution name shall be selected for address matching; If not, the address is matched by the method of "place name + institution name".

For example, in the sentence "The Taiwan industrial park in Gaocheng Town, Gaoqing County, Zibo City, has serious chemical gas emissions. Please investigate". only had the place name, so, we just use the place name "Taiwan industrial park, Gaocheng Town, Gaoqing County, Zibo City" to match the address, if it contains only institution name is the same situation. But if this sentence is changed to "The xx factory in the Taiwan Industrial Park of Gaocheng Town, Gaoqing County, Zibo City, has serious chemical gas emissions. Please investigate.", among them, the xx factory is institution name, the Taiwan Industrial Park of Gaocheng Town, Gaoqing County, Zibo City is place name, because the institution name does not contain the place name, so, we should use the "place name + institution name" to match the address.

Finally, we used geocoding to convert address information to coordinate, and located the address matching result on the map to realize the spatialization of atmospheric pollution public opinion information by making heat map.

4 Spatialization and Comparative Analysis of Atmospheric Pollution Information in Shandong Province

4.1 Spatialization of Public Opinion Information

This paper selected 5,028 effective public complaints from February 2017 to October 2018 on the public complaint platform of Shandong provincial environmental protection department, processed and analyzed the data according to the above data processing flow. Among them, in the data preprocessing, the Chinese word segmentation efficiency was improved by removing special symbols, spaces and stop words. After part of speech tagging, the uniqueness of the complaint pollution site was ensured by repeating search the combined place names, and set filtering combination condition to get the structured address information. Finally, the structured address information was converted into coordinates by address matching, which realizes the spatialization of the complaint location.

In the obtained address information and address matching result, we extracted 100 pieces of data for manual verification, judged the accuracy of address extraction and address matching. The result of verifying is that the accuracy of the address extraction to the city level was 86% and the village level was 79%, the accuracy of the address matching was 92%.

In order to display the distribution of complaints about atmospheric pollution in the cities of Shandong Province from a macro perspective, statistics were carried out by districts and counties, we used the nuclear density analysis function of GIS to make the heat map of atmospheric pollution complaints distribution in Shandong Province, the distribution is shown in Fig. 2. It can be seen from the spatialization results that the key areas of public complaints are Zibo, Weifang, Jinan, Laiwu and Zaozhuang. The overall situation in the eastern coastal areas is better, but there are also some places where public complaints are concentrated. Such as the urban area of Qingdao, Yantai and other cities.

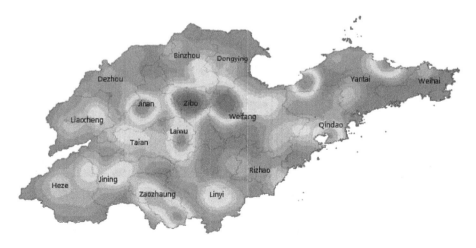

Fig. 2. Nuclear density analysis

4.2 Comparative Analysis of Public Opinion Macro Spatialization Results and Published Data

In order to verify the consistency between the results of the spatial analysis of public opinion and the data published by authoritative organizations or departments, we obtained the atmospheric quality data in 2017 and January to October of 2018 released by the Shandong Provincial Department of Ecology and Environment (as shown in Table 3 and Fig. 3). It can be seen from the comparison with Table 3 and Fig. 3 that the results of public opinion data analysis are basically consistent with those published by authoritative organizations, only the situation in northwestern and southwestern of Shandong province is deviated. The reason may be related to the public's awareness of rights protection and the state of regional economic development.

Table 3. 2017 atmospheric quality ranking of Cities in Shandong Province

City	Provincial ranking	National ranking	City	Provincial ranking	National ranking
Weihai	1	71	Jinan	10	339
Yantai	2	123	Zaozhuang	11	345
Qingdao	3	157	Zibo	12	346
Rizhao	4	256	Laiwu	13	351
Jining	5	315	Binzhou	14	352
Taian	6	316	Dezhou	15	366
Dongying	7	319	Heze	16	367
Weifang	8	325	Liaocheng	17	372
Linyi	9	329			

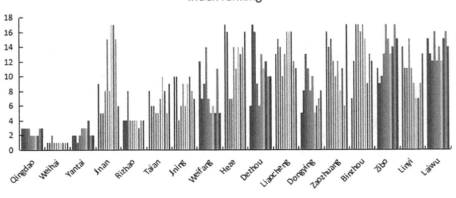

Fig. 3. Shandong Province January–October 2018 comprehensive atmospheric index ranking

4.3   City-Level Comparative Analysis of Public Opinion Information

According to the complaint categories, the content of public complaints mainly focuses on the exhaust emission of industrial and mining enterprises, especially in some cities with poor atmospheric quality, emissions account for a large proportion. In Zibo City, where complaints are more serious, there are many factories and enterprises that produce serious exhaust pollution. The main components of the exhaust gas are: sulfur dioxide, carbon monoxide and nitrogen dioxide. Figure 4 shows the monthly trend data of the Zibo city Atmospheric Quality Index (AQI), SO2, CO, and NO2 obtained from the Shandong Provincial Department of Ecology and Environment and synchronized with the public opinion data. It can be seen from Fig. 4 that the various pollution indicators of atmospheric quality in Zibo City shows certain cyclical changes, especially in December to January of the following year, the most serious atmospheric pollution in January, through comparative analysis with public complaint data. It also reflects that the public complaints are also the most concentrated during this period.

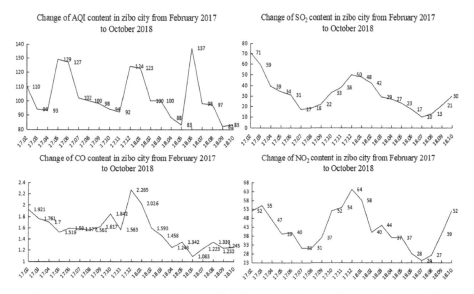

Fig. 4. Atmospheric quality data of Zibo City from February 2017 to October 2018

In addition, most of the atmospheric quality data were released by the state come from national monitoring stations, which are sparsely distribution. Therefore, it is unable to effectively find some local atmospheric pollution events. As shown in Fig. 5, the monitoring stations in Yantai City are mainly distributed in the northeastern coastal area. The atmospheric quality status of Yantai City monitored by these stations ranked second in Shandong Province and ranked 123 in the country. Figure 6 shows the distribution of public complaint information in the Yantai City, and there are cases where chemical pollution and dust pollution complaints are concentrated in Laizhou.

Fig. 5. Distribution of atmospheric quality monitoring stations in Yantai City

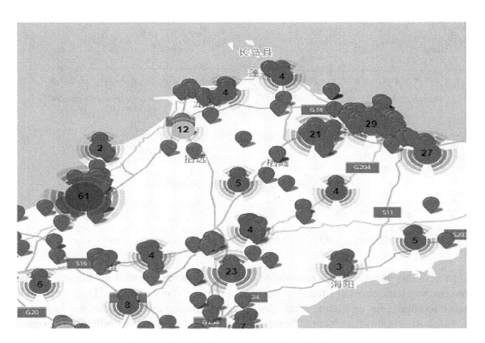

Fig. 6. Yantai City complaint data distribution

This indicates that there are illegal emissions of toxic and harmful gases and illegal mining incidents. Therefore, by spatializing the public opinion information of public complaints, it can effectively reflect the local atmospheric pollution situation at a smaller time and space scale, and can achieve a strong complement to the results of the national control monitoring station.

5   Conclusion

For national monitoring sites cannot effectively monitor regional air pollution events at smaller spatial and temporal scales, this paper proposed a spatialization method based on NLP for atmospheric quality public opinion data, through Chinese word segmentation, part of speech tagging and address matching, realized the spatial localization and transformation of public complaint data by means of heat map. Through further comparison and analysis with the pollution result data released by the national authority, it shows that the proposed method can realize the dynamic monitoring of air pollution status from a finer scale, and can effectively improve the public's participation in air quality monitoring, which can provide a useful supplement for atmospheric environmental monitoring at national control stations.

Acknowledgments. This work was supported in part by a grant from the Major Science and Technology Innovation Projects of Shandong Province (2019JZZY020103) and the National Science Foundation of China (41471330).

References

1. Greene, J.S., Kalkstein, L.S., Ye, H., et al.: Relationships between synoptic climatology and atmospheric pollution at 4 US cities. Theoret. Appl. Climatol. **62**(3–4), 163–174 (1999). https://doi.org/10.1007/s007040050081
2. Pantavou, K., Lykoudis, S., Psiloglou, B.: Air quality perception of pedestrians in an urban outdoor Mediterranean environment: a field survey approach. Sci. Total Environ. **574**, 663–670 (2017)
3. Song, J.G., Guo, M.Y., Yin, G.B.: Urban air quality management satisfaction evaluation method and case study. Environ. Pollut. Control **33**(09), 81–86 (2011)
4. Chen, S.: The development of government service under the era of "Internet +" [J/OL]. Electr. Technol. Softw. Eng. (22), 19 (2018)
5. Wang, M.M., Wang, J.L.: Spatialization of township-level population based on nighttime light and land use data in Shandong Province. J. Geo-Inf. Sci. **21**(05), 699–709 (2019)
6. Yin, Y.J., Liu, H., Ye, L.: Application of data cleaning and spatial visualization in floating car data processing. Geosp. Inf. **17**(05), 116–119+6 (2019)
7. Xu, Y.L.: Review of natural language processing based on deep learning. In: The 22nd Network New Technology and Application Annual Conference Proceedings. China Computer User Association Network Application Branch, Beijing Union University Beijing Information Service Engineering Key Laboratory, p. 4 (2018)
8. Li, S.: Research and development of natural language processing. J. Yanshan Univ. **37**(05), 377–384 (2013)

9. Ma, L.B., Gong, J.Y.: Application of spatial information natural language query interface. Editor. Board Geomatics Inf. Sci. Wuhan Univ. (03), 301–305 (2003)

10. Matci, K., Dilek, A.U.: Address standardization using the natural language process for improving geocoding results. Comput. Environ. Urban Syst. S0198971517300455 (2018)

11. Xu, P.L., Wang, Y., Huang, Y.K.: Chinese place-name address matching method based on large data analysis and Bayesian decision. Comput. Sci. **44**(09), 266–271 (2017)

12. Cheng, C.X., Yu, B.: A rule-based segmenting and matching method for fuzzy Chinese addresses. Geogr. Geo-Inf. Sci. **27**(03), 26–29 (2011)

13. Li, X.F., Song, Z.L., Chen, X.Y.: Research and implementation of fuzzy matching for K-tree address. Bull. Surv. Mapp. (09), 126–129+155 (2018)

14. Xu, M.H.: User comments data model and information processing. Inf. Technol. Inf. (03), 147–149 (2019)

15. Xu, W.: Research and implementation for Chinese Lexicon analysis system based on neural network. Harbin Institute of Technology (2017)

16. Zong, C.Q.: Statistical Natural Language Processing, 2nd edn, p. 135. Tsinghua University Press, Beijing (2013)

17. Chegn, Y.S., Shi, Y.T.: Domain specific Chinese word segmentation. Comput. Eng. Appl. **54**(17), 30–34+109 (2018)

18. Zheng, J.: Principles and practice of NLP Chinese natural language processing. Publishing House of Electroni2017 Air Quality Ranking of Cities in Shandong Provinces Industry, Beijing, p. 196 (2017)

19. Wang, X., Sun, W.W., Hui, Z.F.: Research on Chinese semantic role labeling based on shallow parsing. J. Chin. Inf. Process. **25**(01), 116–122 (2011)

Extension and Expression of OWL Modeling Primitives Based on Spatio-Temporal Ontology of Ocean Flow Field

Min Ji[1], Qingsong Shi[1(✉)], Fengxiang Jin[2], Ting Li[1], and Yong Sun[1]

[1] College of Geomatics, Shandong University of Science and Technology,
Qingdao 266590, China
1021428690@qq.com
[2] College of Surveying and Geo-Informatics, Shandong Jianzhu University,
Jinan 250101, China

Abstract. For the issue that it was difficult to describe and express the dynamic, multi-dimensional and time-varying characteristics of ocean flow field by existing ontology language, by introducing spatio-temporal ontology, the paper constructed a three-tier architecture which was composed of ocean flow field concept ontology, spatio-temporal object ontology and spatio-temporal process ontology. And based on this, the paper proposed a spatio-temporal ontology model for the dynamic variability characteristics of ocean flow field, and realized the semantic expression of ocean flow domain knowledge. Based on the semantic analysis of ocean flow field spatial primitives and temporal primitives, the paper established the criteria for determining the topological relationship, the sequential relationship, the distance relationship and the temporal relationship between ocean flow field objects. In order to give the formal expression of the spatial-temporal ontology for ocean flow field, based on Web Ontology Language (OWL) syntactic rules, the paper constructed spatio-temporal predicates and terms, extended OWL spatio-temporal modeling key nodes, and formed Ocean Current Spatio-temporal Web Ontology Language (OST OWL). And taking the cold vortex as an example, the spatio-temporal process has modeled. It showed that OST OWL can express ocean flow field more effectively.

Keywords: Ocean flow field · Ontology modeling · Spatio-temporal primitives · OST OWL

1 Introduction

With the continuous development of ocean stereo detection technology, human beings have acquired a large amount of ocean observation data. However, due to the characteristics of multi-source and heterogeneous ocean observation data, semantic ambiguity and difficulty of sharing ocean data were caused. Ontology as a formal, explicit specification of a shared conceptualization, it is conducive to the formal expression and reasoning of knowledge [1]. For this reason, domestic and foreign scholars have proposed some ontology models for expressing Marine domain knowledge, such as the

© Springer Nature Singapore Pte Ltd. 2020
Y. Xie et al. (Eds.): GSES 2019/GeoAI 2019, CCIS 1228, pp. 321–335, 2020.
https://doi.org/10.1007/978-981-15-6106-1_24

Bay knowledge ontology, Marine disaster domain ontology, Marine ecology ontology, Maritime accident ontology, etc. [2–6]. The construction of these Marine domain ontology models provides important theoretical support for the integration and sharing of Marine domain knowledge. However, due to the inherent multi-dimensional and time-varying characteristics of ocean flow field, the existing ontology models in ocean domain fail to effectively describe and define this spatio-temporal feature. Moreover, although OWL, the most widely used ontology description language, can better express the concepts and relationships defined in the ontology model, it has obvious deficiencies in rule representation and spatio-temporal expression. In order to realize the spatio-temporal semantic description and formal expression of ocean flow field, it is imperative to conduct spatio-temporal ontology modeling of ocean flow field and expand spatio-temporal modeling key nodes.

At present, most of the existing OWL extensions were based on probability theory and fuzzy theory. For example, Ding [7] realized the probability extension of OWL by transforming OWL ontology into a Bayesian network. Gao [8] introduced fuzzy description logic as a theory combining with OWL, and carried out theorems and constraints of some concepts. Li [9] used the theoretical system of fuzzy logic to extend the rules of Horn clause in OWL, and proposed OWL expression based on the extension of fuzzy logic. Starting from the limitation of fuzzy knowledge, Stoilos [10] created the fuzzy OWL (f-owl) and proposed the basic fuzzy extended grammar and semantics of OWL. Due to the lack of formal expression of the spatio-temporal ontology in these OWL extensions, to this end, based on the spatio-temporal ontology modeling of the ocean flow field, through the analysis of the spatial and temporal primitives of the ocean flow field, the OST OWL is constructed based on the OWL syntax rules, thereby implementing the extension of the OWL spatio-temporal modeling key nodes, it provides theoretical and methodological support for the spatio-temporal semantic sharing and knowledge reasoning of ocean flow field.

2 Spatio-Temporal Ontology Modeling of Ocean Flow Field

Spatio-temporal ontology is a normative system to describe the concepts of spatio-temporal objects and the relationships between them, a method to classify the characteristics of spatio-temporal objects, and a tool to share and reuse standard spatio-temporal knowledge [11]. Based on the idea of spatio-temporal ontology database model [12] and fuzzy spatio-temporal ontology [13], this paper constructs the spatio-temporal ontology of ocean flow field from the three-tier architecture of ocean flow field concept ontology, ocean flow field spatio-temporal object ontology and ocean flow field spatio-temporal process ontology. Among them, the concept ontology of ocean flow field is the abstraction of the ocean flow field. obtains the concept knowledge of the ocean flow field domain, and realizes the expression of properties, relations and instances according to the concept; the spatio-temporal object ontology of ocean flow field is based on the instance of ocean flow field concept to study the knowledge of spatial and temporal domain; the spatio-temporal process ontology of ocean flow field is based on the spatio-temporal objects of ocean flow field, to study the process changes of the objects, and to realize the expression of the life cycle of the spatio-temporal objects of ocean flow field.

2.1   Construction of the Ocean Flow Field Concept Ontology

The concept of ocean flow field has divided into five aspects: the nature, location, origin, scale and flow state of ocean flow. Warm Current, Cold Current and Natural Current can be classified according to the nature of ocean flow field, Nearshore Current, Boundary Current and Ocean Currents can be divided according to the position of ocean flow field, and the divided ocean flow field can be subdivided according to different factors, so as to realize the definition of the ocean flow field concept hierarchy. On this basis, by adding the properties of concepts, the instance of concepts and the relationships between concepts, the construction of ocean flow field concept ontology is realized.

Definition 1. The basic structure of ocean flow field concept ontology can be defined as a quintuple:

$$O_{stc} = (C, P^C, R, I, A) \tag{1}$$

In the formula, C denotes the concept set, including all the concepts in the field of ocean flow field, such as wind driven current, geostrophic current, etc. P^C denotes the property set of concept, which represents the properties of ocean flow field concept, such as temperature, direction, etc. and $P^C = (c, p, v)$, where $c \in C$ is the concept of ocean flow field concept ontology, p is the property contained in c, and v is the corresponding property value of p. R denotes the relationship set between concepts, referring to the semantic relationship between concepts, such as kind-of relationship, part-of relationship, etc. I denotes the instance set, that is the set of objects in each concept set. For example, instances of warm current include Japan warm current, north Pacific warm current, etc. A denotes an axiom set used to constrain concepts and relationships.

Above, the concept ontology of ocean flow field has formally defined in the form of quintuple, and the concept ontology of Geostrophic Current can be expressed as follows: $O_{Geostrophic\ Current} = (\{Geostrophic\ Current\}, \{Geostrophic\ Current, [Flow\ Direction, Flow\ Rate], [Northern\ hemisphere\ to\ the\ right, Decreases\ with\ depth]\}, \{[Ocean\ Flow\ Field, kind-of], [Wind\ Driven\ Current, disjoint]\}, \{Kuroshio\}, \{Ocean\ Flow\ Field\ contains\ Geostrophic\ Current, Geostrophic\ Current\ and\ Wind\ Driven\ Current\ do\ not\ intersect\})$. Figure 1 is the Ontology Model Diagram of "Geostrophic Current" and "Wind Driven Current".

2.2   Construction of the Ocean Flow Field Spatio-Temporal Object Ontology

The spatio-temporal object ontology of ocean flow field is a spatio-temporal ontology modeling of ocean flow field objects, which is mainly used to express the spatio-temporal properties and spatio-temporal relationships of ocean flow field objects.

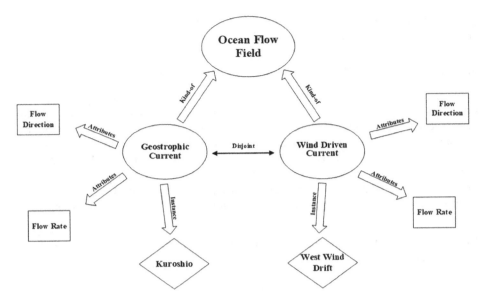

Fig. 1. Ontology model diagram of "geostrophic current" and "wind driven current"

Among them, spatial properties include: spatial location, spatial form and spatial scale; temporal properties including: year, month and day, etc.; spatial relations include: topological relations, sequence relations and distance relations; temporal relations include: disjointing, meeting, overlapping, etc.

Definition 2. The basic structure of ocean flow field spatio-temporal object ontology can be defined as a quad:

$$O_{sto} = (C_o, P_o, S_o, T_o) \tag{2}$$

In the formula, C_O denotes the object set, corresponding to the instance set in the ocean flow field concept ontology; P_O denotes the property set, which can be expressed as $P_O = (P^S, P^T)$, where P^S denotes the spatial properties of the object and P^T denotes the temporal properties; S_O denotes the spatial relation set, which can be expressed as $S_O = (t^r, s^r, d^r)$, where t^r is the topological relation, s^r is the sequence relation, and d^r is the distance relation; T_O denotes the set of temporal relations, mainly including: disjointing, meeting, overlapping, etc.

2.3   Construction of the Ocean Flow Field Spatio-Temporal Process Ontology

Due to the dynamic, periodic and time-varying characteristics of ocean flow field, the construction of spatio-temporal object ontology of ocean flow field can not effectively describe the periodic process change of ocean flow field. Therefore, it is necessary to introduce the ontology of ocean flow field spatio-temporal process and to analyze the ocean flow field life cycle process in order to realize the spatio-temporal process expression of ocean flow field.

Definition 3. The spatio-temporal process ontology of ocean flow field can be defined as a quad structure:

$$O_{stp} = (O, S, P, I) \tag{3}$$

In the formula, O denotes the spatio-temporal entity of ocean flow field; S denotes the set of property states of ocean flow field spatial-temporal entities, covering all possible states of spatio-temporal entity objects; P denotes the rule state sequence of process change, which is the constraint set of spatio-temporal process change and expresses the relationship between states. I denotes the set of instances of the corresponding sequence. In the ocean flow field, most of the ocean phenomena follow the life cycle process of generation, expansion, stability, weakening and extinction. In the ocean flow field, most of the ocean phenomena follow the life cycle process of generation, expansion, stability, weakening and extinction. The Spatial-temporal Process Ontology of Ocean Flow Field is shown in Fig. 2.

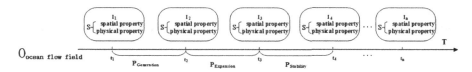

Fig. 2. The spatial-temporal process ontology of ocean flow field

3 Spatio-Temporal Primitives Analysis of Ocean Flow Field

Spatial and temporal primitive analysis of objects mainly includes two aspects: spatial and temporal. There are two kinds of spatial primitives between objects: point-based and region-based, which reflect the spatial feature information by describing the spatial relationship between objects; and temporal primitives between objects are also two kinds: instants-based and intervals-based, which are mainly used to represent the characteristics of object state changes [14].

3.1 Ocean Flow Field Spatial Primitives

1) Spatial Relations

Topological Relationship (TR) has used to describe topological invariants of spatial objects in the case of translation, scale transformation, etc. [15]. which is the most common expression of spatial relation in ocean flow field.

Topological spatial relations of ocean flow field include three kinds of relationships between ocean flow field and land, such as $CL_{Disjoint}$, CL_{Touch} and $CL_{Contain}$, and the relationships among four kinds of ocean flow field objects, such as $C_{Disjoint}$, C_{Touch},

$C_{Overlap}$ and $C_{Contain}$. Based on part-of (x, y) theory and topological intersection model theory, two objects A and B were selected, and each object has divided into two parts: boundary X^- and internal X^+. The discrimination of spatial topological relation of ocean flow field in this paper was shown in Table 1:

Table 1. Discrimination of topological relations in ocean flow field

Relational name	Formal discrimination	Description
Disjoint	Disjoint(A,B):= Part-of(C,A) $\wedge$ Part-of(C,B)=0	Non-existent object C, making C $\subseteq$ A, C $\subseteq$ B
Touch	Touch(A,B):= Part-of(A$^-$,A) $\wedge$ Part-of (B$^-$,B)=1	Two objects are intersected only by boundary line
Overlap	Overlap(A,B):= Part-of(C,A) $\wedge$ Part-of(C,B)=C	Existing object C, making C $\subseteq$ A, C $\subseteq$ B
Contain	Contain(A,B):= Part-of(B,A$^+$)	B $\subseteq$ A$^+$

Sequential Relationship (SR) has used to describe the relative position information between spatial objects [15]. "Projection-based" azimuth model has adopted to classify the sequential relationships between ocean flow field objects, including eight sequential relationships: East, West, South, North, NE, SE, NW, and SW. The sequential relationship has determined by comparing the longitude and latitude values of the objects. In this paper, longitude and latitude were instantiated as coordinate axes, with the prime meridian as the longitudinal axes of longitude and latitude coordinate axes, the north latitude as positive, the south latitude as negative, the equator as the horizontal axes of longitude and latitude coordinates, the east longitude as positive, and the west longitude as negative. Set the longitude and latitude coordinates of object A to be (X_A, Y_A), and the longitude and latitude coordinates of object B to be (X_B, Y_B), The discrimination of spatial sequential relation of ocean flow field in this paper was shown in Table 2:

Table 2. Discrimination of sequential relationship in ocean flow field

Relational name	Formal discrimination	Description
East	$X_A > X_B, Y_A = Y_B$	Object A is positioned East of Object B
West	$X_A < X_B, Y_A = Y_B$	Object A is positioned West of Object B
South	$X_A = X_B, Y_A < Y_B$	Object A is positioned South of Object B
North	$X_A = X_B, Y_A > Y_B$	Object A is positioned North of Object B
NE	$X_A > X_B, Y_A > Y_B$	Object A is positioned NE of Object B
SE	$X_A > X_B, Y_A < Y_B$	Object A is positioned SE of Object B
NW	$X_A < X_B, Y_A > Y_B$	Object A is positioned NW of Object B
SW	$X_A < X_B, Y_A < Y_B$	Object A is positioned SW of Object B

Distance Relationship (DR) has used to describe the relative position between spatial entities and to reflect the proximity between adjacent objects [16]. The distance relationship between ocean flow field objects can be divided into qualitative (Fig. 3) distance and quantitative distance, in which the qualitative distance includes three elements: target object A, reference object B and reference frame. With reference object B as the center of the circle, several concentric circles are divided, the qualitative distance between A and B can be divided into five relationships according to the location of A's concentric circle: VeryNear, Near, Moderate, Far and VeryFar. The quantitative distance can be expressed by "numerical value + length unit", such as "five miles apart".

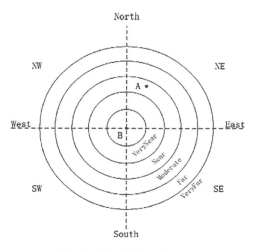

Fig. 3. Qualitative distance

2) Spatial Location (SL)

The expression of ocean flow field spatial location can be divided into point-based longitude and latitude coordinates and area-based "node" longitude and latitude coordinates. Point-based longitude and latitude coordinates are used to express the longitude and latitude coordinates of the object by taking the center point or critical point of the object as the mark point of the object. The longitude and latitude coordinate set based on the region instantiates the ocean flow field region into a polygon, determines the longitude and latitude of each vertex of the polygon, and expresses it in the form of a set. For example, "an anticyclonic vortex located at 119°E and 20°N on the southwest side of Taiwan island", in the example, the spatial position of the vortex is represented by the longitude and latitude coordinates of the central point of the vortex.

3.2   Ocean Flow Field Temporal Primitives

Time is an abstract concept, which has the characteristics of dynamic, multi-scale and uncertainty. Dynamicity mainly refers to the dynamic change characteristics, namely the timing characteristics; Multi-scale was shown in the length of time series, such as the change of time series of year, month and day; The uncertainty manifests itself in the imprecision of the timing stages.

1) Time Granularity (TG)

Generally speaking, time has defined as absolute time and relative time [17], and time can be expressed directly or indirectly through time points and time slots of time elements. In the real world, time points and slots are generally instantiated by time granularity (year, month, day, etc.). Time granularity can express both time scale and absolute time.

The ocean flow field was a dynamic change process, which can be abstracted as a process from generation to extinction. Based on the analysis of life evolution process of process objects by Xue [18], we divide the life cycle of ocean flow field into five stages: Generate, Expand, Stabilize, Weaken and Die.

According to the characteristics of ocean flow field, the expression of time granularity has divided into four forms: 1) Absolute Time: such as November 23, 2019; 2) Relative Time, such as 1 week before, 2 months after; 3) Specific Time, such as 2019–2020; 4) Duration, such as 1 year, 2 weeks. Its concrete manifestation was shown in Table 3:

Table 3. Ocean flow field time granularity expression

Time granularity	Description	Instance
Year	Indicates the year	2019 Year
Years	Indicates the number of years	10 Years
	N years ago	10 Years Ago
	N years later	10 Years Later
Month	Indicates the month (Value 1–12)	5 Month
Months	Indicates the number of months	5 Months
	N months ago	5 Months Ago
	N months later	5 Months Later
Day	Indicates the day (Value 1–31)	28 Day
Days	Indicates the number of days	28 Days
	N days ago	28 Days Ago
	N days later	28 Days Later

2) Temporal Relationship

Temporal Relationship (TempR) has used to describe the relationship between time periods (time points). There were three kinds of time relations between time points: Before, After and Equal. As for the expression of time slot, according to the 13 kinds of time expression relationships proposed by Allen [19], this paper divides seven kinds of relationships among time slots belonging to the ocean flow field: TDisjoint, TMeet, TOverlap, TStart, TContain, TEnd and TEqual. Let's say there are two time periods T1 and T2, where t_s^1 represents the starting time of time period T1, t_e^1 the ending time of time period T1, t_s^2 the starting time of time period T2, and t_e^2 the ending time of time period T2, then the temporal relationship of ocean flow field was shown in Table 4:

Table 4. Discriminating temporal relationship of ocean flow field

Relational name	Formal discrimination	Description
TDisjoint	$t_e^1 < t_s^2$	T1 end time earlier than T2 start time
TMeet	$t_e^1 = t_s^2$	T1 end time is equal to T2 start time
TOverlap	$(t_s^1 < t_s^2) \cup (t_s^2 < t_e^1 < t_e^2)$	T1 start time earlier than T2, and T1 end time between T2 start time and T2 end time
TStart	$t_s^1 = t_s^2$	T1 start time at the same time as T2 start time
TContain	$(t_s^1 > t_s^2) \cup (t_e^1 < t_e^2)$	T1 start time is later than T2 start time, T1 end time is earlier than T2 end time
TEnd	$t_e^1 = t_e^2$	T1 end time equals T2 end time
TEqual	$(t_s^1 = t_s^2) \cup (t_e^1 = t_e^2)$	T1 start time is equal to T2 start time, T1 end time is equal to T2 end time

3.3   Spatial-Temporal Relationship of Ocean Flow Field

At present, the study of spatio-temporal model, mainly in the following two approaches [14]: 1) to make orthogonal combination of spatial model and temporal model; 2) to attempt to establish a unified model of spatio-temporal. The former is more closely combined with the existing research and easier to realize, the latter is more perfect than the former, but it is difficult to set up.

Because the ocean flow field was a dynamic entity, there were complex spatial and temporal properties, and the spatial properties change with the temporal properties. A single analysis of spatial properties or temporal properties can not reflect the temporal and spatial characteristics of the ocean flow field, so the establishment of the temporal and spatial model of the ocean flow field has essential. Based on the knowledge of spatial and temporal of ocean flow field, this paper combines 4 spatial topological relationships, 8 spatial sequential relationships, 6 spatial distance relationships and 7 temporal relationships of ocean flow field object, and obtains 1344 possible spatial and temporal relationships in ocean flow field. Setting two ocean flow field objects A and B, then the spatio-temporal relationship between ocean flow field objects can be expressed as A{(topological relationship, sequential relationship, distance relationship), temporal relationship}B. Table 5 lists the spatio-temporal relationships among some ocean flow field objects:

Table 5. Spatio-temporal relationship between ocean flow field objects

Time	Space						
	A{D}B	A{M}B	A{O}B	A{S}B	A{C}B	A{End}B	A{E}B
A{d, East, Far}B	A{(d, East, Far), D}B	A{(d, East, Far), M}B	A{(d, East, Far), O}B	A{(d, East, Far),S}B	A{(d, East, Far), C}B	A{(d, East, Far), End}B	A{(d, East, Far), E}B
A{t, East, Far}B	A{(t, East, Far), D}B	A{(t, East, Far), M}B	A{(t, East, Far), O}B	A{(t, East, Far), S}B	A{(t, East, Far), C}B	A{(t, East, Far), End}B	A{(t, East, Far), E}B
A{o, East, Far}B	A{(o, East, Far), D}B	A{(o, East, Far), M}B	A{(o, East, Far), O}B	A{(o, East, Far), S}B	A{(o, East, Far), C}B	A{(o, East, Far), End}B	A{(o, East, Far), E}B
A{c, East, Far}B	A{(c, East, Far), D}B	A{(c, East, Far), M}B	A{(c, East, Far), O}B	A{(c, East, Far), S}B	A{(c, East, Far), C}B	A{(c, East, Far), End}B	A{(c, East, Far), E}B

4 OWL Extension and Formal Description

4.1 OWL Extension

OWL is a knowledge representation language composed of syntactic rules, vocabulary and semantics [20]. To realize the extension of OWL, new syntactic rules, vocabulary or semantics need to be constructed. Following the OWL syntactic rules, this paper establishes an Ocean Flow Field Spatio-temporal Ontology Language Layer on the basis of OWL Layer, constructs new spatio-temporal predicates and terms, and realizes the spatio-temporal formal expression of ocean flow field objects. The naming space of ocean flow field object spatio-temporal has defined as "ost-owl", the naming space of ocean flow field object spatial has defined as "ost-Spatial", and the naming space of ocean flow field object temporal has defined as "ost-Time".

1) Spatio-temporal predicates extension

Description Logic is a concept-based knowledge representation method that is a decidable subset of first-order predicate logic [21, 22]. OWL is an ontology description language based on descriptive logic, its class constructor and axioms have corresponding representations with descriptive logic. Based on this, through the existing ocean flow field spatio-temporal primitives, implement OWL spatio-temporal predicate extension.

This paper mainly expresses the spatial topological relationship between objects by defining "ost-Spatial:TR". For example, "ost-Spatial:Disjoint" declares that the spatial topological relationship between objects is phase separation. Similarly, "ost-Spatial:

SR" is defined to express the order relationship between objects, and "ost-Spatial:DR" is defined to express the distance relationship between objects. The following table lists some spatial predicates, where θ represents azimuth, d represents qualitative distance, and the result of qualitative distance is i(i=1,2,...5) represents the hierarchical relationship divided by reference objects. The extension of OWL spatial predicates was shown in Table 6:

Table 6. OWL spatial predicates expansion

Keyword	Syntax	Description	Formal expression

ost-Spatial:Disjoint	$A \cap B = \emptyset$	Declare that object A is disjoint from object B	`<ost-owl:Class rdf:ID="A">`
			`<ost-Spatial:Disjoint rdf:resource ="#B"/>`
			`</ost-owl:Class>`

ost-Spatial:Contain	$A \subseteq B$	Declares that object A is contains with object B	`<ost-owl:Class rdf:ID="A">`
			`<ost-Spatial:Contain rdf:resource="#B"/>`
			`</ost-owl:Class>`

ost-Spatial:West	$\theta.(A, B) = 180°$	Declares that object A is west of object B	`<ost-owl:Class rdf:ID="A">`
			`<ost-Spatial:West rdf:resource ="#B"/>`
			`</ost-owl:Class>`

ost-Spatial:SE	$\theta.(A, B) \subseteq (270°, 360°)$	Declares that object A is southeast of object B	`<ost-owl:Class rdf:ID="A">`
			`<ost-Spatial:SE rdf:resource="#B"/>`
			`</ost-owl:Class>`

ost-Spatial:VeryNear	$d.(A, B) = 1$	Declare that object A is very near from object B	`<ost-owl:Class rdf:ID="A">`
			`<ost-Spatial:VeryNear rdf:resource="#B"/>`
			`</ostowl:Class>`

ost-Spatial:VeryFar	$d.(A, B) = 5$	Declare that object A is very far from object B	`<ostowl:Class rdf:ID="A">`
			`<ost-Spatial:VeryFar rdf:resource="#B"/>`
			`</ost-owl:Class>`

...	...	...	...

The expansion of OWL temporal predicate is based on the temporal relationship between ocean flow field objects, and describes the possibility of temporal relationship between ocean flow field objects by defining "ost-owl:Restriction+ost-Time:onProperty+ost-Time:TempR". The specific expansion of temporal predicates was shown in Table 7:

2) Terminological lexical expansion

In OST OWL, besides expressing the temporal-spatial relationship between ocean flow field objects in the form of predicates, there are specific terms to describe the spatial and temporal knowledge. Such as "ost-Spatial:longitudeValue", "ost-Spatial:latitudeValue" and "ost-Spatial:llCoordinates" were defined to represent the longitude coordinates, latitude coordinates and longitude-latitude coordinates set of the object respectively; "ost-Spatial:SS" was defined to represent the spatial scale of the object; "ost-Time:TG" was defined to represent the time granularity of the object; "ost-Time:

Table 7. OWL temporal predicates expansion

Keyword	Description	Formal expression

ost-Time:TDisjoint	Declares that the life cycle of object A is disjoint from that of object B	`<ost-owl:Class rdf:ID="A">`
		`  <ost-owl:Restriction>`
		`    <ost-Time:onProperty rdf:resource="#Lifecycle">`
		`    <ost-Time:TDisjoint rdf:resource="#B"/>`
		`  </ost-owl:Restriction>`
		`</ost-owl:Class>`

ost-Time:TMeet	Declares that the life cycle of object A is meets from that of object B	`<ost-owl:Class rdf:ID="A">`
		`  <ost-owl:Restriction>`
		`    <ost-Time:onProperty rdf:resource="#Lifecycle">`
		`    <ost-Time:TMeet rdf:resource="#B"/>`
		`  </ost-owl:Restriction>`
		`</ost-owl:Class>`

ost-Time:TOverlap	Declares that the life cycle of object A is overlap from that of object B	`<ost-owl:Class rdf:ID="A">`
		`  <ost-owl:Restriction>`
		`    <ost-Time:onProperty rdf:resource="#Lifecycle">`
		`    <ost-Time:TOverlap rdf:resource="#B"/>`
		`  </ost-owl:Restriction>`
		`</ost-owl:Class>`

...	...	...

Lower" and "ost-Time:Upper" were defined to represent the upper limits of the temporal and the lower limits of the temporal respectively. In addition, "ost:value" was defined to represent attribute values; "ost:maxValue" and "ost:minValue" were defined to represent the maximum and the minimum respectively, etc.

4.2 Example Verification

In literature [23], vortex changes in spring and winter of 2013 in Bashi Strait were studied. The central position of a cold vortex in this sea area was 26°N, 133°E, and its diameter was about 150 km–200 km. After nearly 5 months, the central position moved to 22°N, 123°E on the east side of Bashi Strait. Therefore, the cold vortex has defined as Bashi east cold vortex, and the cold vortex has modeled with OST OWL as follows:

1) Determine the research field and scope

The above is an example of a cold vortex, so the research field is the field of ocean flow field.

2) Concept abstraction and hierarchical structure determination

According to the concept ontology of ocean flow field, the vortex is a subclass of the ocean flow field, the cold vortex is a subclass of vortex, the Bashi east cold vortex is an instance of cold vortex, which is expressed as follows:

```
<ost-owl:Class rdf:ID="Vortex">
        <ost-owl:subClassOf rdf:resource="#OceanFlowField"/>
</ost-owl:Class>
<ost-owl:Class rdf:ID="ColdVortex">
<ost-owl:subClassOf rdf:resource="#Vortex"/>
</ost-owl:Class>
<ColdVortex rdf:ID="BashiEastColdVortex"/>
```

3) Define the attribute characteristics and attribute values of the concept

The attributes of the Bashi East Cold Vortex have shown as spatial attributes and temporal attributes, in which the spatial attributes include the spatial position and spatial scale, and the time attributes have shown as the duration of five months, which were expressed as follows:

```
<ost-owl:Class rdf:ID="BashiEastColdVortex">
        <ost-owl:Restriction>
                <ost-Spatial:onProperty
rdf:resource="#SpatialScale"/>
                <ost-Spatial:Diameter ost:maxValue=200km/>
        </ost-owl:Restriction>
        <ost-owl:Restriction>
                <ost-Spatial:onProperty
rdf:resource="#SpatialScale"/>
                <ost-Spatial:Diameter ost:minValue=150km/>
        </ost-owl:Restriction>
</ost-owl:Class>
        <ost-owl:Class rdf:ID="BashiEastColdVortex">
    <ost-owl:Restriction>
                <ost-Time:onProperty
rdf:resource="#Duration"/>
                <ost-Time:Months ost:value=5/>
        </ost-owl:Restriction>
        </ost-owl:Class>
```

4) Spatio-temporal process expression

The spatio-temporal properties of the ocean flow field have changed with the change of time. Set the cold vortex five months ago to BashiEastColdVortex_A, five months later to BashiEastColdVortex_A'. Through the change of the longitude and latitude position of the cold vortex, it can be determined that A is located in the northeast direction of A', and it is a disjoint relationship, which is expressed as follows:

```
<ost-owl:ObjectProperty rdf:ID="Duration">
                        <rdfs:domain                rdf:re-
source="#BashiEastColdVortex_A"/>
                        <rdfs:range>
```

```
                                            <ost-owl:Restriction>
                                                <ost-Spatial:NE
rdf:resource="#BashiEastColdVortex_A"'/>
                                                <ost-Spatial:Disjoint
rdf:resource="#BashiEastColdVortex_A"'/>
                                            </ost-owl:Restriction>
                                    </rdfs:range>
            </ost-owl:ObjectProperty>
```

5   Conclusion

OWL as one of the ontology description languages, has strong ability of expression, logical judgment and reasoning. However, OWL is often suitable for describing the logical framework and hierarchical structure of "static" knowledge, and there are still deficiencies in the expression of spatio-temporal "dynamic" knowledge like ocean flow field.

Based on the modeling of ocean flow field spatio-temproal ontology, this paper constructs the OST OWL, OST OWL is based on the domain knowledge of ocean flow field, and based on the syntactic rules of OWL language, expands the spatio-temporal primitives of ocean flow field from vocabulary and semantics, and realizes the expression of the spatio-temporal process of ocean flow field. OST OWL not only inherits the strong expression ability of OWL language, but also has a good formal expression ability for the spatio-temporal process of ocean flow field. Due to the fact that this paper does not consider the characteristics of ocean flow field, such as boundary ambiguity and regional ambiguity, the follow-up research still needs to carry out in-depth expansion of OST OWL from the syntactic rule level. At the same time, it is necessary to study the expression of spatio-temporal knowledge in other fields of ocean.

Acknowledgments. This work was support in part by a grant from the Major Science and Technology Innovation Projects of Shandong Providence (2019JZZY020103) and the National Science Foundation of China (41471330).

References

1. Studer, R., Benjamins, V.R., Fensel, D.: Knowledge engineering, principles and methods. Data Knowl. Eng. **25**(1), 164–197 (1998)
2. Du, Y., Zhang, D., Su, F., et al.: Geospatial data organization of the bay based on the geo-ontology—a case study of the Liaodong Bay. J. Geoinf. Sci. **10**(1), 7–13 (2018)
3. Zhu, G., Shi, S., Xu, X., et al.: Research and application of Marine disaster domain ontology. Mar. Inf. (2), 5–8 (2009)
4. Xiong, J.: Research and application of marine ecology ontology modeling. Ocean University of China, Qingdao (2010)
5. Yun, H., Xu, J., Guo, Z., et al.: Modeling of marine ecology ontology. J. Comput. Appl. **34**(4), 1105–1108 (2014)

6. Shao, B.: Design and implementation of maritime accident determination system based on ontology. Dalian Maritime University, Dalian (2010)
7. Ding, Z., Peng, Y.: A probabilistic extension to ontology language OWL. In: Proceedings of the 37th Hawaii International Conference on System Sciences (2004)
8. Gao, M., Liu, C.: Extending OWL by fuzzy description logic. In: Proceedings of the 17th IEEE International Conference on Tools with Artificial Intelligence (2005)
9. Li, M.: OWL rules extension and application research of its reasoning. Tianjin University, Tianjin (2006)
10. Stoilos, G., Stamou, G., Pan, J.Z.: Fuzzy extensions of OWL logical properties and reduction to fuzzy description logics. Int. J. Approx. Reason. **51**, 656–679 (2010)
11. Hou, J., Kong, Y.: Spatio-temporal philosophical thought to geographical information science. Geomat. Spatial Inf. Technol. **36**(1), 18–21 (2013)
12. Chen, X., Li, S., Zhu, J.: Database model based on spatio-temporal ontology. Geogr. Geo-Inf. Sci. **26**(5), 1–6 (2010)
13. Cheng, H.: Research on representation of fuzzy spatio-temporal knowledge with ontology and construction method based on Petri Net. Northeastern University, Shenyang (2013)
14. Liang, X.: Spatio-temporal analysis and reasoning of cyber physical systems. Guangdong University of Technology, Guangzhou (2014)
15. Huo, L.: Research on some problems of complex spatial relations models and spatial description logic. Jilin University, Jilin (2013)
16. Liang, H., Cui, T., Guo, J.: Expression and organization of geospatial spatial relations based on topic maps. J. Tianjin Normal Univ. (Nat. Sci. Ed.) **37**(2), 50–56 (2017)
17. Wei, H.: Research and application of spatio-temporal GIS modeling. Information Engineering University, Zhengzhou (2007)
18. Xue, C., Zhou, C., Su, F., et al.: Research On Process-Oriented Spatio-Temporal Data Model. Acta Geodaetica et Cartographica Sin. **39**(1), 95–101 (2010)
19. Allen, J.F.: Maintaining knowledge about temporal intervals. Commun. ACM **26**(11), 832–843 (1983)
20. Liu, J., Zhang, X., Sun, W.: OWL extension method. Libr. Inf. Sci. **56**(15), 93–98, 125 (2012)
21. Baader, F., Horrocks, I., Lutz, C., Sattler, U.: An Introduction to Description Logic. Cambridge University Press, Cambridge (2017)
22. Fu, H., Deng, L.: Web page clustering based on ALCIF description logic. Mod. Comput. **4**, 41–45 (2019)
23. Wang, S.: Study on small-scale flow characteristics of ocean based on satellite and onsite observation. Dalian Ocean University, Dalian (2017)

Research on Domain Ontology Modeling and Formal Expression of Ocean Flow Field

Min Ji[1], Shenglu Zang[1(✉)], Yong Sun[1], Ting Li[1], and Yaru Xu[2]

[1] Shandong University of Science and Technology, Qingdao 266590, China
1657302996@qq.com
[2] Jinan Real Estate Measuring Institute, Jinan 250001, China

Abstract. With the continuous acquisition of multi-source marine data, the issue of semantic heterogeneity among ocean flow field phenomena had become the main difficulty for data sharing and analyzing of ocean flow field. For the reason that there was still no complete semantic shared knowledge system for ocean flow field at present, based on extensive reference to marine expertise and related standards, with the idea of ontology, the paper established the hierarchical system of ocean flow field domain knowledge. Through the semantic analysis of the relationship between concepts and attributes of ocean flow field domain ontology and the space-time relationship between ocean flow field instances, the paper proposed an ontology expression model based on a six-tuple, and constructed a basic structure that integrated concepts, attributes, relationships and instance sets together. And through extended modeling key words for Web Ontology Language (OWL), taking the western boundary current as an example, the paper gave the formal expression and description for semantic information based on OWL. The paper's research could provide a theoretical basis for knowledge sharing and data mining in the field of ocean flow field.

Keywords: Ocean flow field · Ocean ontology · Semantic analysis · Knowledge description · OWL

1 Introduction

With the continuous development of three-dimensional ocean exploration technology, mankind had acquired an unprecedented ocean big data set. Because the ocean flow field is an important carrier of material and energy exchange in the ocean, the semantic information sharing and migration rule analysis of ocean flow field based on those heterogeneous ocean data is of great significance to human deep understanding of the ocean.

To solve the problem of semantic heterogeneity in Marine domain knowledge, many scholars had carried out a series of researches with the help of ontology. It was Findler and Malyankar who first established the ontology of coastal entities (such as shoreline, tidal surface, etc.) when hosting the project of expression and distribution of spatial knowledge. Bermudez et al. [1] constructed 28 ocean ontologies including most of the key concepts in Marine science in the Marine metadata interoperability project.

© Springer Nature Singapore Pte Ltd. 2020
Y. Xie et al. (Eds.): GSES 2019/GeoAI 2019, CCIS 1228, pp. 336–351, 2020.
https://doi.org/10.1007/978-981-15-6106-1_25

Based on thesaurus, Zhou [2] constructed the Marine domain ontology covering Marine physics, Marine biology, Marine chemistry and other contents. Xiong [3], Yun et al. [4] conducted ontology modeling research in the field of Marine ecology, and proposed the Marine ecological knowledge organization model. Zhang et al. [5] studied the characteristics of Marine culture and used OWL_DL description language to realize the ontology model description of Marine culture domain. Jia Haipeng et al. [6] proposed an ontology model based on Hozo's role theory and established the ocean carbon cycle ontology. Abidi et al. [7] conducted ontology modeling for Marine biological knowledge and data management. Li [8] and Zhang [9] established the Marine knowledge ontology based on ocean remote sensing data sources by applying hybrid ontology. Based on offshore, shore-based and remote sensing observation data, Li [10] established ocean observation ontology and constructed a set of data standardization service supporting NetCDF format. Lu [11] took ontology as the semantic basis of the multidimensional model of data warehouse and built the multidimensional data model based on the ocean hydrological ontology. For the problem of data organization in the polar Marine environment monitoring system, Zhang [12] adopted the idea of ontology to construct an ontology model framework of Marine environment data including spatial relations.

Inspired by the above knowledge ontology modeling in the Marine domain. To realize knowledge sharing and cognitive consistency in the field of ocean flow field. The paper based on the semantic analysis of the properties and spatio-temporal relations of ocean flow field, with the idea of ontology, proposed the ontology structure of ocean flow field based on six-tuple, and gave the Ontology modeling and formal expression of ocean flow field with OWL. The paper's research could provide some references for the expression and mining of knowledge system of ocean flow field.

2 Application of Ontology in the Field of Ocean

2.1 Ontology and Geographical Ontology

Ontology originated from the field of philosophy, and was first proposed by Goclenius, a German scholar in the 17th century. For the reason that there was no unified definition and fixed application field, in 1995 Gruber from Stanford university came up with a widely accepted definition of "Ontology is an accurate description of concepts and is used to describe the essence of things". Ontology can be divided into domain ontology, general ontology, applied ontology and representation ontology. Subsequently, ontology was introduced into artificial intelligence, information system, knowledge system and other fields by the information academia, and many scholars also gave a series of definitions. In 1998, Studer et al. argued that ontology was a clear formal specification of Shared conceptual model, and formed a unified view of ontology. Ontology construction unifies the description of concepts, attributes and relational sets, making it possible to share domain knowledge.

For the reason that geographical objects have the obvious spatio-temporal characteristics, people had gradually introduced ontology into the field of geographic information science, and created a series of concepts and definitions of geographical ontology. In 1998, Mark et al. [13] believed that the core of geographic ontology was the semantic theory of spatial information. Bishr et al. [14], 1998, held that geographic ontology was a geographic spatial model that could build semantic consistency. In 2005, Jing [15] also proposed the definition of semantic field of geographic ontology. In 2007, Li [16] proposed a three-layer overall framework of geographical ontology.

2.2 Marine Domain Ontology

As a part of the world of geography information, ocean has both geographical spatial characteristics and physical features, so ontology also had been gradually introduced into the ocean. In order to solve the integration and sharing of heterogeneous Marine data, from 2004 to 2008, Su Fenzhen, Du Yunyan, Yang Yiaomei, Xue Cunjin and other scholars studied the organization and modeling of various types of Marine data, and proposed the framework method for the integration and sharing of Marine big data. In 2008, Zhang Feng et al. established the mapping relationship between ontology and data sources by studying the semantic heterogeneity of ocean multi-source data. In 2008, Du [17] took Liaodong Bay as an example and established the gulf knowledge ontology. In 2017, Zhang et al. [18] formulated a series of rules of ocean concepts knowledge. In 2009, Zhu [19] established a Shared Marine disaster ontology. In 2010, Xiong Jing developed an optimized information retrieval system based on the ocean ecological ontology. In 2010, Shao [20] constructed the maritime accident ontology. In 2014, Jia Haipeng established the knowledge sharing system of Marine carbon cycle based on ontology. In 2018, Li et al. [21] carried out the construction and formal expression of the local ontology of the ocean flow field phenomenon and the ontology of the ocean space-time process based on the semantic analysis of the spatial and temporal features of the ocean flow field.

3   Establishment of Knowledge Conceptual System of Ocean Flow Field

Ocean current is the main dynamic flow field in the ocean, which is the result of the comprehensive action of turbulence, fluctuation, periodic tidal current and stable "constant current", and it has different temporal and spatial scales and periodic changes. In order to effectively share and organize knowledge in the field of ocean flow field, this paper established a conceptual hierarchical system of knowledge in the field of ocean flow field based on a large number of references to professional literatures and classification standards in the field of ocean flow field, and according to classification standards such as nature, origin, scale, state and region with the idea of ontology. The specific hierarchical conceptual system of ocean flow field was shown in Fig. 1.

(1) Classification by nature: sea water temperature is the basic index to study the properties of water masses and describe the movement of water masses. According to the index of water temperature, ocean currents could be divided into three categories: warm current, cold current and neutral current.

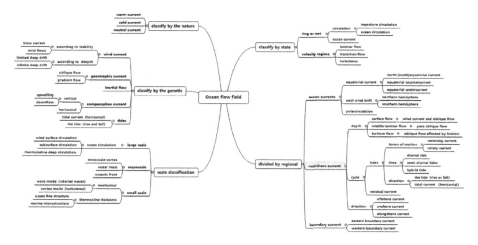

Fig. 1. Concept layered system of ocean flow field

(2) Classification by cause: the causes of seawater flow mainly include the driving effect of wind over seawater(shown in Fig. 2), the horizontal pressure gradient force and coriolis force caused by the superposition of seawater movement and earth movement, the inertial force caused by the wind stress on seawater and the state of seawater itself, the compensation effect caused by seawater flowing at other depths or horizontal positions, and the tide-generating force on seawater on the earth surface. Therefore, ocean currents could be divided into wind currents, geostrophic currents, inertial currents, compensating currents and tides. Among them, the wind current was divided into blowing current and wind current according to stability, and the drift was divided into finite and infinite deep drift according to depth. Geostrophic flow was divided into inclined flow and gradient flow. The compensation flow was divided into vertical and horizontal directions, and the vertical direction includes upward flow and downward flow. Tides were classified into the tides and tidal currents depending on whether the water is moving horizontally or vertically.

(3) Classification by scale: under the influence of the distribution pattern of land and sea, the continuity of sea water, the airflow on the sea surface, and the unequal distribution of heat and salt in sea water, the phenomena of ocean flow field also present different scales. Therefore, ocean flow phenomena could be divided into three scales: large scale, medium scale and small scale. Among them, large-scale refers to thousands of meters in space and months in duration. And ocean circulation could be further subdivided into wind surface circulation, subsurface circulation and thermohaline deep circulation. Mesoscale ocean flow field phenomena could be divided into mesoscale vortices, water masses and ocean peaks. Small scale flow field phenomenon was relatively large ocean flow field and mesoscale flow field, from the superficial, the changes of various physical quantities (temperature, salinity, density and velocity) from the vertical scale of 100 meters to the molecular dissipation scale were in the category of small scale, whose spatial order

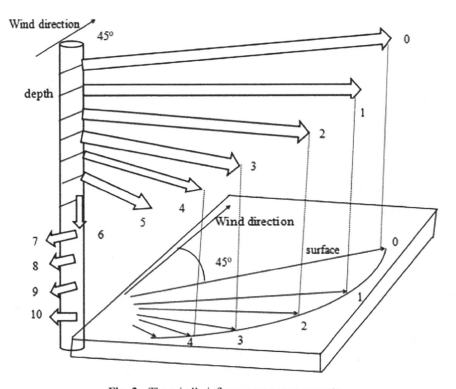

Fig. 2. The wind's influence on ocean currents

is within 100 meters and time span is within a few hours. From the perspective of mechanics, the process of ocean flow field at small scale is divided into two parts: wave mode and vortex mode.

(4) Classification by state: firstly, the flow state of seawater could be divided into circulation and general ocean current according to whether it was a loop, and circulation could be further divided into near-shore circulation and ocean circulation. secondly, according to the Reynolds number, ocean currents could be divided into laminar flow, transitional flow and turbulent flow. The smaller the Reynolds number the more significant the viscous force, the larger the inertial force. When the Reynolds number is less than 2300, laminar flow, also known as steady flow and laminar flow, is formed. When the Reynolds number is more than 4000, the inertial force has a greater influence on the flow field than the viscous force, forming turbulence. Transition flow is a transition state.

(5) Classification by region: according to the position of ocean currents in land-sea relations. Ocean currents could be divided into three categories: nearshore currents, boundary currents and oceanic currents. Among them, the ocean current was divided into equatorial current, westerly drift, polar circulation; Inshore current could be divided into surface flow, middle layer flow and bottom flow according to depth, sea tide and residual flow according to periodicity, and sea tide could be

further classified according to movement form, time and direction, while inshore current could also be classified into offshore flow, shoreward flow and coastal flow according to direction. Boundary flow could be divided into eastern boundary current and western boundary current.

The establishment of the conceptual hierarchy system of ocean flow field phenomena clarified the hierarchical relations, inclusion relations, partial and overall relations among concepts, and provided the framework foundation for the deep semantic analysis of ocean flow field.

4 Semantic Analysis of Ocean Flow Field

4.1 Analysis of Ontology Properties of Ocean Flow Field

There was no obvious boundary between attribute and concept [22]. Some elements could be used as both concepts and attributes. In this paper, concepts that describe the nature or relationship of a certain phenomenon or thing were summarized as category of attribute. Since the ocean flow field has both vector and field characteristics, there is no clear space-time boundary. Before the attributive analysis of ocean flow field domain ontology, attribute types of ocean flow field domain concept were classified into six categories based on the hierarchy system of ocean flow field domain concept established in the previous section. The set of properties of the specific ocean flow field was shown in Fig. 3.

(1) Genetic properties: in physical oceanography, most definitions of various ocean current phenomena were based on their genetic differences, and the main factors affecting the flow of sea water were wind, sea temperature, sea density, sea salinity, etc.

(2) Force properties: force is the source of the motion of the seawater, including horizontal pressure gradient force, wind stress, gravity, tide generating force, inertia force, friction force and coriolis force, etc.

(3) Spatial properties: it mainly focused on the scale, distribution, location and shape of ocean currents.

(4) Time attributes: it described the variation of the duration current phenomenon, from the beginning, formation, development and extinction, and paid attention to the lifecycle of the phenomenon.

(5) Motion attributes: it was divided into horizontal motion state and vertical motion state,with the equation of current motion, analyzing the motion state of ocean current in three-dimensional space. Motion properties included velocity, flow direction, amplitude and axis.

(6) Functional attributes: It described the influence of ocean current on the environmental and economic development of adjacent regions.

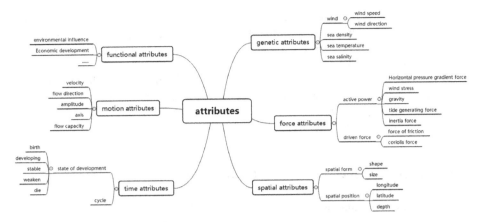

Fig. 3. Concept layered system of ocean flow field

4.2   Asemantic Relation Analysis of Ocean Flow Field

Concepts, attributes and instances constitute the basic elements of ocean flow field description The semantic relations of ocean flow field included inter-concept relations, inter-attribute relations, inter-instance relations, relations between concept and attribute, relations between concept and instance. The semantic description of specific relations was shown in Table 1 and Table 2.

(1) Relations among concepts of ocean flow field
     The relations among the concepts of ocean flow field mainly included equivalence relation, hyponymy and juxtaposition relation. Equivalence relation refers to the relation, in which two or more concepts are same in nature. In the conceptual hierarchy system of ocean flow field, inertial flow was also called residual flow, and equatorial flow was also called information wind current, etc., whose concepts point to the same place in essence and their relation was equivalent. The hyponymy refers to the relation between the parts and the whole and inclusion relation. In the first relation, there was no transfer relationships between their properties, which was represented by component-of. In the second relation, there was a transitive relationship on the attribute, represented by class-of. The hyponymy in the conceptual hierarchy of ocean flow field domain knowledge belonged to this inclusion relation. The relation between the concepts of the same level was defined as juxtaposition.

(2) Relations among ocean flow field properties
     The relations among ocean flow field properties mainly included: containment relation, correlation relation and time relation. Among them, the inclusion relation was used to represent the hierarchical relation in the attribute; Correlation means that there is an correlation between the two attributes. For example, large-scale ocean flow field phenomenon in space generally corresponds to long-period ocean flow field phenomenon in time. Therefore, there was a correlation between the

spatial range attribute of ocean flow field and the periodic attribute in time. Temporal relationships were used to represent the temporal sequence between the phase properties of the lifecycle.

Table 1. Semantic relations between properties and concepts of ocean flow fields.

Relation	Class	Represent	Implication
among conceptions	equivalent	equivalent	the two concepts are essentially equivalent
	hyponymy	component-of	for example, the "vortex core" of the ocean is part of the "vortex" of the ocean
		subclass-of	for example, "ocean circulation" includes "wind surface circulation"
	juxtaposition	partner-of	the relation of the same level of species concepts within the same genus concept. Such as "blow current" and "wind flow"
among attributes	time relation	before	the sequence in which the property appears
		after	
		during	
	contain	kind-of (A,a1)	the complex attribute (a1) under the large category attribute (A), for example, the motion attribute includes the complex attributes such as "flow direction", "velocity", "flow amplitude" and "flow axis"
	related	related	there is an association between the two properties, such as when the "flow" is falling, the general "water color" is light, "transparency" is large
between conception and attribute	attribute of concept	attribute-of	for example, "velocity" is a property of "wind current"
between conception and instance	instance of concept	instance-of	for example, the kuroshio is an instance of western boundary current

(3) Relations between concepts and attributes: it is an ownership relation among numerous attributes and concepts of the ocean flow field domain ontology knowledge system.

(4) Relation between concepts and instances: the instances of concepts represent the correspondence between concrete concepts and abstract generalizations.

(5) Inter-instance relations: the relations between ocean flow field instances mainly included temporal relations and spatial relations. Instances of ocean flow field concepts occur in chronological order, such as La Niña, which mostly occurs in years after El Niño Phenomenon. Spatial relations represent the spatial distribution characteristics of ocean flow field instances, including topological relations, position relations, distance relations, etc. The relations among specific ocean flow field instances was described in Table 2.

Table 2. Relationships among instances.

Class		Represent	Implication
spatial relation	position relation	restricted-south	the two examples are projected onto the two dimensional plane in the directions of southeast, northwest, northeast, southeast, northwest and southwest
		restricted-north	
		restricted-east	
		restricted-west	
		south-east	
		north-east	
		south-west	
		north-west	
		above	the spatial position of two instances is only inconsistent with the y-coordinate
		below	
	topotaxy	overlap	there is an object z, which is part of x and part of y, and x and y are not included. Now the relation between x and y was overlap
		disjoint	if neither x nor y has a partial z, they are separate
		touch	only the boundary part of the two objects overlap, there is no internal overlap
		contain	one contains the other entirely
	distance	far	a pairwise comparison of three or more instances
		near	
		withinKilometers-of	the distance between two instances
		withinMeters-of	
time relation		before	the sequence of the occurrence or existence of one instance and the existence of another
		after	
		during	

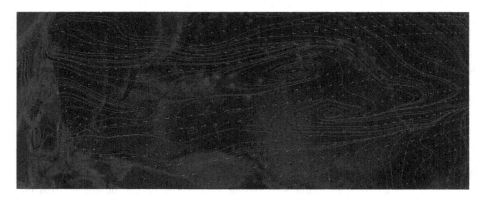

Fig. 4. Symbolization of ocean flow field

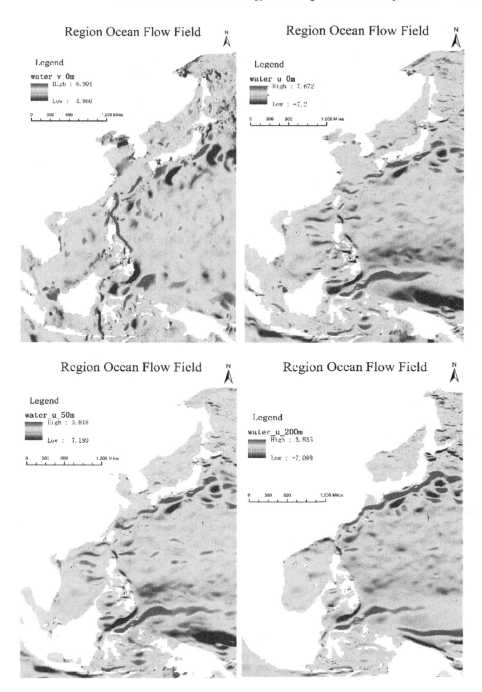

Fig. 5. Schematic diagram of ocean flow field

The boundary of ocean flow field is fuzzy, and its properties and relations are multi-dimensional. Within a certain range, it has visual properties such as direction and velocity, as shown in Fig. 4. There are also drivers, social and environmental benefits and other potential attributes. The maintenance of ocean flow field in different regions and depths and its influence have their own characteristics, as shown in Fig. 5. The ocean flow field shows different forms in different time and space scales, but it is still interrelated. For example, many vortexes are generated in the western boundary flow of the Pacific Ocean, the strength of the east-west equatorial ocean current will lead to the state change of the western boundary flow, and the ocean front will be formed in the transition zone of water body with significantly different characteristics. Based on the complexity, diversity and overall unity of Marine flow field information, this paper conducts ontology modeling of Marine flow field.

5   Modeling and Formal Expression of Ocean Flow Field Domain Ontology

5.1   Modeling of Domain Ontology for Ocean Flow Field

After the establishment of conceptual hierarchy system of ocean flow field domain and the analysis of semantic relations, this article proposed an ontology expression model of ocean flow field in the form of six-tuple based on the ontology idea, which was in the form of O = (C, P, I, R, T, M). Where, C represented the concept set of ocean flow field; P represented the set of attribute vectors of set C; I represented the set of instances of set C; R represented the set of relation of set C, such as semantic relation among concepts and spatial and temporal relation between instances; T represented the set of evolution of an instance corresponding to a concept in set C; M represented the force driving factors for the evolution of examples in T. Considering that the ocean flow field has the characteristics of multi-dimensionality, spatio-temporal dynamics and boundary fuzziness. this paper extended ontology structures such as concepts, attributes, relationships and constructed the basic structure of ocean flow field domain ontology based on the six-tuple model, which was shown in Fig. 6. Because the ocean flow field concept corresponds to not only the instance, but also the instance corresponds to its unique evolution set and driving force elements. So the instance, its evolution set and the driving factors in the evolution process are uniformly organized when describing a collection of instances.

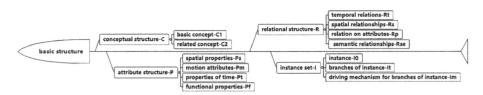

Fig. 6. Symbolization of ocean flow field

Therefore, the model structure of ocean flow field domain ontology could be further expressed as: O = {C(C1, C2); P(Ps, Pt, Pm, Pf); R(Rt, Rs, Rp, Rse); I(I0, It, Im)}. C1 said basic concept, C2 said related concepts. Ps standed for spatial attribute, Pt for time attribute, Pm for motion attribute, Pf for function attribute. Rt referred to temporal relation, Rs to spatial relation, Rp to attribute relation and Rse to semantic relation. I0 represented a specific instance, It represented the instance evolution set, and Im represented the instance evolution drive. The ontology model could be used to express the ontology structure of the entire ocean flow field, or to represent the ontology structure of a specific subclass of the ocean flow field. The overall structure of the ocean flow field domain ontology could be expressed as:

Oocean flow filed = {{all concepts in Marine flow fields}, {{sea-land position, longitude, latitude, depth}, {formation time, end time, lifecycle}, {velocity, flow direction, flow rate, amplitude}, {the influence of climate, economy and environment} ...}, {{before, after, during}, {contained, adjacent, disjoint}, {azimuth relation, distance relation, topological relation}...}, {{instance 1, evolution set of instance 1, evolution driver of instance 1}}, {instance 2, evolution set of instance 2, evolution driver of instance 2},...}}.

5.2   Formal Expression of Ocean Flow Field Domain Ontology Based on OWL

OWL, as one of the most widely used languages for semantic description of ontology, has obvious advantages in the description and expression of domain concepts and conceptual hierarchy relations. However, the ocean flow field has not only a clear conceptual hierarchy, but also a complex spatial-temporal relation. In order to effectively describe and express the ocean flow field domain ontology model, the expansion of OWL language was realized by adding spatial relationship primitives, such as touch, disjoint, after, etc. The semantic description and formal expression of OWL were illustrated as following taking western boundary current as an example.

(1) Conceptual hierarchy description of western boundary current
   In the domain knowledge system of ocean flow field, the west boundary flow is a subclass of the boundary flow divided by the regional location of sea current in the land and sea relation, and the western boundary current is paratactic to the eastern boundary flow. The formal expression of its conceptual hierarchy was as follows:

```
<owl:ObjectProperty rdf:ID="Land and sea area location">
<rdfs:domain rdf:resource="#BoundaryCurrent"/>
...
<owl:Description rdf:about="# WesternBoundaryCurrent"/>
<owl:Description rdf:about="# EasternBoundaryCurrent"/>
...
</owl:ObjectProperty>
```

(2) Attribute description of western boundary current
   The western boundary current has motion properties, function properties, time properties and spatial properties, so the property description of the west boundary

flow includes the property structure description and the property constraint description. According to attribute types, attribute constraints could be divided into numerical constraints, object constraints, etc. For example, velocity in motion attributes is numerical constraints, while direction needed to be defined with some references other than numerical values, which could be defined as object class attribute constraints. For example, the flow direction of the western boundary current was from low latitude to high latitude, and its key expression was as follows:

```
<owl:onProperty rdf:resource="#direction"/>
<owl:hasValue rdf:resource="#zonal(from low latitude to high lati-
tude)"/>
```

(3) Description of the relation structure of western boundary current

The relation structure of western boundary flow included time, space, attribute and semantic relations. At the semantic level, the west boundary current was contained in the boundary flow. In the spatial relation, the west boundary flow was adjacent to equatorial current, but not connected to the eastern boundary flow. In the temporal trend, the western boundary current occurred after the trade(wind) current. The specific OWL was described as follows:

```
<owl:Class rdf:ID="WesternBoundaryCurrent">
<rdfs:subClassOf rdf:resource="#BoundaryCurrent"/>
<rdfs:touch rdf:resource="#TradeWindDrift"/>
<rdfs:disjoint rdf:resource="#EasternBoundaryCurrent"/>
<rdfs:after rdf:resource="#TradeWindDrift"/>
</owl:class>
```

(4) Instance description of western boundary current

Instance description mainly included attributes, instance evolution set, instance driving force and so on. Taking kuroshio as an example, its motion attributes included direction and speed, and its spatial attributes included longitude, latitude and depth. The evolution set included the black tide in the south China sea, ryukyu current, east China sea black tide, tsushima warm current, kuroshio continuation, etc. Its evolution collection could be expressed as follows:

```
<WesternBoundaryCurrent rdf:ID="Kuroshio" >
<owl:unionOf rdf:parseType="Collection">
<owl:Class rdf:about="#South China Sea Branch of Kuroshio"/>
<owl:Class rdf:about="#Ryukyu Current"/>
...
<owl:Class rdf:about="# Kuroshio Extension"/>
```

The driving force of the case evolution set was essentially a special constraint, whose expression form was the same as that of the set of evolution. The driving factors of the evolution process of the instance were given by enumeration. In addition, our study expressed the specific constraint of driving force factors according to the expression form of the above attribute constraints.

The formalized expression of the ocean flow field domain ontology based on OWL stored the semantic information as standardized formalized code, and provided strong support for the ocean flow field information query and semantic sharing. To facilitate model and information management, in this paper, ocean flow field domain knowledge was constructed with Protege ontology modeling tool. The specific ontology structure was shown in Fig. 7. Through this system, the target concept and hierarchical structure could be quickly queried, located and displayed.

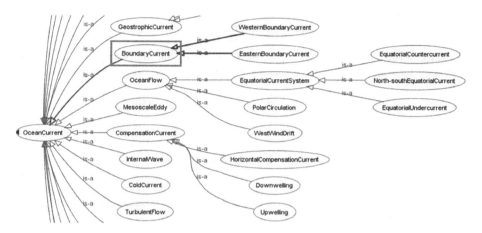

Fig. 7. Ocean flow field domain ontology constructed based on Protégé.

6   Conclusion

Considering the importance of ocean flow field in oceanography research, semantic information sharing and deep rule analysis had also become the focus of people's attention. Therefore, this paper, using the ontological idea, proposed the ontology structure of ocean flow field in the form of six tuples based on the domain knowledge system of ocean flow field and semantic analysis in space-time. In addition, OWL language modeling primitive was extended, taking western boundary current as an example, formal description and expression of ocean flow field domain ontology was carried out. This study could provide methodological support for the integration and sharing of spatio-temporal data of ocean flow field. The analysis of its concept, attribute, spatio-temporal and semantic relationship could be used for knowledge inference in the field of ocean and ocean current field. Combined with natural language and descriptive logic, semantic extraction could be carried out on the latent features of ocean flow field phenomena, so as to achieve the application of big data information mining of ocean flow field and trend prediction based on temporal and spatial characteristics.

Ocean flow field is the result of interaction of various factors and has typical spatio-temporal distribution characteristics. Since this paper is a knowledge system of ocean flow field domain established from the perspective of semantic analysis, the organization and analysis of spatio-temporal process data of ocean flow field are few. The relationship between time and space proposed in this paper can solve the basic expression of time and space and is a preliminary study on the spatio-temporal characteristics of ocean flow field. Future research could be focus on ontology-based spatio-temporal data organization of ocean flow field, owl-based extension of ocean flow field spatio-temporal information expression, ontology-based ocean flow field knowledge inference and spatio-temporal prediction model.

Acknowledgments. This work was supported in part by a grant from the Major Science and Technology Innovation Projects of Shandong Province (2019JZZY020103) and the National Science Foundation of China (41471330).

References

1. Bermudez, L., Graybeal, J., Arko, R.: A marine platforms ontology: experiences and lessons. In: Proceedings of the ISWC 2006 Workshop on Semantic Sensor Networks, Athens, GA, USA. Aachen, Germany (2006)
2. Zhou, J.G., Liu, B.S.: Ontology construction in marine domain based on thesaurus. J. Ningbo Univ. (Sci. Technol. Ed.) (25), 1:10–1:12 (2012)
3. Xiong, J.: Research and application of modeling method of marine ecological ontology. Ocean University of China, Qingdao (2010)
4. Yun, H.Y., Xu, J.L., Guo, Z.B., et al.: Ontology modeling of marine ecology. Comput. Appl. **34**(4), 1105–1108 (2014)
5. Zhang, J.X., Liu, Y.: Ontology modeling research in the field of marine culture in the context of big data. J. Southeast Univ. **16**(S2), 74–76 (2014)
6. Jia, H.P., Xiong, J., Xu, J.L., et al.: Research and application of role theory in ocean carbon cycle ontology construction. J. Ocean Univ. China **13**(6), 979–986 (2014)
7. Abidi, S.R., Abidi, S.S.R., Kwan, M., et al.: Ontology modeling for oceans knowledge and data management. In: International Conference on Intelligent Computational Systems (2011)
8. Li, H.T.: Research on marine environment information integration method and development of new generation MAPGIS platform software. Ocean University of China (2007)
9. Zhang, F.: Research on ontology-based ocean data integration method. Ocean University of China, Qingdao (2008)
10. Li, J.: Key technologies and development of ocean data sharing platform. Tianjin University, Tianjin (2008)
11. Lu, Q.: Design of marine environment data warehouse based on domain ontology. Northeast University, Shenyang (2009)
12. Zhang, Q.: Research on the representation and the construction of marine environment data ontology based on spatial relations. Shanghai Ocean University, Shanghai (2018)
13. Smith, B., Mark, D.M.: Ontology and geographic kinds. In: Poiker, T.K., Chrisman, N. (eds.) Proceeding of International Symposium on Spatial Data Handling (SDH 1998), pp. 308–320. International Geographical Union, Vancouver (1998)
14. Bishr, Y.: Overcoming the semantic and other barriers to GIS interoperability. Int. J. Geograph. Inf. Sci. **12**(4), 299–314 (1998)

15. Jing, D.S.: Research on semantic expression and service of geospatial information based on ontology. University of Chinese Academy of Sciences (Institute of Remote Sensing Application), Beijing (2005)
16. Li, H.W.: Study on geographic information service based on ontology. The PLA Information Engineering University, Zhengzhou (2007)
17. Du, Y.Y., Zhang, D.D., Su, F.Z., et al.: Gulf spatial data organization method based on geographic ontology – a case study of Liaodong Gulf. Geo-Inf. Sci. 10(1), 7–13 (2008)
18. Zhang, Y., Su, F.Z., Wang, Q., et al.: Reef ontology construction and its application in the spatial data organization. Geogr. Geo-Inf. Sci. 33(3), 52–58 (2017)
19. Zhu, G.Q., Shi, S.H., Xu, X.C., et al.: Ontology research and application in the field of Marine disasters. Mar. Inf. (02), 5–8 (2009)
20. Shao, B.: Design and implementation of maritime accident determination system based on ontology. Dalian Maritime University (2010)
21. Li, T., Fu, Y., Ji, M., et al.: Semantic analysis of ocean flow field based on ontology. J. Geo-Inf. Sci. 20(10), 1373–1380 (2018)
22. Wang, G.H., Huang, Q., Qin, C., et al.: Research on semantic analysis methods in ontology construction. Libr. Inf. Work 7(57), 106–111 (2013)

Ecology, Environment and Social Sustainable Development

Study on Space-Time Development of Urban Areas in Shenyang Using Landsat Remote Sensing Data

Zhiwei Xie, Yuntao Ma[✉], Guoqing Su, and Jiaqiang Shan

School of Transportation Engineering, Shenyang Jianzhu University,
Shenyang 100168, China
mayuntao7@163.com

Abstract. As the urbanization process continues to accelerate, this paper studies the development of time and space in the built-up area of Shenyang City based on Landsat image data to analyze the urban expansion, study the urban spatial layout and rationally plan urban development. The study first prepro-cesses the images. The multiresolution segmentation of the study area is carried out by using the classification network evolution algorithm, and the unsuper-vised decision tree classification method is used to extract the built-up area of Shenyang City. Finally, using the landscape index to study the urban devel-opment and changes in Shenyang. According to the experimental results, the Percentage of Landscape (PLAND) of urban built-up areas in Shenyang from 2007 to 2014 increased by 7.00%. However, it merely increased by 1.17% from 2014 to 2017. The overall expansion rate of the built-up area in Shenyang City showed a trend of "first fast and then slow".

Keywords: Remote sensing · Urban built-up area · Landscape index · Urban expansion

1 Introduction

Urban expansion is the result of social and economic development and is increasingly becoming a major problem facing cities. Remote sensing technology has the charac-teristics of multi-temporal, wide coverage and high efficiency. It has unique advantages in using multi-temporal images to describe the temporal and spatial trends of urban expansion, and provides a basis for predicting the future urbanization process. Therefore, researching urban spatiotemporal development based on remote sensing data and predicting urban development direction is the basis for urban scientific development and sustainable development.

The study of urban time and space development includes two aspects: using remote sensing data to extract urban built-up areas and using landscape indices to analyze urban development changes. In terms of urban built-up area extraction, domestic and foreign scholars have also proposed a variety of effective means. For example, Jin combines object-oriented and deep learning, grasps the characteristics of textures of different objects, and forms a deep learning model to guide object classification [1].

© Springer Nature Singapore Pte Ltd. 2020
Y. Xie et al. (Eds.): GSES 2019/GeoAI 2019, CCIS 1228, pp. 355–367, 2020.
https://doi.org/10.1007/978-981-15-6106-1_26

This method solves the problem that the object-oriented method does not describe the texture features comprehensively. Chen used the high-scoring satellite data of China to formulate the classification rules for buildings in built-up areas from the aspects of spectral features, geometric features, texture features and edge density based on object-oriented methods [2]. According to the above analysis, the object-oriented classification method can extract the urban built-up area better.

With the rapid development of remote sensing and geographic information technology, the theory of landscape pattern and landscape index has been introduced on the road of using urban remote sensing data to study urban time and space development gradually. Landscape ecologists proposed a quantitative analysis of the landscape pattern of the landscape index. The landscape index is formed in the process of establishing the pattern and process of landscape spatial analysis and the infiltration of other theories into landscape ecology. Forman has divided the landscape indices describing plaque into two categories, namely landscape indices describing the shape of plaques, such as shape indices and landscape indices describing plaque inlays, such as relative abundance, dominance and fractal dimension [3]. Cui used the landscape index such as Percentage of Landscape (PLAND) to study the relationship between land use and landscape pattern in the American white moth epidemic area, and achieved good results [4]. Pan used Percentage of Landscape (PLAND) and Class Area (CA) indexes to study the change of landscape pattern of panda habitat [5]. In summary, we know CA and PLAND have been used more in the analysis of urban space-time development. Therefore, this study selected two basic landscape indices, CA and

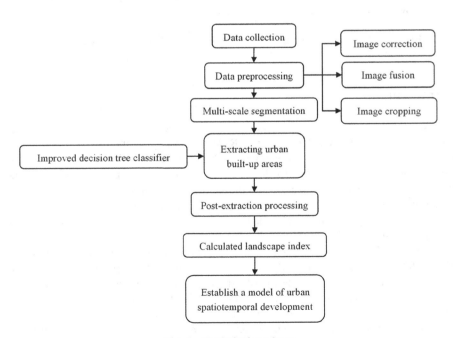

Fig. 1. Technical roadmap

PLAND. According to calculate and reasonably analyze the trend of urban development and change was finally obtained.

In order to accurately grasp the urban development and rationally plan the urban construction, this paper uses Landsat data to study the development of the built-up area in Shenyang. In this study, the remote sensing data is preprocessed by image fusion, cropping, etc., and then the improved decision tree classification method is used to extract the built-up area. The technical flow chart of the research is shown in Fig. 1.

2 Data Preprocessing

2.1 Image Fusion

In this paper, the Gram-Schmidt Pan Sharpening method is used to fuse the corresponding multi-spectral band and the full-color band in 2014 and 2017 respectively. The resulting image has both multi-spectral features and improved image resolution. Since the 2007 and 2010 data uses the Landsat 5 satellite imagery, there is no panchromatic band, so the resolution is 30 m.

2.2 Image Registration

Image registration refers to the process of geometrically calibrating two images with overlapping regions acquired at different times, different sensors or under different conditions. Image registration technology is the basis of remote sensing image processing, image mosaic, change monitoring, multi-sensor image fusion and other fields.

2.3 Radiation Correction

Radiation correction refers to the correction of systematic, random radiation distortion or distortion caused by external factors, data acquisition and transmission systems, eliminating or correcting the process of image distortion caused by radiation errors.

3 Extracting Urban Built-Up Areas by Unsupervised Decision Tree Classification

Based on the pixel classification method, only the built-up area is extracted by spectral information, and the precision is low. Object-oriented is the smallest classification unit after segmentation. Considering multiple features of features in classification, the classification accuracy is higher than that based on pixel classification, and the interpretation efficiency is higher than visual interpretation. Object-oriented classification breaks through the traditional pixel-based classification method. Instead of using pixels as the minimum unit, the image is segmented into objects and classified using spectral and spatial features between the objects. The object-oriented classification method is used to extract the built-up area with higher precision and more reliability.

3.1 Automatically Select Training Samples

In order to improve the automation of decision tree classification, this aper proposes a method of unsupervised decision tree classification. The method firstly performs unsupervised classification on the image, and then uses the algorithm to randomly generate training samples for built-up area in the built-up area classification results in the Third Ring Road of the city. The accuracy of the samples is analyzed by manual interpretation. The sample accuracy can meet the classification requirements.

3.2 Fractal Net Evolution Approach

In this paper, the fractal net evolution approach is used to segment the image. This method is a segmentation algorithm widely used at present, and it is also the basis and core content of object-oriented image analysis technology. After the split is performed, the image is divided into several image objects consisting of homogeneous pixels, and the subsequent feature information extraction will be based on the object. The scale in multi-scale segmentation determines the level of the smallest polygon generated. However, this size is not suitable for extracting all image information, that is, a segmentation scale cannot accurately describe all image information to be extracted. Different scales of feature information need to adopt different segmentation scales. An image usually has a variety of information to be extracted. Therefore, it is necessary to construct a multi-scale image segmentation scale network to select the different scales of different features.

3.3 Decision Tree Classification

The classification decision tree model is a tree structure that describes the classification of instances. The Cart decision tree belongs to one of the supervised classification methods. The principle is to select one or more attribute combinations from a plurality of prediction attributes as a column variable of the tree node, and divide the test variable into each branch, and repeat the process to establish a large enough The classification tree is then pruned by the pruning algorithm to obtain a series of nested classification trees. Finally, a series of classification trees are tested by test data to select the optimal classification tree. The core of the algorithm is to determine a decision tree branch criterion, using the Gini coefficient in economics as the selection of the best test variable and the segmentation threshold criterion. When constructing the decision tree, it can make full use of various kinds of feature categories in a reasonable range. Characteristics. In this paper, the decision tree classifier is used to divide the ground objects in the study area into built-up areas, vegetation and water bodies. Sample points will be generated from the results of unsupervised classification.

4 Research on the Development of Built-Up Area Using Landscape Index

The landscape pattern plays a vital role in the study of landscape ecology. All landscape studies are inseparable from its pattern analysis. The landscape index refers to the highly concentrated landscape pattern information, which reflects the simple quantitative indicators of its structural composition and spatial configuration, and is suitable for quantitative analysis of the spatial analysis method of the relationship between landscape pattern and ecological process. It is used to describe the landscape pattern, and then to establish the connection between the landscape structure and the process or phenomenon to better explain and understand the landscape function. The calculation of the landscape pattern index includes three types: plaque level, plaque type level and landscape level. The first two types of indexes are for individual plaques or different types of plaques, while the landscape level index is for research. A description of the overall characteristics of the scope. The landscape pattern index has a scale effect, and the ecological significance between the indices is basically the same. Any two indexes show a strong correlation, and the information is more repetitive.

Landscape indices are diverse and some landscape indices are highly correlated. Therefore, when selecting the landscape index at two levels of plaque type and landscape level, the following principles should be followed:

(1) The selected landscape index can better reflect the structure, characteristics and process of the landscape;
(2) The selected landscape indices have relative independence and low redundancy;
(3) It can characterize the state of various ecological land landscape patterns.

According to the above three criteria, the selected indicators have plaque type area (CA) and plaque type index (PLAND) at the plaque type level.

Class area (CA) is equal to the sum of the areas of all plaques in a plaque type (m^2), divided by 10,000 and converted to ha (ha); that is, the total area of a plaque type. Unit: ha, range: CA > 0. CA measures the components of the landscape and is the basis for calculating other indicators. It has important ecological significance.

Percentage of Landscape (PLAND) is equal to the total area of a plaque type as a percentage of the total landscape area, which is the proportion of the plaque's landscape area. Unit: percentage, range: 0 < PLAND ⩽ 100. When the value tends to 0, it indicates that the plaque type in the landscape becomes very rare; when the value is equal to 100, the whole landscape is composed of only one type of plaque. PLAND measures the composition of the landscape, which calculates the relative proportion of a patch type to the area of the landscape.

5 Experiment and Analysis

5.1 Data Source and Pretreatment

In this paper, the image data of Landsat5 and Landsat 8 are used, the central meridian is 123 °E, and the spatial resolution of the data is 30 m. In this paper, we first select 10 control points to register images by polynomial registration, and then perform "radiation calibration" and "atmospheric correction" on the results after registration. The data after pre-processing are shown in Fig. 2. The following figures are false color images.

Fig. 2. Preprocessed image

5.2 Selection of Sample Points

This paper first uses the unsupervised classification method to classify images. Then randomly generate sample points in the center of each class of result. Sample point results are shown in Fig. 3.

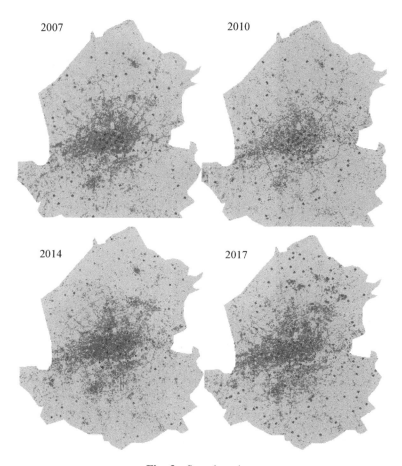

Fig. 3. Sample points

5.3   Multiresolution Segmentation

This article uses eCognition software to segment images multiple scales. Multiresolution segmentation requires three parameters, namely segmentation scale parameter, shape factor parameter, and compactness factor parameter. After trial and error, the landsat5 image segmentation in 2007 and 2010 was scaled using the scale 50, the shape factor was set to 0.1, and the compactness factor was set to 0.5. The landsat8 image segmentation in 2014 and 2017 was scaled using the scale 500, and the shape factor was set to 0.1. The tightening factor is set to 0.5. After the segmentation is completed, the segmentation effect is shown in Fig. 4.

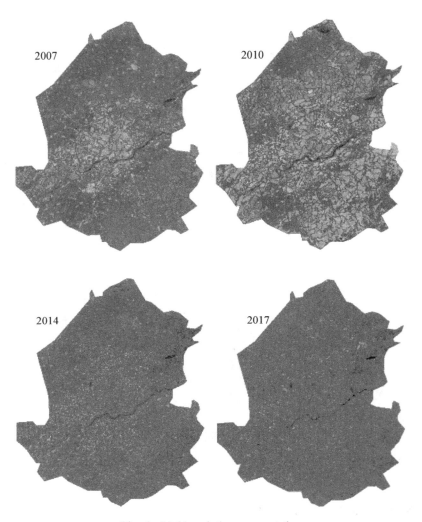

Fig. 4. Multiresolution segmentation

5.4 Decision Tree Classification

This paper classifies decision trees in the main urban area of Shenyang. The classification effect is shown in Fig. 5. The green part is vegetation, the blue part is the built-up area, and the pink part is water.

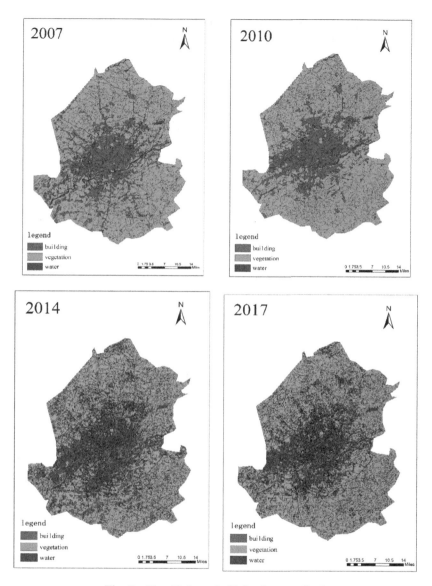

Fig. 5. Classified result (Color figure online)

5.5   Post-classification

According to the definition of urban built-up area in the Urban Planning Law of the People's Republic of China, the built-up area is the area where urban construction is connected, infrastructure and public facilities arrive, and it is the core part of the urban planning area. Therefore, the types of vegetation, rivers, and water bodies within the core area of urban buildings are also within the scope of built-up areas. In this paper, the three images in 2007, 2010, 2014 and 2017 are filled by manual visual interpretation, and the holes in the main urban area are filled into the building category as shown in Fig. 6.

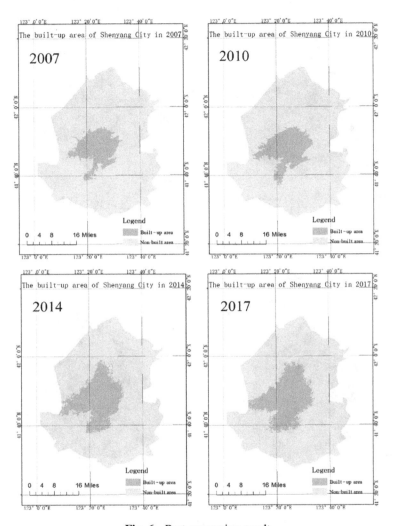

Fig. 6. Post processing result

5.6   Precision Analysis

The extraction accuracy evaluation of the built-up area of this paper is based on the Landsat image of Shenyang area. The method of manual interpretation and visual interpretation is used to map out the boundary of the built-up area in Shenyang, 2007, 2010, 2014 and 2017, and the results of this paper are compared and analyzed.

According to the classification method of the "built-up area-non-built-up area" two-point method, the confusion matrix is obtained, and the obtained user precision, drawing precision, overall precision and Kappa coefficient are shown in Table 1. It can be seen from Table 1 that the average kappa coefficient extracted from the built-up area is 0.88, which has high precision.

Table 1. Decision-making tree classification method extraction accuracy from 2007 to 2017

Year	2007		2010		2014		2017	
	B	N	B	N	B	N	B	N
B	358.33	19.99	426.55	15.60	575.99	42.01	627.26	29.87
N	86.10	2956.49	82.86	2715.90	130.70	2492.21	153.90	2609.88
UA	0.81		0.84		0.82		0.80	
PA	0.95		0.96		0.93		0.95	
OA	0.97		0.92		0.90		0.95	
Kappa	0.85		0.75		0.74		0.84	

(1) B = Built-up area, (2) N = Non-built area, (3) UA = User's Accuracy, (4) PA = Producer's Accuracy, (5) OA = Overall Accuracy, (6) Kappa = Kappa Coefficient

5.7 Analysis of Urban Time and Space Development by Using Landscape Index

This paper uses Fragstats software to calculate the landscape index and select two landscape indices, CA and PLAND. The calculation results of the landscape index can be obtained as shown in Table 2 and Table 3.

Table 2. CA value statistics table of Shenyang urban built-up area from 2007 to 2017

CA	Built-up area	Non-built area
2007	37774.88	304348.72
2010	44176.60	297947.00
2014	61732.32	280398.04
2017	65727.48	276396.12

Table 3. PLAND value statistics table of Shenyang urban built-up area from 2007 to 2017

PLAND	Built-up area	Non-built area
2007	11.04	88.96
2010	12.91	87.09
2014	18.04	81.96
2017	19.21	80.79

According to the calculation results and statistical tables, the curve of the CA value in the built-up area of Shenyang City from 2007 to 2017 is shown in Fig. 7 where blue is the built-up area and green is the non-built area. The unit of CA value is hectare.

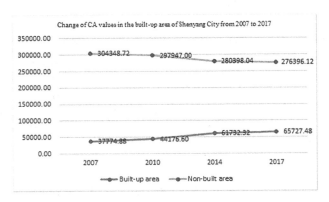

Fig. 7. Statistical analysis of CA values in the built-up area of Shenyang City from 2007 to 2017 (Color figure online)

According to the statistical line chart of the change of plaque type area in the built-up area of Shenyang City from 2007 to 2017, the slope of the curve was larger and changed rapidly from 2007 to 2014, and the CA value in the built-up area increased rapidly. Between 2014 and 2017, the slope of the curve is small and the change is slow. The CA value in the built-up area is still growing, but the growth rate is obviously slower. Overall, the CA value in the built-up area of Shenyang City has gradually increased from 2007 to 2017.

The curve of the PLAND value in the built-up area of Shenyang City from 2007 to 2017 is shown in Fig. 8, where blue is the built-up area and green is the non-built area.

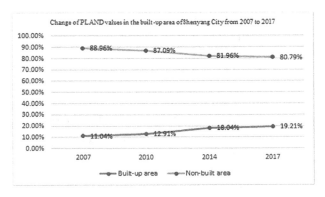

Fig. 8. Statistical analysis of PLAND values in the built-up area of Shenyang City from 2007 to 2017 (Color figure online)

According to the CA value change statistical line chart of Shenyang City built in 2007–2017, similar to the CA value change statistical line chart of the built-up area of Shenyang City, the slope of the curve is larger and the change is faster between 2007 and 2014. The PLAND value of the built-up area is increasing fast. Between 2014 and 2017, the slope of the curve is small and the change is slow. The CA value in the built-up area is still growing, but the growth rate is obviously slow. In general, from the period of 2007 to 2017, the PLAND value of the built-up area in Shenyang is gradually increasing. It can be seen that the expansion rate of the built-up area in Shenyang City showed a trend of "first fast and then slow".

6 Summary

This paper uses Landsat image to extract the built-up area of Shenyang City, and based on the calculation of the landscape index, the development of built-up area in Shenyang has shown a trend of "first fast and then slow". The landscape index only calculates the characteristics of a landscape pattern of a phase and neglects the process of studying the temporal and spatial development of the city. Therefore, in the study of urban expansion, the landscape expansion index can also be considered.

Acknowledgments. This study was supported by Scientific Research Project of Education Department of Liaoning Province of China: Analysis and Research on Polycentric Urbanization in Liaoning Province with Traffic Big Data (No. lnqn201917).

References

1. Jin, Y.T., Yang, X.F., Gao, T., Guo, H.M., Liu, S.M.: The typical object extraction method based on object-oriented and deep learning. J. Remote Sens. Land Resour. **30**(01), 22–29 (2018)
2. Chen, K.X., Zhang, F., Du, Z.H., et al.: Analysis of urban expansion and driving forces in Jiaxing City based on remote sensing image. J. Zhejiang Univ. (Sci. Ed.) **06**, 709–715 (2016)
3. Forman, R.T.T.: Land mosaics: the ecology of landscapes and regions. J. Landscape Urban Plann. **36**, 229–231 (1996)
4. Cui, M.R., Hu, S.F., Zhou, R.L.: Analysis of land use and landscape patterns in the American white moth disease area in Liaoning Province. J. Anhui Agric. Sci. **47**, 78–82 (2019)
5. Pan, X., Li, W.Q., Huo, L., Zhang, Z.Q., Peng, Y.: Analysis on change of landscape pattern and optimization of panda habitat. J. Sci. Surv. Mapp. **41**, 61–65 (2016)

Ecological Environment Evaluation of Liuxi River Basin During Rapid Urbanization

Yunwen Xie[1,2], Xulong Liu[1,3(✉)], Zhifeng Wu[2], Ji Yang[1,3], and Jingjun Zhang[4]

[1] Key Laboratory of Guangdong for Utilization of Remote Sensing and Geographical Information System, Guangdong Open Laboratory of Geospatial Information Technology and Application, Guangzhou Institute of Geography, Guangzhou 510070, China
lxlong020@126.com
[2] School of Geographical Sciences, Guangzhou University, Guangzhou 510006, China
[3] Southern Marine Science and Engineering Guangdong Laboratory (Guangzhou), Guangzhou 511458, China
[4] Xinhua College of Sun Yat-Sen University, Guangzhou 510520, China

Abstract. This article takes Liuxi River Basin in Guangzhou as the research object, selects Landsat remote sensing data of 1993, 2000, 2011 and 2018 as data source, to access the five indices, including vegetation coverage, DEM, land use, soil index and slope. The comprehensive index method is used to obtain the ecological environment comprehensive index of this study area. The environmental quality is divided into 4 grades. The ecological environment of the basin and its dynamic changes have been comprehensively evaluated in the past 25 years. The results show that: In the Liuxi River Basin in 2018, the percentage of excellent natural ecological environment quality was 4.80%, the percentage of the areas with good quality was 44.53%; the percentage of the areas with general quality was 45.97% and the percentage of the areas with poor quality was 4.68%. Overall, the good and the general ecological environment quality is the majority. From the perspective of time from 1993 to 2018, the percentage of the area with excellent environmental quality increased by 1.69%; the percentage of the area with good environmental quality gradually decreased, with a total of 32.88%, but still occupied the largest proportion; The percentage of area with general environmental quality went up and down, with a total of 29.64%; the percentage of the area with poor environmental quality rose by 1.54%, the overall change was not big. The downstream ecological environment is showing a trend of deterioration. In upstream, the excellent ecological environment is the majority. The overall ecological environment of the river basin is developing for the better.

Keywords: Natural ecological environment · Remote sensing · Evaluation · Liuxi River

© Springer Nature Singapore Pte Ltd. 2020
Y. Xie et al. (Eds.): GSES 2019/GeoAI 2019, CCIS 1228, pp. 368–380, 2020.
https://doi.org/10.1007/978-981-15-6106-1_27

1   Introduction

Ecological environment evaluation is a qualitative or quantitative analysis of the quality level of the ecological environment in the study area from a specific perspective [1]. Scientific understanding, reasonable evaluation and optimized regulation of the regional ecological environment are not only the basic work of regional development planning and ecological environmental protection, but also an important criterion for objectively measuring the healthy and sustainable development of the region [2]. Applying remote sensing technology to ecological environment evaluation can save the manpower and material resources required by the field monitoring and investigation, and achieve rapid monitoring of the ecological environment [3]. The watershed is a comprehensive ecological system including nature, society, and economy, and is an independent and complete ecological landform unit, which has the characteristics of its own environment within the region. Almost all ecological and environmental problems and the irrational use of resources can fall into the scope of a specific river basin. Therefore, solving the environmental problems from the perspective of the river basin is a way that can achieve sustainable development more effectively in a comprehensive and systematic manner [4].

Liuxi River is the main source of water supply for Huadu District and Baiyun District and the main drinking reserve water for downtown Guangzhou as well. It's a part and parcel of Guangzhou city water supply security system. In the situation of rapid urbanization, the impact of human activities on drinking water sources is becoming more serious. Watershed ecological environment quality evaluation is an important method to judge the safety of drinking water sources. Environment protection administration monitoring data indicates that the overall index of the main downstream water quality is at the level of category 5, far from meeting the requirements for water environment function zones. In early 2013, the operation based on Guangdong's water clearing plan of the Guangdong environment protection bureau focused on improvement and protection of the water quality of the Liuxi River; the regulations on the protection of Liuxi River Basin in Guangzhou have been implemented since June 1, 2014. The comprehensive plan for Liuxi River Basin in Guangzhou (2015–2030) proposes to slow down the hydrological effects of urbanization, reduce the interference of human activities, and establish an ecological environment system with rich cultural deposits, harmonious landscapes, and a harmonious relationship between people and water. A lot of attention is paid to the water quality of the Liuxi River by the public. Therefore, in order to strengthen the water environmental protection and management of Liuxi River Basin, enhance the ability to prevent floods and droughts, and systematically and comprehensively protect the drinking water sources in Guangzhou, it is of strategic significance for study personnel to understand the ecological environment and changes in the region in a timely manner.

Most of the existing studies on ecological environment assessment take administrative units as the evaluation units, and there are few studies on river basins. The advantages of remote sensing technology have not been fully reflected. The study of Liuxi River Basin mostly focuses on water quality, water environment, flood forecasting, etc., and its overall ecological environment assessment is relatively insufficient. Based on this, this article takes the Liuxi River Basin in Guangzhou as the research object, and uses Landsat remote sensing data from 1993, 2000, 2011, and 2018 as the data sources to assess vegetation coverage, DEM, land use, soil index, and slope. The comprehensive index method is used to obtain the comprehensive index of the ecological environment in the study area. According to its quality status, it is divided into 4 grades of excellent, good, average, and poor. The ecological environment status of the river basin and the ecological environment changes in the past 25 years are comprehensively evaluated. This study provides technical support for timely understanding of the ecological environment in the region and formulating scientific ecological environment protection measures.

2　Data and Methods

2.1　Study Area

The Liuxi River originated in the northeast of Conghua District, Guangzhou, the area between Dalingtou and Guifeng Mountain which is at the junction of Lutian County and Longmen County (Fig. 1). Flow through Huadu District, Luogang District, Baiyun District. After passing through the river network of the Pearl River Delta, and finally injected into the South China Sea, 157 km in total. Located between 113°10′ 12″ ~ 114°2′00″E and 23°12′30″ ~ 23°57′36″N, Watershed area of 2290 km^2, 2225.72 km^2 is in Guangzhou and 33.28 km^2 is outside Guangzhou, is an important ecological barrier in the north of Guangzhou, There are Liuxihe Forest Park, one of the top ten forest parks in the country, and Liuxihe Reservoir, the only large reservoir in the city. The average rainfall for many years is 1823.6 mm, which belongs to the subtropical monsoon climate zone with mild climate and abundant rainfall. The upstream channel of Liuxi River, which is about 10 km above Liangkou Town, passes through the deep mountain canyon. The average slope of the river bed is 1/1250, and the water flow is turbulent. In the middle and lower reaches of Wenquan Town and below, the average slope of the river bed is reduced to 1/2500, the water flow is relatively gentle. The population of the basin is about 300,000. Since 1995, the urbanization of the Liuxi River Basin has entered a stage of rapid development, and the area of cultivated land has decreased sharply. The ecological environment of the Liuxi River Basin has suffered to varying degrees under the background of urbanization.

The Liuxi River originates in the northeast of Conghua District, Guangzhou and flows through Huadu District, Luogang District, Baiyun District. After passing through the river network of the Pearl River Delta, it finally empties into the South China Sea, with a length of 157 km in total. Located between 113°10′12″ and 114°2′00″E and 23° 12′30″ and 23°57′36″N, it has a watershed area of 2290 km^2, with 2225.72 km^2 in Guangzhou and 33.28 km^2 outside Guangzhou. It is an important ecological barrier in the north of Guangzhou. There are Liuxi River Forest Park, one of the top ten forest parks in China, and Liuxi River Reservoir, the only large reservoir in the city. The average rainfall for many years is 1823.6 mm, which means it belongs to the subtropical monsoon climate zone with a mild climate and abundant rainfall. The upstream channel of the Liuxi River, which is about 10 km away from Liangkou Town, passes through the deep mountain canyon. The average slope of the river bed is 1/1250, and the water flow is turbulent. The average slope of the riverbed was reduced to 1/2500, the water flow is relatively gentle. The population of the basin area is about 300,000. Since 1995, the urbanization of Liuxi River Basin has entered a stage of rapid development, and the area of cultivated land has decreased sharply. The ecological environment of Liuxi River Basin has suffered to varying degrees under the background of urbanization.

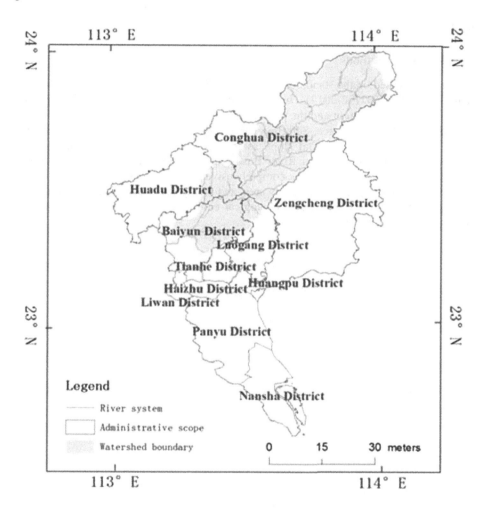

Fig. 1. Geographical location of the study area

2.2   Data Pre-processing

The main data used in this research comes from the geospatial data cloud (http://www. gscloud.cn/), including Landsat images and DEM data covering Liuxi River Basin. In order to form a uniform data format, preprocessing was performed under the support of

ENVI5.3: the Landsat image obtained was subjected to projection conversion, radiation calibration, and atmospheric correction processing. The purpose of image atmospheric correction is to eliminate the effects of atmospheric conditions and lighting conditions on surface reflectivity. The corrected image basically removes the influence of water vapor particles and other factors in the air, the spatial difference of the types of ground features has been improved [5], and the reflection spectrum curve has become normal. Then the Liuxi River boundary is used as the ROI for stitching and cropping, and finally the processed remote sensing image map is obtained.

2.3   Research Methods

This article is based on the standards of Technical Specification for Eco-Environmental Status Evaluation (HJ192-2017), combining the actual situation of rapid development of Liuxi River urbanization and the principles of easy availability of ecological factors, scientificity and independence, Selecting the five basic factors of vegetation coverage, DEM, land use, soil index, and slope as the ecological factors for evaluating the ecological environment of Liuxi River Basin, and normalizing the above-mentioned five ecological factor indicators, the comprehensive index method was used to evaluate the ecological environment of the study area, and finally the ArcGIS software was used to generate the hierarchical map of ecological environment assessment. Technology roadmap is shown in Fig. 2.

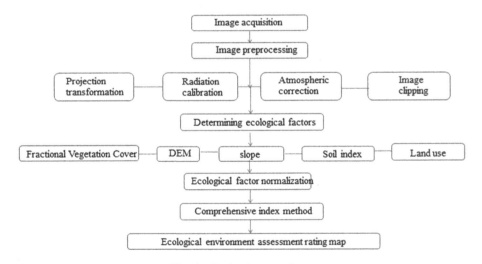

Fig. 2.   Technology roadmap

2.4   Remote Sensing Ecological Index

In this paper, vegetation coverage, DEM, land use, soil index, and slope are selected as ecological factors to evaluate the ecological environment of Liuxi River Basin. Fractional Vegetation Coverage (FVC) is often used to study vegetation changes, ecological environment research, soil and water conservation, and climate. This study uses remote sensing estimation method [6], the sensor can detect vegetation information mainly because the green plant canopy has spectral characteristics with high reflectivity in the red and near-infrared bands. Based on this, the ecological status of green vegetation on the regional scale is evaluated [7]. The calculation formula of vegetation coverage is as follows:

$$FVC = (NDVI - NDVI_{min})/(NDVI_{max} - NDVI_{min}) \tag{1}$$

NDVI is a normalized index, and $NDVI_{max}$ and $NDVI_{min}$ are the maximum and minimum values in the region, respectively. Due to the inevitable noise in the image, $NDVI_{max}$ and $NDVI_{min}$ are not necessarily the maximum NDVI value and the minimum NDVI value. You need to set the confidence value (empirical value) to select. The NDVI calculation formula is as follows:

$$NDVI = (TM4 - TM3)/(TM4 + TM3) \tag{2}$$

Among them: TM4 and TM3 are the reflectance of Landsat ETM near infrared and red band, respectively.

The soil index adopts the widely used bare soil vegetation index model (GRABS) [8]. The specific algorithm is as follows:

$$GRABS = VI - 0.09178BI + 5.58959 \tag{3}$$

BI is the soil brightness index (the first component), and VI is the greenness index (the second component); BI has a considerable impact on the vegetation index, because the main part of the information change of the bare soil is caused by their brightness; There is great correlation between VI and different vegetation cover. Since the first and second components are concentrated >95% of the information, they can well reflect the differences in spectral characteristics such as vegetation conditions and soil moisture content, organic matter content, particle size, mineral composition and surface roughness, therefore, the bare soil vegetation index formed by the linear combination of BI and VI can well reflect the bare condition of the soil.

The change of the terrain gradient has a significant impact on the growth and distribution of surface vegetation and the soil anti-erosion ability. For soil erosion, under the same vegetation coverage, the area with greater terrain slope has a worse surface soil anti-erosion ability and the smaller the terrain slope, the stronger the resistance to erosion, and the larger the index value [9]. The slope index is obtained on the basis of the DEM image. The obtained two-scene DEM images are stitched, the boundary vector data of the Liuxi River administrative region is projected and transformed, and the Liuxi River Basin boundary is used as the ROI for cropping to obtain the Liuxi River Basin DEM data. The slope index can be directly generated through the Topographic Modeling module using ENVI.

Land use is classified using unsupervised classification. First, by setting reasonable parameters, a training template file (SIG) and a spectral clustering image are obtained, and then the training template file is edited, evaluated, and adjusted to form a classification template that finally meets the requirements. Using this classification template as a training sample for supervised classification and supervised classification using the minimum distance method, the land use map of the study area can be obtained. Formed six first-class categories including cultivated land, forest land, grassland, waters, urban and rural industrial and mining residents' land, and unused land. Finally, 120 points were randomly selected for verification, and the overall accuracy reached 92%, and the classification accuracy reached practical requirements.

2.5   Normalization of Ecological Factors

In the study, the indicators are standardized. The data of each participating factor after normalization is a set of values that reflect its attribute characteristics. The values of all indicators are within the range of [0, 10]. According to the scores of each indicator and its contribution to the quality of the ecological environment, the principle of uniform order is adopted. That is, according to the magnitude of their positive impact on the ecological environment, they are divided into several levels from high to low, and the greater their contribution to environmental quality, the smaller the code value. The vegetation coverage, DEM, soil index, slope, and land use are divided into 10 levels according to Table 1 and re-assigned. The larger the code, the better the ecological environment. The land use treatment method is used to understand the use of each land type and its impact on the ecological environment, and then the analytic hierarchy process is used to determine the weight of each land use type, and then they are re-encoded.

Table 1. Comparison of ecological factor normalization code values

Coded value	Fractional vegetation coverage	Soil index	Slope	DEM
1	0–10	−439.04 to −394.58	78.3–87.0	929.12–1039.91
2	10–20	−394.58 to −350.12	69.6–78.3	818.33–929.12
3	20–30	−350.12 to −305.65	60.9–69.6	707.54–818.33
4	30–40	−305.65 to −261.19	52.2–60.9	596.75–707.54
5	40–50	−261.19 to −216.73	43.5–52.2	485.96–596.75
6	50–60	−216.73 to −172.26	34.8–43.5	375.18–485.96
7	60–70	−172.26 to −127.80	26.1–34.8	264.39–375.18
8	70–80	−127.80 to −83.34	17.4–26.1	153.60–264.39
9	80–90	−83.34 to −38.87	8.7–17.4	42.81–153.60
10	90–100	−38.87 to 5.58	0.0–8.7	−67.97 to 42.81

The assessment method used in this study is the comprehensive index method, and the calculation formula is as follows:

$$E = W_1 * S_1 + W_2 * S_2 + W_3 * S_3 + W_4 * S_4 + W_5 * S_5 \qquad (4)$$

Among them, E is the result of ecological environment assessment, S1 is vegetation coverage, S2 is DEM, S3 is slope, S4 is land use, and S5 is soil index. W_i (i = 1,2,3,4,5) is the weight of each coefficient, which is determined by the analytic hierarchy process. The analytic hierarchy process is that several factors that affect the target are divided into several criterion layers. Comparison between each other, the qualitative indicators are blurred to get the weight of each factor. W_1, W_2, W_3, W_4, and W_5 take 0.4534, 0.2473, 0.1442, 0.1009, and 0.0542, respectively. Because the threshold of the ecological environment assessment results is not uniformly and clearly defined [10]. Referencing to the ecological environment assessment in the "Technical Specifications for the Evaluation of Ecological Environment Conditions (HJ192-2017)", and combines the relevant literature and the specific conditions of the Liuxi River Basin, the environmental quality of the study area is divided into four grades: excellent, good, average, and poor (Table 2).

Table 2. Classification of ecological environment assessment results in Liuxi River Basin

Grade	Score	Description of the ecological environment of the Liuxi River Basin
Excellent	(9–10)	High vegetation coverage, reasonable ecological structure, stable ecosystem, strong self-recovery ability, and no damage to natural ecological environment
good	(6–9)	High vegetation coverage, reasonable ecological structure, stable ecosystem, strong self-recovery ability, and basically no damage to natural ecological environment
General	(4–6)	Low vegetation coverage, general ecological structure, relatively unstable ecosystem, self-recovery ability, natural ecological environment is destroyed
Low	(1–4)	Vegetation coverage is low, ecological structure is poor, eco-system is extremely unstable, self-recovery is poor, and natural ecological environment is seriously damaged

3 Results

3.1 Spatial Change Analysis of Ecological Environment

The area with excellent ecological environment quality in Liuxi River Basin is 317 km^2, which is mainly located upstream, accounting for 13.8%. The ecological environment in this part of the region is basically not damaged, the ecological structure is reasonable and stable, and the ecosystem's own functions and self-restoration ability are very strong. The area of good quality is 1209 km^2, accounting for 52.79%, and the ecological structure is basically reasonable and stable. The area of general quality is 700 km^2, accounting for 30.56%, and the ecological environment is basically damaged, but the eco-system's own functions and it's self-recovery ability is weak; the area with poor quality is 30 km^2, which is mainly located downstream, accounting for 1.3%. This part of the ecological environment is severely damaged and the ecological structure is unreasonable and therefore, the ecosystem functions and self-recovery capabilities are weak. Generally speaking, the quality of the ecological environment of the river basin is mainly moderate and good. The results of the ecological evaluation of the Liuxi River also show that the environmental quality of the river basin is quite different. The ecological environment in the Conghua section is better, but it is obviously worse in the middle and lower reaches, and the organic pollution is serious.

3.2   Ecological Time Change Analysis

From the perspective of time change, the proportion of areas with excellent environmental quality increased by 4.46% overall from 1993 to 2018; the proportion of areas with good environmental quality decreased by 4.93% overall, with the maximum specific gravity. The area proportion of general environmental quality continued to rise, but the rate slowed down, with a total increase of 8.42%, which indicated that the area of general environmental quality was getting larger and larger; the percentage of areas with poor environmental quality was decreasing overall, but the overall change was not big, with a total decrease of 3.83%, as shown in Table 3. The areas with good ecological environment quality were an absolute majority, the proportion of area with excellent environmental quality was still rising and the proportion of area with excellent environmental quality was decreasing, so the overall ecological environment of the river basin tended to develop.

Table 3. Changes in ecological environment quality in 1993, 2000, 2010 and 2018

Evaluation grade	Percentage (%)				Change
	years : 1993	2000	2011	2018	
Excellent	8.34	17.57	10.32	13.8	rise4.46%
Good	64.39	52.66	54.54	52.79	decline11.6%
General	22.14	25.36	29.42	30.56	rise8.42%
Low	5.13	4.41	6.72	1.3	decline3.83%

Through comparison of the results of the evaluation of the ecological environment of Liuxi River Basin in 1993, 2000, 2011, and 2018, as shown in Fig. 3, it can be seen clearly that the areas with poor ecological environment in the downstream region increased before 2011. While the upstream Liuxi River Forest Farm, Dalingshan Forest Farm, etc. were mostly with excellent ecological environment quality, and the ecological environment was better. The reason is that the leading industries in the upper reaches of Liuxi River Basin are different from those in the middle and lower reaches. Although the upstream industries have developed rapidly in recent years, they are still dominated by agriculture and tourism. Generally speaking, the degree of pollution of water bodies is positively related to living standards and the local GDP [11]. According to the monitoring data of the environmental protection department from 2009 to 2011, the Liuxi River was classified as Class III when it came out of Conghua District, and it has become inferior to Class V in the middle and lower reaches. The "Southern Daily" and "Guangzhou Daily" and other media repeatedly exposed water pollution events in the Liuxi River. However, the situation was different since 2018, the environmental quality of the downstream in Baiyun and Huadu section have obviously improved, which showed that the ecological environment downstream showed a trend of

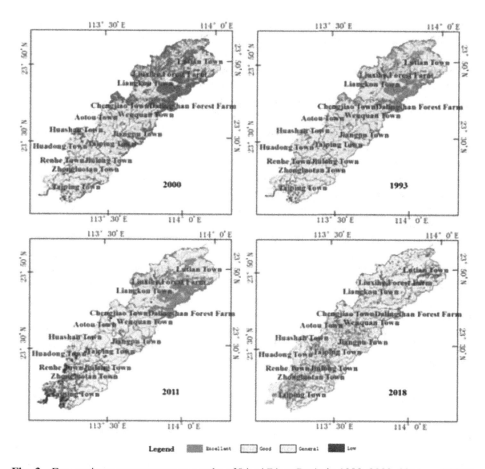

Fig. 3. Eco-environment assessment results of Liuxi River Basin in 1993, 2000, 2011, and 2018

improving. According to the information released by the Guangdong Province on the water quality of key rivers in the fourth year of 2018, the water quality of the Baiyun section of Liuxi River has improved. In the meantime, the quality of the ecological environment of upstream has declined.

3.3   Analysis of Driving Factors of Ecological Environment Change

Analyze the driving factors in terms of land use change, leading industries, and population changes. According to the land use classification results of this study, in terms of construction land, since 2013, the construction land in Taihe Town and Baiyun District has increased rapidly. Due to the development of urbanization and the increase in construction land, the ecological environment quality in the southeast of Baiyun District has declined. In Huashan Town and Huadu Town of Huadu District, the area of cultivated land has decreased significantly, converted into construction land, which has led to a decline in the quality of the ecological environment in this region. The forest land has not changed significantly from 1993 to 2018. It can be seen that forest land is not the main factor affecting the ecological environment of the river basin. In terms of industries, because the middle and lower reaches are dominated by industries and agriculture, and the upper reaches are dominated by tourism, the ecological environment of the lower and middle reaches is generally worse than the upper reaches. From the perspective of the entire river basin, the middle and upper reaches are most affected by domestic pollution. The lower reaches of the Liuxi River are most affected by industrial pollution [12]. The reasons for the deterioration of the upstream ecological environment in recent years are mainly rapid population growth and increased pollution. This year, the upstream ecological environment has deteriorated. In recent years, the total resident population of forest farms, Lutian Town and Liangkou Town in the upper reaches of the river basin has increased to more than 70,000, resulting in a significant increase in domestic sewage and garbage discharged by residents [13]; the direct cause of the impact on the ecological environment of the river basin is the delayed construction of sewage treatment facilities and regional soil and water loss caused by vegetation destruction. The main source is the unreasonable industrial structure and layout of the river basin [14]. The upstream region has good agricultural development conditions and superior tourism resources. It mainly develops agriculture and tourism, while the middle and lower reaches should strictly restrict the development of industries and other organic pollution industries. Eco-tourism and characteristic agriculture for its economic development should be the focused on [15], in order to seek a balance between development and protection.

4   Conclusion

This study takes the Liuxi River Basin as the research area, and fully understands the selection of indicators for eco-environmental assessment and the concepts of each influencing factor and selects the vegetation coverage, DEM, land use, slope index, and soil index as the five impact factors to study the current status of the ecological environment in Liuxi River Basin and its ecological environment change trends and characteristics in the past 25 years. The following conclusions are drawn:

(1) From the perspective of spatial distribution, the area of the ecological environment with excellent water quality in 2018 is mainly located upstream, accounting for 13.8% of the total area, and the area with good evaluation quality accounts for 52.79%. In general, the assessed ecological environment of the Liuxi River is dominated by areas with general and good conditions. The reason that the quality of the environment in the lower reaches of the river is generally worse than that of the middle and upper reaches is that their leading industries are different. The leading industries of the middle and upper reaches are mainly tourism, while those of the downstream areas are mainly industries and agriculture.

(2) In terms of time change, the proportion of the areas with excellent environmental quality in each year increased by 13.8%; the proportion of the areas with good environmental quality decreased by 11.6%; the proportion of the areas with

general environmental quality continued to rise but the rising has slowed down, with a total increase of 8.42%; the area with poor environmental quality declined, but the overall change was not big. The excellent and good ecological environment quality of the basin has an absolute advantage and is still improving, and the proportion of area with general and poor environment quality are declined in the meantime, so the overall ecological environment of the basin is developing.

(3) The areas with poor ecological environment in the downstream region increased before 2011. The reason is that the reduction of cultivated land and construction land in the lower reaches has increased since 2013. Since 2018, the area with poor environment quality in the lower reaches of the basin has gradually decreased, which indicates the ecological environment is better and better. Besides, the overall environmental quality of the upper reaches of the basin in 2018 has suffered due to the increase in the total population of the upper reaches.

Acknowledgments. This research is jointly supported by GDAS' Project of Science and Technology Development (2016GDASRC-0211, 2017GDASCX-0601, 2018GDASCX-0403), duced Talents Team of Southern Marine Science and Engineering Guangdong Laboratory (GML2019ZD0301) and 2019 science and Technology Innovation Project of Provincial Special Fund for Economic Development (2019B1).

References

1. LiuPang, R.C., Zongming, W.: Nanchong River Natural ecological environment quality evaluation of remote sensing. J. Appl. Ecol. **29**(10), 3348–3350 (2018)
2. Wang, L., Zhang, J., Chen, W., et al.: Spatiotemporal Manas river basin ecological environment quality points different evaluation. Arid Study Area 1–10 (2019)
3. Xu, H.: Typical red soil erosion dynamic change of the surface area of the South bare soil analysis-taking Fujian Ting County as an example. Geogr. Sci. **33**(4), 489–496 (2013)
4. Zhao, Y., Zhang, L.: Fragile ecology quantitative assessment method. Geogr. Sci. **18**(1), 73–79 (1998)
5. Lai, Z., Xia, S., Cheng, J.: Research on application model of high-resolution remote sensing satellite data in urban ecological environment assessment. Prog. Geogr. Sci. **19**(4), 361–364 (2000)
6. Zhang, Y., Yuan, J., Liu, B.: Progress vegetation cover and management factor in the soil erosion prediction model. J. Appl. Ecol. **13**, 1033–1036 (2002)
7. Zhang, W., Xie, Y., Liu, B.: China erosivity spatial variation characteristics. J. Mt. Sci. **21**(1), 33–40 (2003)
8. Yang, C.J., Liu, J., et al.: Remote sensing and GIS support China reforestation project at the Grassland. J. Remote Sens. **6**(3), 206–210 (2002)
9. Gong, C., Chen, Q., Yin, Q., et al.: Remote sensing research base quality evaluation of coastal environment- to Shanghai-Nanhui East Beach as an example. Mar. Environ. Sci. **30**(5), 712–718 (2011)
10. Li, H., Shi, Z., Guo, Y.: Evaluation of ecological environment quality in Fujian Province based on remote sensing and GIS technology. Remote Sens. Technol. Appl. **21**(1), 49–54 (2006)

11. Zhang, Y.: Study on the water quality change law of Conghua section of Liuxi River. J. Eco-Environ. **21**(11), 1902–1904 (2012)
12. Chen, Q.: Impact of human activities on eutrophication of Liuxihe Reservoir in the past 50 years. Jinan University, Guangzhou (2014)
13. Chen, X.: Study on legislation of local watershed management—taking legislation of Liuxi River Managementing Guangzhou as an example. Guangdong University of Foreign Studies, Guangzhou (2013)
14. Chen, J., Wu, E.: Water resources protection of Liuxi River for 20 years. Guangzhou Environ. Sci. **17**(3), 41–42 (2002)
15. Yang, Z., Gao, L., Chen, J., et al.: Based on the geochemical characteristics of the river riparian zone-groundwater interaction boundary identification-taking Guangzhou flow River. Sci. Environ. **39**(6), 1843–1847 (2019)

Analysis of Urban Greenness Landscape and Its Spatial Association with Urbanization and Climate Changes

Jingli Wang[1(✉)], Chengjie Yang[1,2], Zongyao Sha[2], and Dai Qiu[2]

[1] School of Transportation Engineering, Shenyang Jianzhu University,
Shenyang 100168, China
cejlwang@sjzu.edu.cn
[2] School of Remote Sensing and Information Engineering, Wuhan University,
Wuhan 430079, China

Abstract. Urban green space and urban landscape patterns are of great significance to the sustainable development for urban ecosystems. Spatial statistics such as Moran's I indicator can only reveal the spatial autocorrelation of urban green space or urban landscape itself, while the dynamic changes of urban green space and urban landscape in spatial correlation and spatial heterogeneity must be investigated under spatio-temporal contexts, with possible driving factors such as urban climate and urbanization process taken into account. The purpose of this paper was to study the dynamics of urban greenness patterns using urban landscape indices as well as the spatial association of the indices with the climate changes and urbanization process. Wuhan, a key megacity in Central China was selected as the case study. To this end, we mapped the urban greenness through NDVI of the city from 2000 to 2018 using Landsat imagery. The dynamics of the urban green space indicated by two landscape indices, namely Percent of landscape (PLAND) and landscape shape index (LSI), was analyzed for the study region. Time-series analysis using Mann-Kendall and Sen's slope were adopted to reveal the temporal changes of the landscape pattern and Pearson's correlation analysis was performed to explain its association with the climate changes and urbanization process. Results indicated that urban greenness not only significantly decreased but also fragmented.

Keywords: Greenness landscape pattern · Climate · Urbanization · Mann-Kendall

1 Introduction

Urban green space is an important part of urban ecosystem, and it is the center of material circulation and energy flow [1]. It plays an important role in regulating energy balance [2], biochemical cycle [3] and water cycle [4], and is a sensitive indicator of environmental impacts of climate and human factors. It is also a resource that has made important contributions to human social and economic activities [5]. The land use and land cover of the city have changed significantly many parts of the world in the past few decades. On the one hand, continuous effort such as artificial urban greening and

© Springer Nature Singapore Pte Ltd. 2020
Y. Xie et al. (Eds.): GSES 2019/GeoAI 2019, CCIS 1228, pp. 381–393, 2020.
https://doi.org/10.1007/978-981-15-6106-1_28

afforestation have helped the green space recovery; on the other hand, the rapid development of urbanization and industrialization has led to the disappearance of greenness in some areas [6]. Mapping the dynamic trend of urban green space and understanding the driving factors of change and the landscape pattern of green space can provide a scientific basis for making decisions on sustainable urban development.

In order to study the urban green space, the traditional method of measuring the urban green space distribution requires to carry the instrument on-site survey, which is time-consuming and labor-intensive. The powerful aerospace technology and sensor technology has greatly facilitated the acquisition of high-resolution remote sensing images. Due to its unique spectral characteristics, vegetation can be inverted in remote sensing images through a variety of vegetation indices, so the vegetation index, including NDVI, has been widely applied [7].

As an important restoration organization of urban ecosystems, urban green space cannot neglect the protection of urban ecological environment, such as effectively alleviating urban "heat island effect" [8] and improving urban microclimate. However, due to the development of the modern economy, the urban development process is happening at an accelerating rate, resulting in a decrease in urban green space. Therefore, more and more scholars have carried out a series of studies around urban green space and using vegetation coverage as a monitoring indicator. Miller et al. examined vegetation spatial autocorrelation [9], but did not discuss its spatial heterogeneity. Wu et al. studied the factor of urbanization intensity; according to the level of urban development, the global cities were divided into five intensity levels to explore the impact of urbanization on the growth of 14 plantations [10]. Riaz et al. collected the multi-temporal Landsat satellite remote sensing map from 1992 to 2015, and calculated the NDVI vegetation index to obtain the urban vegetation coverage of Pakistan Sargodha over the years to study the future sustainable development trend of the city [11]; in their work, although multi-temporal monitoring of green space changes was realized, the factors contributing to or causing the changes in urban vegetation cover were not well investigated. It is necessary to analyze the correlation between multivariate factors and urban green space dynamics through time series model, and to understand the change relationship.

In this study, we attempted to introduce a landscape pattern index instead of NDVI as an indicator of the dynamic change of urban green space. Taking Wuhan, China as an example, we select urban climate and urbanization process to quantify the influencing factors of urban green space dynamics. Through the 19-year time series, explore the correlation between influencing factors and green space changes. Based on the calculation results of the model, the influence of factors is effectively evaluated to provide an analysis basis for decision-making.

2   Methods

2.1   Study Area

Wuhan is a mega city located in the Central China region. It is located in the eastern part of the Jianghan Plain and is the "heart" of Chinese geography. Its east-west direction spans 134 km, and the north-south direction is about 155 km apart, with an area of about 847 km². Due to the intersection of the Yangtze River and Hanshui. In the "8 + 1" urban circle planned by the state, Wuhan is the center of the urban circle and is of great significance to the development of small and medium-sized cities around it (Fig. 1).

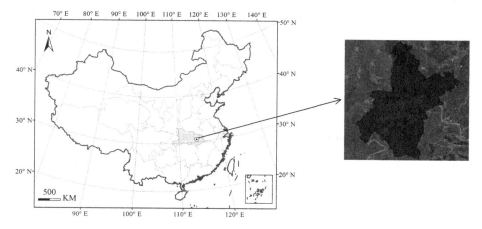

Fig. 1. The study area

2.2 Data Sources

Climate Data. The climate data comes from China's Meteorological Data Sharing Platform during 2000–2018 (http://cdc.cma.gov.cn) [12]. In this study, monthly precipitation and temperature, which have great influence on vegetation, were selected to be variables that can potentially affect urban greenness landscape pattern. Plant growth is a slow process, taking from a few months, to half a year in the study region. Therefore, the aggregated climate data was processed from April to September of each year, which strides over the most important period for the vegetation growth was selected [13]. Specifically, the average monthly precipitation and temperature during those 6 months were processed and averaged to be climate factors which are taken to explain the dynamics of the urban greenness landscape pattern.

NDVI Data. The Normalized Vegetation Index (NDVI), also known as the standardized vegetation index, is an important parameter for estimating vegetation coverage. It is defined as the difference between the reflectance values of Near Infrared (NIR) and Visible red ban (R) and these two bands. The ratio of the sum of the reflectance values:

$$NDVI = (NIR - R)/(NIR + R) \tag{1}$$

The NDVI data is processed under the Google Earth Engine (GEE) [14–16] platform for the Landsat7 data from 2000 to 2012 and the Landsat8 data from 2013 to 2018, providing a data base for the calculation of the green landscape pattern index.

Urbanization. In addition to natural factors, there are also human factors influencing the urban green space landscape pattern [17–19]. Human production and practice activities must be based on land, and it will inevitably affect the green space. The urbanization factor is a comprehensive factor in the human factor, including population density [20], imperviousness [21] and GDP [22]. Urbanized area was extracted based on the NDVI distributions. Data source used is shown in Table 1.

Table 1. Data source used

Dataset	Data source	Temporal scale
Rainfall & temperature	China's Meteorological Data Sharing Platform	2000–2018
NDVI & urbanized area	Landsat7-ETM+/Landsat8-OLI from Google Earth Engine (GEE)	2000–2018

2.3 Methods

The landscape pattern [23] refers to the type, quantity and spatial distribution of landscape units, which is the concrete manifestation of landscape heterogeneity, and also the result of various ecological processes acting on different scales. The landscape pattern index [24] can highly enrich the landscape pattern information, thus reflecting the structural composition and spatial distribution characteristics of the landscape [25]. Based on the Fragstats platform [26–29], this study uses NDVI images as research data, and selects two landscape pattern indices, Percent of landscape (PLAND) and Landscape shape index (LSI), to describe the landscape pattern of Wuhan. Figure 2 is the workflow of study process.

PLAND is an indicator used to measure the proportion of green landscapes in the entire urban landscape. In the case of a certain urban area, the increase or decrease in the area of green space can be reflected by inter-annual dynamics of PLAND. The calculation formula of PLAND is as follows: let a_i is the area of a green patch, i is the number of the patches classified as being urban greenness ($i = 1, 2, \ldots, n$), and A is the summed area of all the patches, then,

$$\text{PLAND} = \frac{\sum_i^n a_i}{A} \times 100\% \tag{2}$$

Given a region that is partly covered by discrete areal green patches, LSI is equivalent to the sum of all patch edge length divided by the length of a square having the same area as the summed area of all the patches. LSI can indicate the fragmentation degree of the landscape of the urban greenness. When it equals to 1, the green landscape is maximally clustered or no fragmentation. When the LSI becomes larger, it indicates a

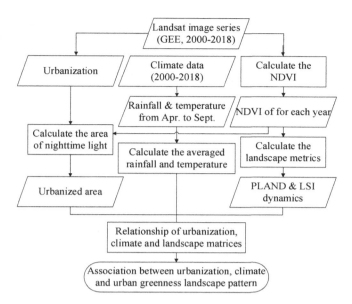

Fig. 2. Workflow of study process

more fragmented green landscape. The calculation formula of LSI is as follows: e_i is the length of a green patch, where i is the number of green patches (i = 1, 2, ..., n), and min_{edge} is the edge length divided by the length of a square having the same area as the summed area of all the patches, then,

$$LSI = \frac{\sum_{i=1}^{n} e_i}{min_{edge}} \qquad (3)$$

Linear regression is often applied to study the time-series trend of a geographical event but it requires that time series data conform to a normal distribution. Instead, Mann-Kendall (M-K) [30] method does not require the sequence distribution and is not sensitive to the outliers. Mann-Kendall trend analysis can complete the test of the significance of the time-series trend. Sen's slope is an index that indicates the magnitude of a time-series trend using the median of the time-series sequences. From 2000 to 2018, time series data is likely to have some outliers due to factors such as the atmosphere and clouds. The combination of Mann-Kendall and Sen's slope [31–34] can enhance the noise immunity in the trend analysis of urban greenness. For the time series X, the trend degree β is used to indicate the time series trend. Provided by M-K testing, a positive β suggests that the time series shows a rising trend, and vice versa. Sen's slope is calculated as:

$$\beta = \text{Median}\left(\frac{x_i - x_j}{i - j}\right), \forall i > j \qquad (4)$$

The Mann-Kendall process is as follows. For a time series vector $X(x_1, x_2, \ldots, x_n)$ where $n = 19$ (number of years), a sequential time series statistic s_k is computed,

$$s_k = \sum_{i=1 \, i=1}^{k} r_i, k = 1, 2, \ldots, n \tag{5}$$

$$r_i = \begin{cases} 1 & x_i > x_j \\ 0 & else \end{cases}, \quad j = 1, 2, \ldots, i \tag{6}$$

A statistic UF_k, calculated from the chronological sequence $X(x_1, x_2, \ldots, x_n)$:

$$UF_k = \frac{|s_k - E(s_k)|}{\sqrt{Var(s_k)}}, k = 1, 2, \ldots, n \tag{7}$$

Where $UF_1 = 0$, $E(s_k)$ and $Var(s_k)$ are the mean and variance of s_k given by,

$$\begin{cases} E(s_k) = \frac{k(k-1)}{4} \\ Var(s_k) = \frac{k(k-1)(2k+5)}{72} \end{cases}, \quad k = 1, 2, \ldots, n \tag{8}$$

In a similar manner, the definition statistic UB_k is computed as the following,

$$UB_{K'} = UF_k, k' + k = n + 1 \tag{9}$$

Clearly, $UB_1 = 0$.

A significant level of $\alpha = 0.05$ was measured in the test, namely $U0.05 = 1.96$ was used to test if a given trend was statistically significant. The two statistic sequence curves of UF_k and $UB_{K'}$ and the ± 1.96 lines were plotted to visually illustrate the result.

3   Results

3.1   Urban Climate States

Figure 3 shows the inter-annual process and changes of precipitation in Wuhan. As can be seen from Fig. 3, the precipitation in growing season has been relatively small and stable over the past 19 years. The average monthly precipitation during the growing season strides over from 77 mm (2001) to 240 mm (2016), with an average of 149 mm/per month during the growing season. Figure 3 shows the inter-annual process and changes in temperature in Wuhan. It can be seen that the temperature has fluctuated around the average of 23.5 °C in the past 19 years.

3.2   Urbanization Process

The urbanized area of Wuhan during the period from 2000 to 2018 can be largely indicated from Landsat NDVI images. The expansion speed and expansion direction of urban construction land revealed partially the intensity of land use and economic development, and reflected the process of urbanization development. The urbanization expansion in Wuhan has been observed during the whole study period. While the urbanization process has supported the improvement of urban functions and facilities, it also brought continuous pressure on the urban green space environment.

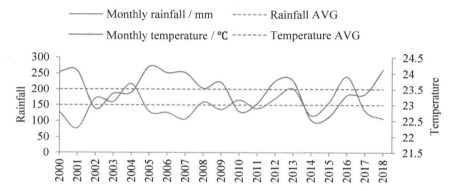

Fig. 3. Monthly average rainfall and temperature for the growing season during 2000–2018

3.3   Greenness Landscape Indices

Figure 4 shows the NDVI data for 2000, 2010 and 2018. Overall, the NDVI showed obvious degradation over the years. From the dynamics of the time series, the PLAND and LSI indicators vary significantly. In general, the PLAND index showed a downward trend, while the LSI index showed an upward trend during 2000–2018, suggesting that the city's greenness space was reduced and that the greenness landscape was more fragmented (Fig. 5). Moreover, the critical inter-annual variation in the two landscape indicators also demonstrated some complicated factors were involved in the dynamics. Table 2 is the example of the greenness landscape indices (years 2000, 2010 and 2018).

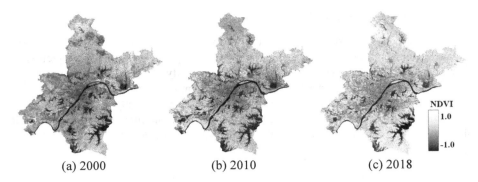

(a) 2000   (b) 2010   (c) 2018

Fig. 4. Example of NDVI maps

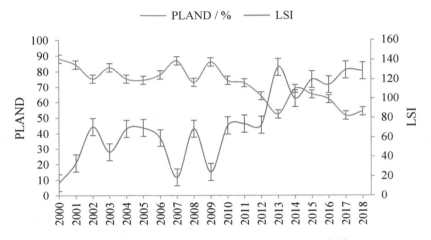

Fig. 5. Dynamics of the urban greenness landscape indices

Table 2. Example of the greenness landscape indices (years 2000, 2010 and 2018)

	2000	2010	2018
PLAND	88.124	74.281	53.952
LSI	13.157	72.194	128.431

3.4 Association Analysis Between Urban Landscape and Influential Factors

Figure 6 shows the dynamics of the examined variables, including temperature, precipitation, urbanization process, PLAND index, and LSI index, using Mann-Kendall (M-K) trend analysis. The M-K curves of temperature and precipitation have UF values located within the significant lines between -1.96 and 1.96, which means that it only fluctuates up and down within the significant horizontal line, indicating that the trend of temperature and precipitation is not statistically significant. The monotonously upward UF and downward UB in the M-K curve for the urbanized area, along with a positive β from Sen's slope (Table 3) indicated that the urbanization process was significantly observed. Similarly, when combine Sen's slope analysis, the result indicated the PLAND index was declined significantly and that the LSI index showed a significant increasing trend over the years.

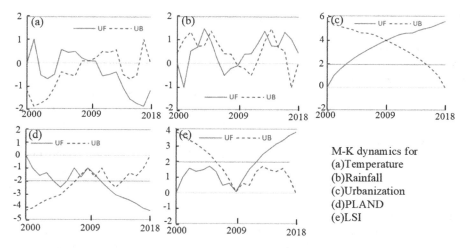

Fig. 6. Time-series trend of the landscape indices for urban greenness and the selected factors through Mann-Kendall (M-K) analysis

Table 3. Sen's slope for the time-series data for the selected variable

Variable	β
Temperature	−0.113
Rainfall	0.119
Urbanization	0.272
PLAND	−0.346
LSI	0.296

Table 4. Correlation coefficient between the landscape indices and the selected climate/urbanization factors

Landscape index	Temperature	Rainfall	Urbanization
PLAND	0.189	0.326	0.765
LSI	0.360	0.358	0.784

In order to further explore the relationship between the considered influential factors and greenness landscape patterns, the correlation coefficients were computed between the selected factors. It can be seen that both PLAND and LSI were mostly correlated to the urbanization process. Climate factors seemed to contribute the landscape LSI as well. For example, Table 4, the coefficient between the rainfall and LSI reached over 0.3. In summary, the climate temperature imposed impact on the LSI index of urban landscape pattern, and that urbanization factor might be the dominant factor determining both LSI and PLAND dynamics.

4   Discussion

As one of the landscapes of urban ecosystems, urban green space is also an important open system. If we can introduce the theory of system science, we can further understand the association rules in urban green space. In addition to its internal role, green space also has material exchange, energy flow and information transmission with the natural environment and human activities of the outside world, which constitutes a complex system. In each system, there are factors that make up the system and characteristics that affect the system. In addition to directly affecting the system, the impact factors also form complex synergies and trade-offs between factors, which in turn enhance or reduce the impact on the system. Synergistic effect refers to a state in which the influence factors appear to increase or decrease simultaneously when acting on the system. When the trade-off effect acts on the system, one influence factor increases while the other influence factor decreases. The role of these two effects depends mainly on the relationship between the influence factors and their associated mechanisms.

Corresponding to this study, it is to discuss the synergy and trade-off effect between precipitation, temperature and urbanization process in the urban green space system. According to the Pearson correlation analysis, the equation coefficient of precipitation and temperature is -25.580, the coefficient of temperature and urbanization is -680.722, and the coefficient of precipitation and urbanization is 0.501. It is obvious from this that temperature and precipitation, temperature and urbanization all show significant trade-off effects, and there is a slight synergy between precipitation and urbanization. It is well known that at high temperatures, the amount of water vapor contained in the air is large, and it is difficult to have precipitation conditions, making it more difficult to form precipitation. And the high temperature is not suitable for long-term settlement of human beings, and thus shows a strong trade-off effect with the degree of urbanization. In ancient times, urban areas with suitable precipitation were indeed more popular with people. After all, water is an important resource for people's lives. However, today's technology is developed, people have more ways to obtain water resources, so the way to collect water resources through precipitation is not so important, so it shows a slight synergy in the urban green space system.

For urban green space systems, precipitation, temperature, and urbanization can be seen as exogenous variables, while PLAND and LSI can be considered as endogenous variables. Through the previous discussion, this study has found that there are certain association rules between exogenous variables and endogenous variables. Unfortunately, there is no specific relationship between variables. The next step in our work is to create a model that expresses the association rules between variables so that we can clearly see the proportion of variables affecting the urban green space system. And we can use this model to verify the trend accuracy of PLAND and LSI to make up for the lack of this research.

Most of our research on urban green space is extremely macroscopic, starting from the perspective of the entire city to analyze the public green space ecosystem. However, we all know that the city's green space can also be divided into green areas in public areas, such as parks, squares and streets, as well as a large number of private green areas, such as residential areas, courtyards and villas. These private green spaces often have obvious regional characteristics - they exist in the more affluent areas of the city, which has a significant impact on the landscape pattern of urban green spaces. Future research can make a more detailed division of urban green space. By analyzing the different roles of green space, it will have different effects on the urban landscape pattern.

5 Conclusions

The research on the changing pattern of urban greenness landscape and its association analysis to the driving factors is of great significance to improve the sustainable development of urban environment. The current study took use of the long-term sequence of NDVI during 2000 to 2018 in Wuhan, China to project the temporal urban landscape indices in terms of the urban greenness space. To this end, two urban greenness landscape indices, namely PLAND and LSI, were selected to quantify urban greenness space in Wuhan. It was found that PLAND showed a significant decline trend, proving a reduced land cover area in terms of the urban greenness. On the contrary, LSI showed an upward trend, proving a significant phenomenon of urban green space fragmentation in Wuhan. Multiple correlation analysis revealed that both PLAND and LSI were closely related to urbanization process and that urban conditions might also update the overall trend of the landscape indices.

Further research is needed in the future. First, the satellite images may contain considerable noise for extracting NDVI and urbanized area. Landsat images ETM+ and OLI have different image sensors. At the same time, the spatial resolution of the ETM and OLI images is 30 m, making small greenness patches, for example the green space along some streets or close to some residential areas, not accurately extracted. Second, the evolution of the landscape pattern in terms of the urban green space could be driven by both natural and anthropogenic factors. However, this paper only took the urbanization expansion as the main indicator as anthropogenic impact, it is preferable that more human factors should be added.

References

1. Wolch, J.R., Byrne, J., Newell, J.P.: Urban green space, public health, and environmental justice: the challenge of making cities 'just green enough'. Landscape Urban Plann. **125**, 234–244 (2014)

2. Merina, A., Gallego, C., Coca, D., et al.: Initial results from Phase 2 of the international urban energy balance model comparison. Int. J. Climatol. **31**(2), 244–272 (2011)
3. Pataki, D.E., Carreiro, M.M., Cherrier, J., et al.: Coupling biogeochemical cycles in urban environments: ecosystem services, green solutions, and misconceptions. Front. Ecol. Environ. **9**(1), 27–36 (2011)
4. Musolff, A., Leschik, S., Reinstorf, F., et al.: Micropollutant loads in the urban water cycle. Environ. Sci. Technol. **44**(13), 4877 (2010)
5. Horwood, K.: Green infrastructure: reconciling urban green space and regional economic development: lessons learnt from experience in England's north-west region. Local Environ. **16**(10), 963–975 (2011)
6. Zhou, X., Wang, Y.C.: Spatial-temporal dynamics of urban green space in response to rapid urbanization and greening policies. Landscape Urban Plann. **100**(3), 268–277 (2011)
7. Liu, Y., Guo, J.: The research of NDVI-based urban green space landscape pattern and thermal environment. Prog. Geogr. **28**(5), 798–804 (2009)
8. Park, J., Kim, J.H., Dong, K.L., et al.: The influence of small green space type and structure at the street level on urban heat island mitigation. Urban For. Urban Greening **21**, 203–212 (2017)
9. Miller, T.F., Mladenoff, D.J., Clayton, M.K.: Old-growth northern hardwood forests: spatial autocorrelation and patterns of understory vegetation. Ecol. Monogr. **72**(4), 487–503 (2002)
10. Wu, S., Liang, Z., Li, S.: Relationships between urban development level and urban vegetation states: a global perspective. Urban For. Urban Greening **38**, 215–222 (2019)
11. Riaz, O., Munawar, H., Khalid, M.: Spatio temporal evaluation of vegetation cover in Sargodha (Pakistan) for sustainable urban future. Eur. J. Sustain. Dev. **6**(2), 33–40 (2017)
12. Sha, Z., Zhong, J., Bai, Y., et al.: Spatio-temporal patterns of satellite-derived grassland vegetation phenology from 1998 to 2012 in Inner Mongolia, China. J. Arid Land **8**(3), 462–477 (2016). https://doi.org/10.1007/s40333-016-0121-9
13. Miao, L., Qiang, L., Fraser, R., et al.: Shifts in vegetation growth in response to multiple factors on the Mongolian Plateau from 1982 to 2011. Phys. Chem. Earth Parts A/B/C **87–88**, 50–59 (2015)
14. Gorelick, N., Hancher, M., Dixon, M., et al.: Google earth engine: planetary-scale geospatial analysis for everyone. Remote Sens. Environ. **202**, 18–27 (2017)
15. Patel, N.N., Angiuli, E., Gamba, P., et al.: Multitemporal settlement and population mapping from Landsat using Google Earth Engine. Int. J. Appl. Earth Obs. Geoinf. **35**, 199–208 (2015)
16. Dong, J., Xiao, X., Menarguez, M.A., et al.: Mapping paddy rice planting area in northeastern Asia with Landsat 8 images, phenology-based algorithm and Google Earth Engine. Remote Sens. Environ. **185**, 142–154 (2016). S003442571630044X
17. Baker, D.: HUMAN FACTORS in the USE of GREEN TRANSPORTATION: sociology for public policy in transportation, urban planning, and environmental quality (2017). https://deepblue.lib.umich.edu/handle/2027.42/139601
18. Shi, F., Li, X., Xu, H.: Analysis of human activities in nature reserves based on nighttime light remote sensing and microblogging data - by the case of national nature reserves in Jiangxi Province. In: ISPRS - International Archives of the Photogrammetry, Remote Sensing and Spatial Information Sciences, vol. XLII-2/W7, pp. 1341–1348 (2017)
19. Zhang, Q., Seto, K.C.: Mapping urbanization dynamics at regional and global scales using multi-temporal DMSP/OLS nighttime light data. Remote Sens. Environ. **115**(9), 2320–2329 (2011)
20. MoLler, A.P., Diaz, M., Flensted-Jensen, E., et al.: High urban population density of birds reflects their timing of urbanization. Oecologia **170**(3), 867–875 (2012). https://doi.org/10.1007/s00442-012-2355-3

21. Ogden, F.L., Pradhan, N.R., Downer, C.W., et al.: Relative importance of impervious area, drainage density, width function, and subsurface storm drainage on flood runoff from an urbanized catchment. Water Resour. Res. **47**(12), 1–12 (2011)
22. Farhani, S., Ozturk, I.: Causal relationship between CO_2 emissions, real GDP, energy consumption, financial development, trade openness, and urbanization in Tunisia. Env. Sci. Pollut. Res. **22**(20), 15663–15676 (2015)
23. Griffiths, G.H., Lee, J., Eversham, B.C., et al.: Landscape pattern and species richness; regional scale analysis from remote sensing. Int. J. Remote Sens. **21**(13–14), 2685–2704 (2000)
24. Zhou, Z.X., Li, J.: The correlation analysis on the landscape pattern index and hydrological processes in the Yanhe watershed, China. J. Hydrol. **524**, 417–426 (2015)
25. Van Der Zanden, E.H., Verburg, P.H., Mücher, C.A.: Modelling the spatial distribution of linear landscape elements in Europe. Ecol. Indic. **27**, 125–136 (2013)
26. Kupfer, J.A.: Landscape ecology and biogeography: rethinking landscape metrics in a post-FRAGSTATS landscape. Prog. Phys. Geogr. **36**(3), 400–420 (2012)
27. Lamine, S., Petropoulos, G.P., Singh, S.K., et al.: Quantifying land use/land cover spatio-temporal landscape pattern dynamics from Hyperion using SVMs classifier and FRAG-STATS. Geocarto Int. **33**(8), 862–878 (2017)
28. Midha, N., Mathur, P.K.: Assessment of forest fragmentation in the conservation priority Dudhwa landscape, India using FRAGSTATS computed class level metrics. J. Indian Soc. Remote Sens. **38**(3), 487–500 (2010). https://doi.org/10.1007/s12524-010-0034-6
29. Xing, X., Zhou, Q., Feng, F.: Investigating landscape fragmentation in suburban area using remote sensing and fragstats: a case study of Shanghai (2012)
30. Atta-ur-Rahman, Dawood M.: Spatio-statistical analysis of temperature fluctuation using Mann-Kendall and Sen's slope approach. Clim. Dyn. **48**(3–4), 783–797 (2017). https://doi.org/10.1007/s00382-016-3110-y
31. Gocic, M., Trajkovic, S.: Analysis of changes in meteorological variables using Mann-Kendall and Sen's slope estimator statistical tests in Serbia. Glob. Planet. Change **100**, 172–182 (2013)
32. Da Silva, R.M., Santos, C.A.G., Moreira, M., Corte-Real, J.: Rainfall and river flow trends using Mann-Kendall and Sen's slope estimator statistical tests in the Cobres River basin. Nat. Hazards **77**(2), 1205–1221 (2015). https://doi.org/10.1007/s11069-015-1644-7
33. Rahman, M.A., Yunsheng, L., Sultana, N.: Analysis and prediction of rainfall trends over Bangladesh using Mann-Kendall, Spearman's rho tests and ARIMA model. Meteorol. Atmos. Phys. **129**, 409–424 (2017). https://doi.org/10.1007/s00703-016-0479-4
34. Wang, X., Shen, H., Zhang, W., et al.: Spatial and temporal characteristics of droughts in the Northeast China Transect. Nat. Hazards **76**(1), 601–614 (2015). https://doi.org/10.1007/s11069-014-1507-7

Urban-Expansion Driven Farmland Loss Follows with the Environmental Kuznets Curve Hypothesis: Evidence from Temporal Analysis in Beijing, China

Ying Tu[1,2], Bin Chen[3], Le Yu[1,2], Qinchuan Xin[4], Peng Gong[1,2],
and Bing Xu[1,2(✉)]

[1] Ministry of Education Key Laboratory for Earth System Modeling,
Department of Earth System Science, Tsinghua University,
Beijing 100084, China
bingxu@tsinghua.edu.cn
[2] Joint Center for Global Change Studies, Beijing 100084, China
[3] Department of Land, Air and Water Resources, University of California,
Davis, CA 95616, USA
[4] School of Geography and Planning, Sun Yat-sen University,
Guangzhou 510275, China

Abstract. Since its Reform and Opening-Up, China has been undergoing an unprecedented urbanization process, primarily manifested by intensive urban sprawl and continuous farmland shrinkage. However, few studies have given heed to the interrelation of these two phenomena at a finer spatiotemporal resolution, and limited researches have well quantified what controls the temporal rates of urban-expansion driven farmland loss. By considering Beijing as a case study, here we quantified the rates, patterns, spatiotemporal dynamics and interactions of urban expansion and farmland loss from 1980 to 2015 using the annual land-use/land-cover data. Additionally, by introducing the Environmental Kuznets Curve hypothesis, we further explored the relationship between urban-expansion driven farmland loss and economic growth. Results showed that rapid urban expansion (1592.57 km^2) and extensive farmland loss (1591.36 km^2) were observed during the study period. Three urban growth modes coexisted, where the edge-expansion was dominant (780.98 km^2, 49.06%), followed by the infilling (675.85 km^2, 42.44%) and the outlying (135.72 km^2, 8.50%). Urban expansion was identified to be the dominant driver of farmland loss (96.13%), leading to a more spatially irregular and fragmented distribution of the farmland extent. Lastly, an inverted U-shape relationship was verified between urban-expansion driven farmland loss and economic growth, which indicated a shift from extensive to intensive and economizing land-use patterns in the future.

Keywords: Urbanization · Cropland loss · Land use · Sustainable development

© Springer Nature Singapore Pte Ltd. 2020
Y. Xie et al. (Eds.): GSES 2019/GeoAI 2019, CCIS 1228, pp. 394–412, 2020.
https://doi.org/10.1007/978-981-15-6106-1_29

1   Introduction

Land-use activities, whether converting natural landscapes for human use or changing management practices on human-dominated lands, have transformed a large proportion of the planet's land surface [1]. These changes in land use and land cover considerably alter the Earth's energy balance and biogeochemical cycles, thus contributing to climate change and in turn affecting land surface properties and the provision of ecosystem services [1–4]. Recent studies have revealed that only 5% of the world's land remains unaffected by humans [5] and 60% of global land changes are associated with direct human activities [6]. Among all these human-related modifications on global lands, one typical manifestation is urbanization, which is characterized by the demographic shift from rural to urbanized areas and urban land expansion [7]. As the most populous country in the world, China has been undergoing an unprecedented urbanization process over the past four decades. The proportion of urban population in China rose from 17.92% to 55.88% during the period 1978–2015, an increase of nearly 600 million people [8]. This increasing trend is expected to continue that nearly 70% of the Chinese population—about one billion—will live in urban areas by 2030, based on the latest forecast by the World Bank [9]. The ongoing rapid urbanization process has greatly reshaped China's land use pattern, primarily characterized by intensive urban sprawl and continuous cropland loss [10, 11]. China's urban and rural area reached 209,950 km^2 in 2017 (approximately 13.6 times of the 15,364 km^2 in 1978) while over 118,205 km^2 of farmland was converted into impervious surfaces between 1978 and 2017 [12]. It raises a critical challenge for sustainable development goals that China needs to use limited croplands to feed more than 1.3 billion people.

A number of studies have been working on quantifying the rates, dynamics and driving forces of urban expansion in China, especially in those megacities such as Beijing, Shanghai, Guangzhou, etc. [13–15], and the urban agglomerations including the Beijing–Tianjin–Hebei agglomeration region [7, 16–19], the Yangtze River Delta [20] and the Pearl River Delta [21, 22]. Besides, relevant researches have been carried out at the national scale [23–27]. In the meantime, the continuous loss of farmlands in China has been another focus of increasing public concerns because of their potential threats on food security issues. Numerous empirical studies have summarized the spatiotemporal patterns of farmland loss across different regions [28–32], explored its transitional mechanisms and driving forces [33–35], revealed its impact on grain production and food security [36–38], as well as the protection policy analysis [39–41]. Literally, many researches indicate that urban expansion is the dominant driver of farmland loss [16, 38, 39, 42–44]. For example, according to Liu's latest study in 75 sample cities of China, more than 10,000 km^2 farmlands were encroached by urban lands during the past four decades, which accounted for 54.95% of the newly-expanded urban lands and became the primary land source of urban expansion [43].

However, current studies normally regard urban expansion or farmland loss as a relatively independent phenomenon while few pay attention to their mutual interactions. Hu, Kong, Zheng, Sun, Wang and Min [8] quantitatively evaluated the interaction between urban expansion and farmland loss in Beijing from 1980 to 2015 by developing two relative indicators. But the time interval of the land use data used in that study is large (10 years) and its spatial resolution is relatively coarse (100 m).

Exploring the precise relationship between urban expansion and farmland loss in China at a finer spatiotemporal resolution still remains an open topic.

In 1955, Simon Kuznets suggested that an inverted U-shape relationship exists between per capita income and income inequality. This relationship is now known as the Kuznets Curve, which indicates that whilst at first income inequality increases with per capita income, after a certain turning point it begins to decline [45]. Since 1991, the Environmental Kuznets Curve (EKC) has become a vehicle for describing the relationship between per capita income and the use of natural resources or the emission of wastes [46]. The EKC hypothesis postulates that as income grows, the use of natural resources and pollutions will initially increase, but subsequently decline beyond a certain turning point if economic growth proceeds well enough. During the past few years, this hypothesis has been analyzed in relation to various environmental indicators like deforestation [47], carbon emissions [48], oil exploitation [49] and biodiversity conservation [50].

With increasing attention being paid to China's environmental problems, proving the existence of an EKC for various regions has raised great interest. In fact, by using farmland conversion rates as the environmental indicators, scholars have verified an EKC pattern between economic growth and farmland conversion in China during the post-reform era [51, 52]. However, limited studies have well quantified what controls the temporal rates of urban-expansion driven farmland loss. Could urban-expansion driven farmland loss be compatible with economic growth? Is there any EKC relationship between economic growth and urban-expansion driven farmland loss in Chinese cities? This study attempts to fill these gaps.

In response to the above analysis, here we used the long-term annual land-use/land-cover (A-LULC) product to quantify the rates, patterns and dynamics of urban expansion and farmland loss, in addition to their interrelationship in Beijing, China for the years 1980–2015. Furthermore, a relative factor, which denotes the contribution of urban expansion on farmland loss, was used as the environmental indicator for EKC analysis. Our results would reveal the real situation of continuous farmland shrinkage consumed by urban expansion and the changes of their relationships along with economic growth, which help formulate China's urban planning and farmland protection polices in the future.

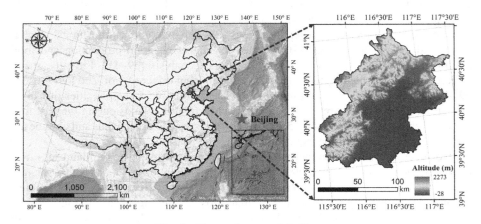

Fig. 1. The study area of Beijing.

2   Study Area and Data Sources

2.1   Study Area

The geographical location and topographic features of Beijing city is shown in Fig. 1. Located in the northwest of the North China Plain of China (between 39°38′–40°51′ N and 115°25′–117°30′ E), Beijing covers an area of 16,410.54 km^2 with an average altitude of 43.71 m. The north-west of the city is dominated by mountains and hills, accounting for 62% of the area, while the central and southeastern are flat plains. As the capital of China, in addition to the political and cultural centre of the country, Beijing has undergone rapid urbanization over the past four decades: the urban population grew from approximately 5.21 million (57.62%) in 1980 to 18.78 million (86.51%) in 2015 [8].

2.2   Collection and Process of Data

The annual land-use/land-cover (A-LULC) data from 1980 to 2015 with a spatial resolution of 30 m was retrieved from our previous study using the time-series Landsat and MODIS satellite images (Xu et al., under review). We validated these data from various spatiotemporal perspectives. At the national scale, by using the global reference sample set [53], the classification accuracies for the years 2013, 2014 and 2015 were 72.18%, 71.60% and 72.21%, respectively. In the city of Beijing, the overall accuracies for farmland and urban land were 72.10% and 91.89%, respectively, where the multi-temporal validation samples were retrieved from our previous study [54]. Besides, a variation sample set was used to detect the correctness of the change year and its accuracy was 75.61%.

The A-LULC data was first divided into six primary classes: cultivated land, woodlands, grasslands, water bodies, construction land, and unused land, and then further divided into 25 sub-classes. In this study, we extracted the cultivated land and construction land in the general categories and defined them as 'farmland' and 'urban land', respectively. Then, all the LULC maps were clipped based on the administrative boundaries of Beijing and then re-projected into Albers Conical Equal Area projection coordinate system.

In addition to these LULC maps above, the Digital Elevation Model (DEM) data was obtained from the Geospatial Data Cloud Platform (http://www.gscloud.cn/) and the gross domestic product (GDP) statistics for the study period was downloaded from Beijing Municipal Bureau of Statistics (http://www.bjstats.gov.cn/tjsj/).

3   Methods

3.1   Rate of Urban Expansion and Farmland Loss

To track changes in the rate, we used two indices to quantify the speed of urban expansion and farmland loss, which are defined as follows:

$$\text{AUER} = \frac{U_{end} - U_{start}}{U_{start}} \times 100 \div t \tag{1}$$

where AUER is the annual urban expansion rate, U_{start} is the urban area at the starting year, U_{end} is the urban area at the ending year, and t is the time interval.

$$\text{AFLR} = \frac{F_{start} - F_{end}}{F_{end}} \times 100 \div t \tag{2}$$

where AFLR denotes the annual farmland loss rate, F_{start} denotes the area of farmland at the starting year, and F_{end} denotes the area of farmland at the ending year.

AUER (AFLR) directly measures the annual area change rate of urban land (farmland), which is effective for comparison of urban expansion (farmland loss) in different periods.

3.2   Spatial Patterns of Urban Expansion and Farmland Loss

Landscape Expansion Index (LEI) for Urban Expansion. Given an urban growth rate, urbanization may assume different urban expansion modes (i.e., infilling, edge-expansion, and outlying) [55]. Any other pattern can be regarded as the variant or hybrid of these three basic forms [56, 57]. An infilling type refers to the one that the gap (or hole) between old patches or within an old patch is filled up with the newly grown patch. Forman [56] discusses the edge-expansion type, defined as a newly grown patch spreading unidirectionally in more or less parallel strips from an edge. If the newly grown patch is found isolated from the old, then it would be defined as an outlying type.

In this study, the landscape expansion index (LEI) [58] was used to detect these urban expansion modes:

$$\text{LEI} = \frac{A_o}{A_o + A_v} \times 100 \tag{3}$$

where A_o is the intersection between the buffer zone and the occupied category, and A_v is the intersection between the buffer zone and the vacant category.

Table 1. Description and functions of landscape metrics [59].

Metric	Description	Unit	Range
Largest patch index (LPI)	Identifies the largest continuous patch for each class in the entire landscape	Percent	$0 < \text{LPI} \leq 100$ LPI increases as the continuity of the landscape increases LPI approaches 0 when the largest patch occupies a decreasing area in the landscape and equals 100 when the largest patch occupies the entire landscape itself
Aggregation index (AI)	Equals the number of like adjacencies involving the corresponding class	Percent	$0 \leq \text{AI} \leq 100$ A larger AI value indicates a more aggregated and compact patch AI equals 0 when the focal patch type is maximally disaggregated and equals 100 when the patch type is maximally aggregated into a single, compact patch

Three thresholds of LEI values were used to determine the urban expansion types [58]: (1) $50 < \text{LEI} \leq 100$, the infilling type; (2) $0 < \text{LEI} \leq 50$, the edge-expansion type; (3) the outlying type if $\text{LEI} = 0$.

Landscape Metrics for Farmland Loss. Two landscape pattern metrics, the largest patch index (LPI) and the aggregation index (AI), were applied to assess the contiguous degrees of the farmland at the landscape level. Their definitions and functions are listed in Table 1. If the farmland is with an increasing LPI and AI, its corresponding landscape will be characterized as more compact and aggregated. Both LPI and AI were calculated using FRAGSTATS software version 4.2.1.

3.3   Interactions Between Urban Expansion and Farmland Loss

Standard Deviation Ellipse Analysis. A common way to measure the trend of a set of points or areas is to calculate the standard distance separately in the x- and y-directions. These two measures define the axes of an ellipse encompassing the distribution of features. The derived ellipse is regarded as the standard deviational ellipse (SDE), because it calculates the standard deviation of the x-coordinates and y-coordinates from the center to define the axes of the ellipse [60], using Eq. (4) [61].

$$\text{SDE}_x = \sqrt{\frac{\sum_{i=1}^n (x_i - \overline{X})^2}{n}}, \ \text{SDE}_y = \sqrt{\frac{\sum_{i=1}^n (y_i - \overline{Y})^2}{n}} \tag{4}$$

where x_i and y_i are the coordinates for feature i, represents the mean center for the features, and n is equal to the total number of features. In detail, the position of the ellipse center is presumed to be the most representative single position of all locations in the area occupied by the urban-expansion driven farmland loss, and its change reflects the transfer of the mean enter of land gravity [62]. The angle of a long half axis deviating clockwise from the y-coordinates is deemed to be the rotation angle of SDE. Another indicator is the oblateness of SDE, which describes the flattening degree. The lower the value, the more obviously an ellipse tends to be a circle. Conversely, the larger the value of oblateness, the more obvious the directionality of the data.

Quantity Assessment. To assess the relative consumption of farmland for urban expansion, one quantitative index is introduced as follows:

$$\text{Contribution} = \frac{AFU}{AF} \tag{5}$$

where Contribution represents the contribution of urban expansion on farmland loss. AFU is the area of farmland consumed by urban expansion and AF is the total area of farmland transfer out.

The value of Contribution ranges from 0 to 1. The higher the value of Contribution, the greater the contribution of urban expansion on the loss of farmland. When the Contribution value reaches 1 (i.e., $AFU = AF$), all lost farmland is converted to urban land.

3.4   The EKC Model

We use the Contribution index in Sect. 3.3 as an environmental indicator for the EKC model. If an inverted U-shape relationship does exist between economic growth and urban-expansion driven farmland loss, then

$$\text{FU}_t = \alpha_0 + \alpha_1 E_t + \alpha_2 E_t^2 + \theta_t \tag{6}$$

where subscript t refers to time 1980 to 2015, FU denotes urban-expansion driven farmland loss (i.e. values of Contribution), and E denotes economic growth such as values of GDP statistics. α_0 is an intercept term while θ is the white noise error. α_1 and α_2 are the primary coefficient and quadratic coefficient of the quadratic regression, respectively. If $\alpha_1 > 0$ and $\alpha_2 < 0$ are obtained in the regression, then urban-expansion driven farmland loss is said to follow with the EKC assumption.

4   Results

4.1   Urban Expansion in Beijing During 1980–2015

Overall, Beijing experienced rapid urban expansion during the study period. The urban land expanded from 1221.14 km^2 in 1980 to 2813.71 km^2 in 2015, an average expansion area of 45.50 km^2 per year. As showed in Fig. 2, the peak rates of urban expansion took place in the year of 1988 and 2012 (AUER = 15.42 and 16.45, respectively). Before 1987, urban land expanded slowly with an average AUER of 0.38, then it underwent three acceleration stages in 1987–1993, 2000–2004 and 2012–2015 (AUER = 6.24, 4.26 and 5.07, respectively), and two deceleration stages in 1993–2000 and 2004–2010 (AUER = 0.51 and 0.93, respectively).

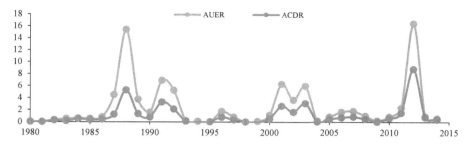

Fig. 2. Rate of urban expansion (AUER) and farmland loss (AFLR) in Beijing from 1980 to 2015.

Figure 3 presents the spatiotemporal distributions of Beijing's urban expansion from 1980 to 2015. Before 1987, urban expansion mainly took place in the southern part of the central city. Then, during 1988–1993, the focus of expansion shifted to the north. After 2000, obvious urban expansion occurred in various regions. As time went by, the newly constructed urban land extended along the boundary of old urban land and was farther and farther away from the city center.

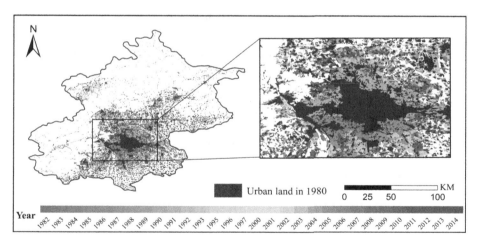

Fig. 3. Spatiotemporal distributions of urban expansion in Beijing from 1980 to 2015. The enlarged area is the central area of Beijing.

We identified three urban growth modes (i.e., infilling, edge-expansion, and outlying) by calculating the LEI values of landscape patches (Fig. 4). The existing urban land in 1980 was mainly surrounded by the infilling mode while the outlying mode was distributed sparsely at outskirts, and edge-expansion mode filled the gaps between them.

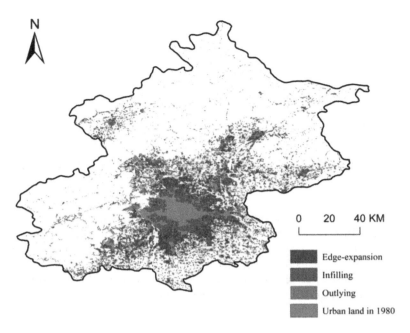

Fig. 4. Spatial distribution of three urban growth modes in Beijing from 1980 to 2015.

As illustrated in Fig. 5, the infilling, edge-expansion and outlying modes of urban expansion coexisted during the study period, and their compositional structure changed over time. The edge-expansion mode played a predominant role, with an area of 780.98 km^2, which accounts for 49.06% of the total expansion area. The infilling mode followed by, which occupied 675.85 km^2 (42.44%) of the land. As for the outlying mode, it was the smallest among all, with only 135.72 km^2 (8.50%). Before 1996, little land of urban expansion was identified as the outlying growth (Fig. 5b). Only in the year of 2000, 2005 and 2010, the proportion of the outlying mode exceeded 30%, but it's still insignificant given the small areas of urban expansion in total during these years.

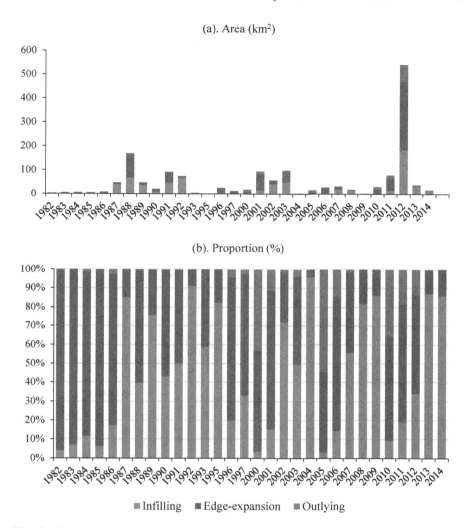

Fig. 5. Three urban growth modes in Beijing from 1980 to 2015: (a) area, (b) proportion.

4.2 Farmland Loss in Beijing During 1980–2015

Under the rapid urbanization process, the area of farmland in Beijing reduced from 4794.90 km^2 in 1980 to 3203.54 km^2 in 2015, with an average loss area of 45.47 km^2 per year. The loss speed of farmland (AFLR) experienced a similar trend of AUER with some rise and fall phases (Fig. 2). The peak rates of farmland loss took place in the year of 1988 and 2012 (AFLR = 5.30 and 8.82, respectively). Before 1987, farmland lost steadily and slowly with an average AFLR of 0.24, then it underwent three acceleration stages in 1987–1993, 2000–2004 and 2012–2015 (AFLR = 2.34, 1.97 and 2.88, respectively), and two deceleration stages in 1993–2000 and 2004-2010 (AFLR = 0.22 and 0.45, respectively). Overall, the average value of AFLR for the whole study period (1.13) was less than that of AUER (2.48), which indicated that urban expansion outpaced farmland loss.

Figure 6 shows the spatiotemporal distributions of farmland loss in Beijing during the study period. During the first decade of 1980–1990, farmland loss was concentrated in the south of central area. After 1990, it gradually moved to the northern part of the city. Since 2000, farmland loss mainly occurred in the suburbs of Beijing. Furthermore, farmland loss exhibited similar spatial distributions compared to urban expansion in the meantime (Fig. 3), which implies that farmland was the primary land source for urban expansion.

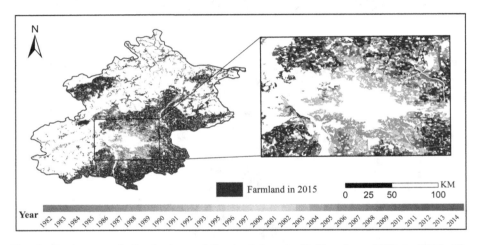

Fig. 6. Spatiotemporal distributions of farmland loss in Beijing from 1980 to 2015. The enlarged area is the central area of Beijing.

Figure 7 displays the changes of landscape metrics of farmland in Beijing. Between 1980 and 2015, the values of LPI and AI decreased by 25.49% and 1.07%, respectively, where LPI reduced from 28.99 in 1980 from 20.71 in 2015 while AI reduced from 96.65 in 1980 to 95.40 in 2015. The decreasing trends of landscape metrics indicated that farmland became more fragment during the study period. In other words, the spatial pattern of the farmland was transformed from centralized to decentralized, which was greatly caused by the expansion of the built-up environment.

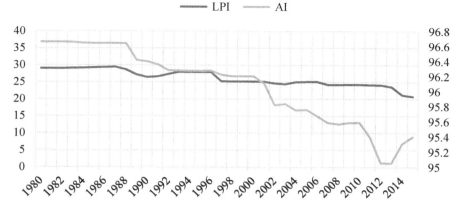

Fig. 7. Changes of landscape metrics (LPI and AI) in Beijing from 1980 to 2015.

4.3   Interrelationship Between Urban Expansion and Farmland Loss

We built the transfer matrix of land use in Beijing between 1980 and 2015 (Table 2). In 2015, 38.03% of the urban land (1070.02 km^2) was built in 1980, while the remaining mainly converted from farmland (54.37%), woodland (4.49%) and grassland (1.77%). As for farmland, only 60.14% (2856.50 km^2) remained unchanged in 2015 while about 32.21% converted to urban land, followed by woodland (4.93%) and grassland (2.14%). In total, approximately 1529.77 km^2 of farmland was occupied by urban land in Beijing, accounting for 96.06% of the newly expanded urban land (1592.56 km^2) and 96.13% of the lost farmland (1591.36 km^2). Moreover, the spatiotemporal distributions of farmland encroached by newly urban land (Fig. 8) were very similar to those of urban expansion (Fig. 3) and farmland loss (Fig. 6). Consequently, we concluded that urban expansion was the dominant driver of farmland loss in Beijing.

Table 2. Transfer matrix of land use in Beijing from 1980 to 2015 (unit: km^2).

	Farm	Wood	Grass	Water	Urban	Others	Total (1980)
Farm	2856.50	234.08	101.69	65.61	1529.77	7.25	**4794.90**
Wood	131.63	5650.49	41.39	21.24	126.42	0.18	**5971.35**
Grass	64.50	144.11	846.19	17.91	49.85	1.41	**1123.97**
Water	46.79	26.97	35.73	181.34	37.22	4.15	**332.20**
Urban	104.10	19.41	18.56	8.75	1070.02	0.30	**1221.14**
Others	0.01	0.08	0.03	0.01	0.43	0.40	**0.96**
Total (2015)	**3203.54**	**6075.12**	**1043.60**	**294.87**	**2813.70**	**13.69**	

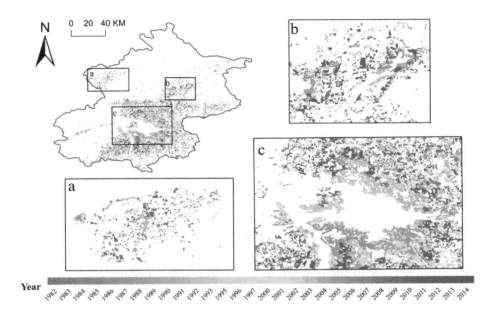

Fig. 8. Spatiotemporal distributions of farmland encroached by newly urban land in Beijing from 1980 to 2015. a–c highlights some areas with intense changes.

As shown in Fig. 9, the center and direction of urban-expansion driven farmland loss migrated during the study period. During 1985–1990, the weighted mean center of SDEs shifted from the northwest to the southeast. Then since 1990, it experienced slightly changes. As for the oblateness of SDEs, it decreased from 0.65 in 1985–1990 to 0.22 in 2010–2015, reflecting a reduced dispersion degree of urban-expansion driven farmland loss in Beijing.

4.4   Urban-Expansion Driven Farmland Loss Keeps to EKC Hypothesis

Based on Eq. (5–6), we further explored the relationships between economic growth and urban-expansion driven farmland loss at the city level (Fig. 10). The R^2 of the two global models were 0.74 ($p < 0.001$) and 0.73 ($p < 0.001$), respectively, and they both exhibited inverted U-shapes (Fig. 10). The turning point, where decreasing urban-expansion driven farmland conversion occurred simultaneously to economic growth, was discovered in the year of 2001 when GDP reached up to 370.80 billion Yuan. As time went by, the contribution of urban expansion on farmland loss initially increased, but subsequently declined when it reached the turning point. In the future, the consumption of urban expansion on farmland will continue to decrease as more attention is paid to the issue of sustainable development.

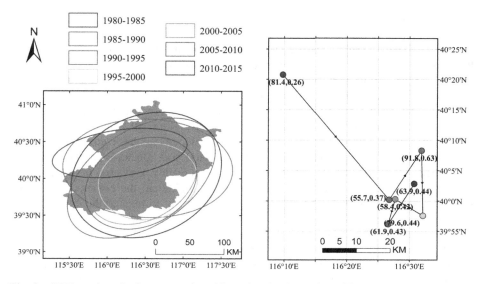

Fig. 9. SDE results of urban-expansion driven farmland loss in Beijing. The distributions of SDEs are shown in the left column and the weighted mean center of SDEs are presented in the right column. The brackets in the right column represent the rotation angle and the oblateness of SDEs, respectively.

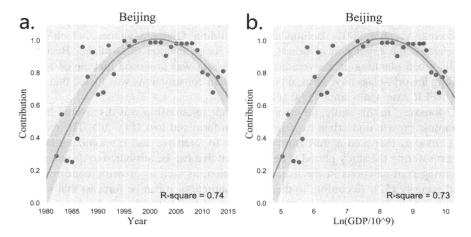

Fig. 10. EKC models between economic growth and urban-expansion driven farmland loss in Beijing. The independent variables used in model (a) and model (b) are year and GDP, respectively. Both models have passed the significance test ($p < 0.001$). Yellow strips denote 95% confidence intervals. (Color figure online)

5 Discussion

This study discovered that Beijing had undergone intense urban expansion and farm-land loss from 1980 to 2015, and their rates fluctuated with some rise and fall phases (Fig. 2). A series of economic reforms and national strategies, in addition to the farmland protection policies, would contribute to the explanation of these findings. After the implementation of the Reformed and Open-Door Policy in 1978, Beijing witnessed a boom in town and village enterprises, development zones and real estate [63]. Although the economy was effectively developed by these enterprises, there followed a bunch of problems that related to land use. Because of its scattered, unplanned spatial distribution, urban land, which was mainly converted from farmland, expanded overly, and ended up idle and wasted. In addition, the construction of the second and third ring roads of Beijing at that time greatly stimulated the expansion of the urban area and its occupation of farmland. As a result, during 1980 to 1993, a considerable area of farmland was occupied (Fig. 8), and enormous amount of urban land left idle due to the transition from intensive to extensive economic growth.

The "overheated" development zones and real estate were noticed by the Chinese government. In 1994, the State Council of China enacted the Regulations on the Protection of Basic Farmland [64], and began to implement a strict farmland protection policy aiming to limit farmland loss occupied by urban expansion. Since then, the expansion of urban land and its occupation of farmland were effectively restrained (Fig. 2, Fig. 8, Fig. 10). In this period, the dominant type of urban expansion was the infilling mode (Fig. 5), which consisted of filling the small gaps in the surrounded urban areas.

However, in 2001, Beijing successfully applied to host the 2008 Olympic Games. Accordingly, after 2001, Beijing started building Olympic venues, which greatly accelerated the expansion of urban land and its occupation of farmland. Besides, in 2001, urban planning of the greater Beijing region was proposed that aimed to make Beijing a world-class city. Thus, a boom in urban constructions was fueled that AUER and AFLR rose again since 2000 (Fig. 2).

Moreover, in this study, an inverted U-shape relationship was discovered between economic growth and urban-expansion driven farmland loss (Fig. 10). The results of EKC models indicated a shift from extensive to intensive and economizing land-use patterns during the study period, a development that can be attributed to the emergence of strong contradictions in the man-land relation, as well as growing concerns about food security [52]. According to the EKC assumption, it can be forecast that with the progress of technology and the transformation of industry, China's land use patterns will be more intensive and efficient in the future. Meanwhile, the government plays an important role in maintaining a dynamic balance between socioeconomic development and resource conservation.

We admitted that there are some limitations enclosed in this study. First, only Beijing was considered as the study area, which restricted the feasibility of the research methods in other regions. For example, can EKC assumption be applied to other Chinese cities or even the entire country? If it does, what are the current stages of urban-expansion driven farmland loss in different cities and what are the driving forces behind them? Second, this study only quantified the loss of farmland, but does not take the compensation of farmland into account, as well as its impact on grain yield, ecosystem, biodiversity, etc. To investigate the current situation of urban expansion and farmland loss in other areas, more experiments need to be conducted in our future work. Besides, some profound mechanism models can be introduced, which reveal the comprehensive impact of urbanization on natural environment.

6 Conclusions

By taking Beijing as the study area, this research quantified the rates, patterns, spatiotemporal dynamics and interactions of urban expansion and farmland loss from 1980 to 2015 using the A-LULC data integrated with geographical information science techniques and landscape ecology approaches. Our results showed the pattern of rapid urban expansion along with continuous farmland loss. We also identified three coexisting urban growth modes for the urbanization process that the edge-expansion mode was dominant, followed by the infilling mode and the outlying mode. In addition, Urban expansion was recognized as the dominant driver of farmland loss, resulting in a more spatially irregular and fragmented distribution of the farmland extent. Last but not least, urban-expansion driven farmland loss was affirmed to follow with the EKC hypothesis. As more attention is paid to China's sustainable development, we forecast that China's land use patterns will be more intensive and efficient in the future.

Acknowledgement. This research was funded by Ministry of Science and Technology of the People's Republic of China under the National Key Research and Development Program (2016YFA0600104), and donations from Delos Living LLC and the Cyrus Tang Foundation to Tsinghua University. We would like to thank the three anonymous reviewers for their constructive comments and suggestions, which greatly help the improvement of this manuscript.

References

1. Foley, J.A., et al.: Global consequences of land use. Science **309**, 570–574 (2005)
2. Turner, B.L., Lambin, E.F., Reenberg, A.: The emergence of land change science for global environmental change and sustainability. Proc. Natl. Acad. Sci. **104**, 20666–20671 (2007)
3. Le Quéré, C., et al.: Global carbon budget 2016. Earth Syst. Sci. Data **8**, 605–649 (2016)
4. Alkama, R., Cescatti, A.: Biophysical climate impacts of recent changes in global forest cover. Science **351**, 600–604 (2016)
5. Kennedy, C.M., Oakleaf, J.R., Theobald, D.M., Baruch-Mordo, S., Kiesecker, J.: Managing the middle: a shift in conservation priorities based on the global human modification gradient. Glob. Change Biol. **25**, 811–826 (2019)
6. Song, X.-P., et al.: Global land change from 1982 to 2016. Nature **560**, 639 (2018)
7. Wu, W., Zhao, S., Zhu, C., Jiang, J.: A comparative study of urban expansion in Beijing, Tianjin and Shijiazhuang over the past three decades. Landscape Urban Plann. **134**, 93–106 (2015)
8. Hu, Y., Kong, X., Zheng, J., Sun, J., Wang, L., Min, M.: Urban expansion and farmland loss in Beijing during 1980–2015. Sustainability **10**, 3927 (2018)

9. Wahnschafft, R., Wei, F.: Urban China: toward efficient, inclusive, and sustainable urbanization the world bank and the development research center of the state council, People's Republic of China world bank, Washington, DC, 583 pages. In: Natural Resources Forum, pp. 151–152. Wiley Online Library (2015)

10. Liu, T., Liu, H., Qi, Y.: Construction land expansion and cultivated land protection in urbanizing China: insights from national land surveys, 1996–2006. Habitat Int. **46**, 13–22 (2015)

11. Long, H., Ge, D., Zhang, Y., Tu, S., Qu, Y., Ma, L.: Changing man-land interrelations in China's farming area under urbanization and its implications for food security. J. Environ. Manag. **209**, 440–451 (2018)

12. Gong, P., Li, X., Zhang, W.: 40-year (1978–2017) human settlement changes in China reflected by impervious surfaces from satellite remote sensing. Sci. Bull. **64**, 756–763 (2019)

13. Kuang, W., Chi, W., Lu, D., Dou, Y.: A comparative analysis of megacity expansions in China and the US: patterns, rates and driving forces. Landscape Urban Plann. **132**, 121–135 (2014)

14. Li, X., Zhou, W., Ouyang, Z.: Forty years of urban expansion in Beijing: what is the relative importance of physical, socioeconomic, and neighborhood factors? Appl. Geogr. **38**, 1–10 (2013)

15. Fei, W., Zhao, S.: Urban land expansion in China's six megacities from 1978 to 2015. Sci. Total Environ. **664**, 60–71 (2019)

16. Tan, M., Li, X., Xie, H., Lu, C.: Urban land expansion and arable land loss in China—a case study of Beijing–Tianjin–Hebei region. Land Use Policy **22**, 187–196 (2005)

17. Kuang, W.: Simulating dynamic urban expansion at regional scale in Beijing-Tianjin-Tangshan Metropolitan Area. J. Geogr. Sci. **21** (2011). Article number: 317. https://doi.org/10.1007/s11442-011-0847-4

18. Sun, Y., Zhao, S.: Spatiotemporal dynamics of urban expansion in 13 cities across the Jing-Jin-Ji urban agglomeration from 1978 to 2015. Ecol. Ind. **87**, 302–313 (2018)

19. Li, X., Yeh, A.G.-O.: Analyzing spatial restructuring of land use patterns in a fast growing region using remote sensing and GIS. Landscape Urban Plann. **69**, 335–354 (2004)

20. Tian, G., Jiang, J., Yang, Z., Zhang, Y.: The urban growth, size distribution and spatio-temporal dynamic pattern of the Yangtze River Delta megalopolitan region, China. Ecol. Model. **222**, 865–878 (2011)

21. Weng, Q.: Land use change analysis in the Zhujiang Delta of China using satellite remote sensing, GIS and stochastic modelling. J. Environ. Manag. **64**, 273–284 (2002)

22. Seto, K.C., Fragkias, M.: Quantifying spatiotemporal patterns of urban land-use change in four cities of China with time series landscape metrics. Landscape Ecol. **20**, 871–888 (2005). https://doi.org/10.1007/s10980-005-5238-8

23. Wang, L., et al.: China's urban expansion from 1990 to 2010 determined with satellite remote sensing. Chin. Sci. Bull. **57**, 2802–2812 (2012). https://doi.org/10.1007/s11434-012-5235-7

24. Schneider, A., Mertes, C.: Expansion and growth in Chinese cities, 1978–2010. Environ. Res. Lett. **9**, 024008 (2014)

25. Zhao, S., et al.: Rates and patterns of urban expansion in China's 32 major cities over the past three decades. Landscape Ecol. **30**, 1541–1559 (2015). https://doi.org/10.1007/s10980-015-0211-7

26. Kuang, W., Liu, J., Dong, J., Chi, W., Zhang, C.: The rapid and massive urban and industrial land expansions in China between 1990 and 2010: a CLUD-based analysis of their trajectories, patterns, and drivers. Landscape Urban Plann. **145**, 21–33 (2016)

27. Liu, F., et al.: Urban expansion in China and its spatial-temporal differences over the past four decades. J. Geogr. Sci. **26**, 1477–1496 (2016). https://doi.org/10.1007/s11442-016-1339-3

28. Zhong, T.-Y., Huang, X.-J., Zhang, X.-Y., Wang, K.: Temporal and spatial variability of agricultural land loss in relation to policy and accessibility in a low hilly region of southeast China. Land Use Policy **28**, 762–769 (2011)

29. Liu, G., Zhang, L., You, H.: Spatiotemporal dynamics of arable land in the Nanjing metropolitan region, China. Environ. Earth Sci. **73**, 7183–7191 (2015). https://doi.org/10.1007/s12665-014-3898-x

30. Chang, H.-S., Chen, T.-L.: Exploring spatial patterns of farmland transactions and farmland use changes. Environ. Monit. Assess. **187** (2015). Article number: 596. https://doi.org/10.1007/s10661-015-4825-7

31. Gong, J., Jiang, C., Chen, W., Chen, X., Liu, Y.: Spatiotemporal dynamics in the cultivated and built-up land of Guangzhou: Insights from zoning. Habitat Int. **82**, 104–112 (2018)

32. Yu, Q., Hu, Q., van Vliet, J., Verburg, P.H., Wu, W.: GlobeLand30 shows little cropland area loss but greater fragmentation in China. Int. J. Appl. Earth Obs. Geoinf. **66**, 37–45 (2018)

33. Song, J., Ye, J., Zhu, E., Deng, J., Wang, K.: Analyzing the impact of highways associated with farmland loss under rapid urbanization. ISPRS Int. J. Geo-Inf. **5**, 94 (2016)

34. Su, S., Jiang, Z., Zhang, Q., Zhang, Y.: Transformation of agricultural landscapes under rapid urbanization: a threat to sustainability in Hang-Jia-Hu region, China. Appl. Geogr. **31**, 439–449 (2011)

35. Xiao, R., Liu, Y., Huang, X., Shi, R., Yu, W., Zhang, T.: Exploring the driving forces of farmland loss under rapidurbanization using binary logistic regression and spatial regression: a case study of Shanghai and Hangzhou Bay. Ecol. Ind. **95**, 455–467 (2018)

36. Yang, H., Li, X.: Cultivated land and food supply in China. Land Use Policy **17**, 73–88 (2000)

37. Shi, W., Tao, F., Liu, J.: Changes in quantity and quality of cropland and the implications for grain production in the Huang-Huai-Hai Plain of China. Food Secur. **5**, 69–82 (2013). https://doi.org/10.1007/s12571-012-0225-9

38. Song, W., Pijanowski, B.C., Tayyebi, A.: Urban expansion and its consumption of high-quality farmland in Beijing, China. Ecol. Ind. **54**, 60–70 (2015)

39. Lichtenberg, E., Ding, C.: Assessing farmland protection policy in China. Land Use Policy **25**, 59–68 (2008)

40. Liang, C., et al.: Farmland protection policies and rapid urbanization in China: a case study for Changzhou City. Land Use Policy **48**, 552–566 (2015)

41. Shen, X., Wang, L., Wu, C., Lv, T., Lu, Z., Luo, W., Li, G.: Local interests or centralized targets? How China's local government implements the farmland policy of Requisition-Compensation Balance. Land Use Policy **67**, 716–724 (2017)

42. Deng, X., Huang, J., Rozelle, S., Zhang, J., Li, Z.: Impact of urbanization on cultivated land changes in China. Land Use Policy **45**, 1–7 (2015)

43. Liu, F., et al.: Chinese cropland losses due to urban expansion in the past four decades. Sci. Total Environ. **650**, 847–857 (2019)

44. Liu, J., Tian, H., Liu, M., Zhuang, D., Melillo, J.M., Zhang, Z.: China's changing landscape during the 1990s: large-scale land transformations estimated with satellite data. Geophys. Res. Lett. **32**, L02405 (2005). https://doi.org/10.1029/2004GL021649

45. Kuznets, S.: Economic growth and income inequality. Am. Econ. Rev. **45**, 1–28 (1955)

46. Grossman, G.M., Krueger, A.B.: Environmental impacts of a North American free trade agreement. National Bureau of Economic Research (1991)

47. Cropper, M., Griffiths, C.: The interaction of population growth and environmental quality. Am. Econ. Rev. **84**, 250–254 (1994)
48. Panayotou, T., Peterson, A., Sachs, J.D.: Is the environmental Kuznets curve driven by structural change? What extended time series may imply for developing countries (2000)
49. Esmaeili, A., Abdollahzadeh, N.: Oil exploitation and the environmental Kuznets curve. Energy Policy **37**, 371–374 (2009)
50. Mills, J.H., Waite, T.A.: Economic prosperity, biodiversity conservation, and the environmental Kuznets curve. Ecol. Econ. **68**, 2087–2095 (2009)
51. Liu, L., Song, M., Yokogawa, H., Qu, B.: Exploring the environmental Kuznets curve hypothesis between economic growth and farmland conversion in China. J.-Fac. Agric. Kyushu Univ. **53**, 321 (2008)
52. Li, Y., Cong, C., Wang, Y., Liu, Y.: Urban-rural transformation and farmland conversion in China: the application of the environmental Kuznets Curve. J. Rural Stud. **36**, 311–317 (2014)
53. Yu, L., et al.: Using a global reference sample set and a cropland map for area estimation in China. Sci. China Earth Sci. **60**, 277–285 (2017). https://doi.org/10.1007/s11430-016-0064-5
54. Li, X., Gong, P., Liang, L.: A 30-year (1984–2013) record of annual urban dynamics of Beijing City derived from Landsat data. Remote Sens. Environ. **166**, 78–90 (2015)
55. Li, C., Li, J., Wu, J.: Quantifying the speed, growth modes, and landscape pattern changes of urbanization: a hierarchical patch dynamics approach. Landscape Ecol. **28**, 1875–1888 (2013). https://doi.org/10.1007/s10980-013-9933-6
56. Forman, R.: Land Mosaics: The Ecology of Landscapes and Regions 1995. Springer, Heidelberg (2014)
57. Wilson, E.H., Hurd, J.D., Civco, D.L., Prisloe, M.P., Arnold, C.: Development of a geospatial model to quantify, describe and map urban growth. Remote Sens. Environ. **86**, 275–285 (2003)
58. Liu, X., Li, X., Chen, Y., Tan, Z., Li, S., Ai, B.: A new landscape index for quantifying urban expansion using multi-temporal remotely sensed data. Landscape Ecol. **25**, 671–682 (2010). https://doi.org/10.1007/s10980-010-9454-5
59. McGarigal, K.: FRAGSTATS help. Documentation for FRAGSTATS 4 (2014)
60. Gong, J.: Clarifying the standard deviational ellipse. Geogr. Anal. **34**, 155–167 (2002)
61. Lefever, D.W.: Measuring geographic concentration by means of the standard deviational ellipse. Am. J. Sociol. **32**, 88–94 (1926)
62. Wang, B., Shi, W., Miao, Z.: Confidence analysis of standard deviational ellipse and its extension into higher dimensional Euclidean space. PLoS ONE **10**, e0118537 (2015)
63. Song, W.: Decoupling cultivated land loss by construction occupation from economic growth in Beijing. Habitat Int. **43**, 198–205 (2014)
64. The State Council of the People's Republic of China: Regulations on the Protection of Prime Farmland (1994)

Spatio-Temporal Evolution of Environmental Efficiency of Construction Land in Yangtze River Economic Zone

Huimin Xu[✉]

Economics School, Wuhan Donghu University, Wuhan 430212, China
xuhuimin1985_2008@163.com

Abstract. With the rapid expansion of construction land in Yangtze River Economic Zone, problems such as shortage of land resources, low utilization efficiency and serious environmental pollution have become increasingly important. This study used Data Envelopment Analysis (DEA) model of super efficiency and Malmquist index to measure the environmental efficiency of construction land in Yangtze River Economic Zone from 2006 to 2016, and analyzed the spatial and temporal evolution characteristics, by using sulfur dioxide, carbon dioxide emissions and PM 2.5 as the environmental cost. It was found that the overall environmental efficiency of construction land in Yangtze River Economic Zone is low, which is smaller than 1. The average efficiency of the eastern coastal economic zones is the highest, which is larger than 1. In addition, from the perspective of the change trend of the environmental efficiency, Shanghai and Jiangsu have witnessed rapid growth, and technological progress is the main driving force. However, some provinces (e.g. Zhejiang, Jiangxi, Hunan and Anhui) also experienced decrease of technical efficiency. The findings provide scientific support for controlling the intensity and allocation of regional construction land under the background of ecological civilization in Yangtze River Economic Zone.

Keywords: Construction land · Environmental efficiency · Regional difference · Super-efficiency DEA · Malmquist index

1 Introduction

1.1 A Subsection Sample

Construction land is a space carrier for human activities. In the past two decades, China's construction land has shown a rapid expansion, resulted in over-occupied agricultural land, and the problem of land environmental pollution has become increasingly prominent [1–3]. With the continuous deepening of ecological civilization construction, other ecological and environmental problems caused by the expansion of construction land, low-efficiency land re-development and carbon emission efficiency have been widely concerned by the society [4, 5].

The utilization efficiency of construction land is a reflection of the utilization status and allocation level of regional construction land. The current research mainly focuses

© Springer Nature Singapore Pte Ltd. 2020
Y. Xie et al. (Eds.): GSES 2019/GeoAI 2019, CCIS 1228, pp. 413–421, 2020.
https://doi.org/10.1007/978-981-15-6106-1_30

on the characteristics, differences and functions of urban construction land use efficiency [6–8]. For example, Zhang Yajie (2015) analyzed the overall characteristics and regional differences in the utilization efficiency of construction land in 41 cities and other areas in the Middle Reaches of the Yangtze River from 2000 to 2012, and the driving efficiency of construction land utilization was explored from the perspective of urbanization level, government intervention level and foreign investment level [9]. Jing Shouwu (2017) constructed a convergence model for the inter-provincial energy environment efficiency and explored the convergence mechanism [10]. In view of the environmental efficiency evaluation of construction land, Wu Zhenhua (2017) used different methods to measure the environmental efficiency of urban construction land in Jianghu, Zhejiang and Wuhan urban circles [11]. Based on the previous researches, the study of environmental efficiency of construction land needs to be further explored. First, the environmental efficiency evaluation indicators for construction land have less consideration for energy consumption and urban residents' living. Second, comparative studies on environmental efficiency differences of construction land in different provinces and major economic regions in Yangtze River Economic Zone should be strengthened. Third, most scholars directly use the mixed Tobit or random effect Tobit method to analyze the factors affecting environmental efficiency, often overlooking the fixed effect characteristics of panel data.

The Yangtze River Economic Zone traverses eleven provincial regions in the three major regions of China's - eastern, central and western regions. The population and economic amount are more than 40% of those of the country. It has important strategic significance in ecological civilization construction. Therefore, in order to observe the differences in environmental efficiency of construction land at provincial and regional scales of the Yangtze River Economic Zone, this paper uses the super-efficient DEA model and Malmquist index to measure and evaluate the temporal and spatial evolution characteristics of the environmental efficiency.

2 Methodology

2.1 Super Efficiency DEA

Andersen and Petersen proposed an effective measure for further distinguishing between effective units in 1993, which was later called the super-efficient DEA model [12]. The model of the super-efficiency DEA is to minimize θ under the following constraint:

$$\begin{cases} \sum_{\substack{i=1 \\ i \neq k}}^{n} X_i \lambda_i \leq \theta X_k \\ \sum_{\substack{i=1 \\ i \neq k}}^{n} X_i Y_i \leq Y_k \\ \lambda_i \geq 0, i = 1, \ldots, n \end{cases} \tag{1}$$

where $X_k = (x_{1k}, x_{2k,...} x_{ak})$, $Y_k = (y_{1k}, y_{2k,...} y_{bk})$, λ_i is the weight of each decision unit, and the optimal solution θ is the efficiency value of the decision unit k. When the efficiency value of the decision unit is less than 1, the decision unit is invalid; when the efficiency value of the decision unit is greater than 1, the decision unit is valid.

2.2 Malmquist Index

In 1953, Malmquist proposed the Malmquist model to study consumption changes over time [13]. On this basis, the Malmquist index model is expressed as:

$$
\begin{aligned}
M_{st} &= \left[\frac{D^s(x^t, y^t)}{D^s(x^s, y^s)} \times \frac{D^t(x^t, y^t)}{D^t(x^s, y^s)} \right]^{1/2} \\
&= \frac{D^t(x^t, y^t)}{D^s(x^s, y^s)} \left[\frac{D^s(x^t, y^t)}{D^t(x^t, y^t)} \times \frac{D^s(x^t, y^t)}{D^t(x^s, y^s)} \right]^{1/2} \\
&= Effch \times Techch \\
&= Sech \times Pech \times Techch
\end{aligned}
\tag{2}
$$

where $D^t(x^t, y^t)$ represents the input distance function of the decision unit with reference to the t-phase technique; $D^s(x^t, y^t)$ represents the input distance function of the decision unit with reference to the s-phase technique. M_{st} represents the degree of change of total factor productivity from the period s to t. If its value is greater than 1, it indicates that total factor productivity is improved during this period. And if a value less than 1, it indicates that total factor productivity decreases during this period. Among them, the change of total factor productivity can be divided into two parts, technological changes (Techch) and the technical efficiency changes (Effch). The technological change is the movement amplitude of the production frontier surface in two periods. The value greater than 1 indicates that the technology can be improved; the technical efficiency change is the change of the distance between the actual output and the specific production frontier surface. A value which is greater than 1 indicates that technical management efficiency has improved. And technical efficiency can be further decomposed into pure technical efficiency (Pech) and scale efficiency (Sech). The former indicates the management efficiency of the decision unit, while the latter is the evaluation of its scale suitability.

3 Indicators and Data

3.1 Input and Output Indicators

The basic idea of environmental efficiency is to maximize the value of output while minimizing resource consumption and environmental pollution, in other words, maximizing the value by minimizing resources input and environmental pollution [14–16]. In view of the comparability, availability and scientific selection of indicators, this paper selected the labor, land, capital and energy inputs as input indicators, the employment of the second and third industries, the area of construction land, and the total investment in fixed assets of the whole society, and the energy consumption deducting the primary industry value; this research will use sulfur dioxide, carbon dioxide emissions and PM 2.5 concentration to represent the environmental cost of using construction land, and use them as an input indicators into the model. The output indicators are divided into economic benefits and environmental benefits, which are expressed by the added value of the secondary and tertiary industries and the green coverage area of the built-up area. It is worth noting that this study assumes when measuring the environmental efficiency of construction land, the statistical scope of relevant indicators should include non-agricultural activities and residents' living activities of which construction carriers is construction land, so carbon dioxide, sulfur dioxide and PM 2.5 data should be included, and that should be the boundary.

3.2   Data

This study selected data from 9 provinces and 2 municipalities in Yangtze River Economic Zone from 2006 to 2016 as the research object, and measured and analyzed the environmental efficiency of construction land in Yangtze River Economic Zone, using static and dynamic methods. The land index data, economic indicators, labor index data, and energy consumption data of this study are all derived from the EPS database, the China Statistical Yearbook, and the provincial statistical yearbooks. The sulphur dioxide data is derived from the China Environmental Yearbook. The ground PM 2.5 data is obtained by combining the Aerosol Optical Depth (AOD) data and using geo-weighted regression to calibrate the global PM 2.5 site observations. According to it, this study employed the PM 2.5 concentration data at resolution of 0.1°*0.1° (http://fizz.phys.dal.ca/~atmos/martin/?page_id=140), using ArcGIS software to obtain the average concentration of PM 2.5 per year in each province of Yangtze River Economic Zone. The carbon emissions of construction land are mainly reflected in the consumption of coal, oil and natural gas. Therefore, the carbon dioxide data is based on CO_2 emission coefficient of the national greenhouse gas inventory guidelines issued by the intergovernmental panel on climate change (IPCC) in 2006, combining with the annual energy consumption of each province, namely the consumption of coal, crude oil, natural gas and coke (e.g. 2.69 tons of CO_2 per ton of coal, 2.76 tons of CO_2 per ton of crude oil, and 2.09 tons of CO_2 per cubic kilometer of natural gas).

4   Result and Analysis

4.1   Environmental Efficiency of Construction Land

In order to comprehensively analyze the environmental efficiency of construction land in different provinces, this paper firstly uses EMS1.3 to calculate the super efficiency value of each province and the efficiency average of each economic zone in Yangtze River Economic Zone. The three division of the economic zone is based on the concept of 13th Five-Year Plan. Overall, the environmental efficiency of construction land for the whole study area was decreasing from 2006 to 2016. From a regional perspective, the overall efficiency average of the eastern coastal economic zone is higher than other two regions in the same period, and the gap is large. The efficiency values of the provinces in the southwestern economic zone and the middle reaches of the Yangtze River are generally low, so the average efficiency value of the region is also lower than that of other economic zones.

Table 1. Regional super efficiency during 2006–2016 (06–16)

Region	06	07	08	09	10	11	12	13	14	15	16
Shanghai	3.10	2.68	2.62	2.50	2.44	2.73	2.53	2.46	2.48	2.08	2.35
Jiangsu	0.81	0.79	0.77	0.74	0.75	0.76	0.78	0.79	0.83	0.85	0.86
Zhejiang	0.85	0.85	0.82	0.78	0.80	0.75	0.74	0.67	0.66	0.63	0.63
Eastern coastal	1.59	1.44	1.41	1.34	1.33	1.41	1.35	1.31	1.32	1.19	1.28
Jiangxi	0.63	0.81	0.83	0.67	0.73	0.73	0.72	0.66	0.65	0.53	0.54
Hubei	0.80	0.67	0.70	0.56	0.52	0.53	0.56	0.57	0.57	0.58	0.58
Hunan	0.64	0.58	0.53	0.52	0.51	0.48	0.48	0.50	0.52	0.54	0.55
Anhui	0.61	0.73	0.74	0.59	0.61	0.63	0.61	0.58	0.61	0.51	0.52
Middle Yangtze	0.67	0.70	0.70	0.58	0.59	0.59	0.59	0.58	0.59	0.54	0.55
Chongqing	0.46	0.52	0.56	0.52	0.56	0.63	0.65	0.64	0.68	0.61	0.62
Sichuan	0.54	0.48	0.46	0.40	0.42	0.42	0.43	0.44	0.45	0.44	0.46
Guizhou	0.57	0.55	0.52	0.49	0.46	0.49	0.52	0.57	0.56	0.56	0.54
Yunnan	0.50	0.49	0.47	0.42	0.42	0.42	0.41	0.43	0.42	0.42	0.41
Southwest	0.52	0.51	0.50	0.46	0.47	0.49	0.50	0.52	0.53	0.51	0.51
Yangtze river economic zone	0.93	0.88	0.87	0.79	0.80	0.83	0.81	0.80	0.81	0.74	0.78

As can be seen from Table 1, Fig. 1 and Fig. 2, the distribution of efficiency values among the provinces within the major economic zones is relatively uneven. In the eastern coastal economic zone, Shanghai has been in an effective state during the study period, and its efficiency value was the highest in Yangtze River Economic Zone all the time. Because of the long-term reform and opening up, the economic and environmental benefits of the construction land for this coastal city have always maintained a high degree of coordinated development. In addition to Shanghai, the environmental efficiency of construction land in Jiangsu has an upward trend, and in the last three years, it has exceeded the average efficiency of Yangtze River Economic Zone. The use of construction land for the provinces in the middle reaches of the Yangtze River and the southwestern still has room for improvement. At present, Chongqing has the highest environmental efficiency in the southwestern economic zone. For the middle reaches of the Yangtze River, Hubei province is the highest, which is the most important force for rise of central China. All the middle reaches of the Yangtze River need to take advantage of geographical regions and national policies, vigorously adjust the industrial structure, and actively develop green industries.

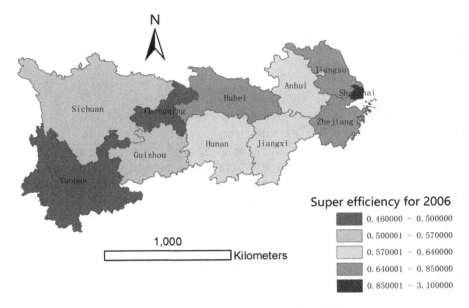

Fig. 1. Regional super efficiency for 2006

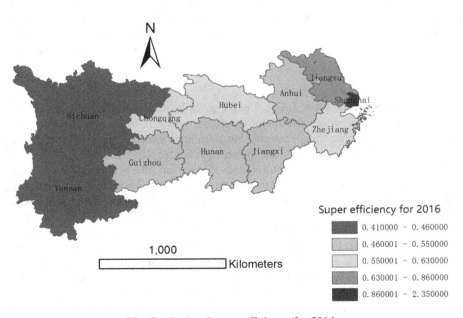

Fig. 2. Regional super efficiency for 2016

4.2   Trends of Environmental Efficiency of Construction Land

As can be seen from Table 2, the Malmquist index of the middle reaches of the Yangtze River is 1.042, the technological progress has increased by 4.6%, and the technical efficiency has decreased by 1.2%. Only for Hubei province, the Malmquist index did not exceed 1, that is, the environmental efficiency of construction land decreased during the study period, mainly due to a slight decrease in technical efficiency, and that was due to the reduction of pure technical efficiency and scale efficiency. Among them, the reduction in scale efficiency is due to the fact that under the recent technology and management level, there may be excessive use of construction land or excessive production of pollution. The reduction in pure technical efficiency is due to the fact that under the current state of the art and scale conditions, the management level has not been improved synchronously, and the resource potential has not been fully exploited. In fact, while most provinces where the Malmquist index has increased, there are also technical efficiencies decreasing in some provinces such as Zhejiang, Jiangxi, Hunan and Anhui. The province with increase of Malmquist index is partly because of technological progress.

Table 2. Annual mean of Malmquist index and its decomposition during 2006–2016

District	Technical efficiency	Technological progress	Pure technical efficiency	Scale efficiency	Malmquist index
Shanghai	1.000	1.102	1.000	1.000	1.102
Jiangsu	1.006	1.080	1.015	0.990	1.086
Zhejiang	0.970	1.089	0.988	0.982	1.057
Eastern coastal	0.992	1.090	1.001	0.991	1.082
Jiangxi	0.985	1.039	0.985	1.000	1.023
Hubei	0.969	1.009	0.972	0.997	0.978
Hunan	0.985	1.042	0.981	1.004	1.026
Anhui	0.985	1.036	0.986	1.000	1.020
Middle Yangtze	0.981	1.032	0.981	1.000	1.012
Chongqing	1.029	1.043	1.021	1.008	1.073
Sichuan	0.983	1.032	0.994	0.989	1.015
Guizhou	0.994	1.038	0.988	1.007	1.032
Yunnan	0.979	1.027	0.978	1.001	1.006
Southwest	0.996	1.035	0.995	1.001	1.032
Yangtze river economic zone	0.990	1.052	0.992	0.997	1.042

5 Conclusion

Based on the panel data of construction land in 9 provinces and 2 municipalities in Yangtze River Economic Zone from 2006 to 2016, this paper uses the Super-efficiency DEA method to measure the environmental efficiency of construction land in each province, three economic zones, and the whole Yangtze River Economic Zone. The following main conclusions are drawn as follows.

The environmental efficiency of construction land in every province and district of Yangtze River Economic Zone is generally low, and there are obvious regional differences. Among them, the eastern coastal economic zones have the highest average efficiency, Shanghai have been in an effective state of environmental efficiency for construction land. The southwestern and the middle reaches of the Yangtze River have the lowest economical areas. During the study period, there were no effective provinces. Even the provinces in the region have not broken through average efficiency of the whole study area over the same period of the year; from the average number of provinces where annual efficiency values exceed the average level, and combining with the average efficiency for every zone, the situation is as follow: the eastern coastal is higher than the middle reaches of the Yangtze River, which is higher than the southwest. In addition, there are some provinces (e.g. Zhejiang, Jiangxi, Hunan and Anhui) with decrease of technical efficiency.

Undoubtedly, the expansion of construction land has seriously threatened the regional ecological environment security and social and economic sustainable development to a certain extent. How to alleviate the negative impact of urban expansion on the ecological environment has aroused widespread concern in the academic community [17]. Compared with previous studies, this paper focuses on the environmental efficiency analysis of construction land, and measures the environmental efficiency of construction land utilization from the perspective of the energy consumption, carbon dioxide, sulfur dioxide and PM2.5 concentration of construction land use. Admittedly, the environmental efficiency of construction land actually includes other aspects of pollution indicators. However, due to the current difficulties in obtaining relevant data, the exactly estimate of environmental efficiency for construction land has to be further studied.

Acknowledgement. This research was supported by Youth Fund of Wuhan Donghu University under grant no. 2018dhsk004.

References

1. Liu, Y.J., Geng, H.: Regional competition in china under the price distortion of construction land: a study based on a two-regime spatial Durbin model. China World Econ. **27**(4), 104–126 (2019)
2. Qu, Y.B., Jiang, G.H., Tian, Y.Y., Shang, R., Wei, S.W., Li, Y.L.: Urban - Rural Construction Land Transition (URCLT) in Shandong Province of China: features measurement and mechanism exploration. Habitat Int. **86**, 101–115 (2019)

3. Yang, Z.W., Chen, Y.B., Qian, Q.L., Wu, Z.F., Zheng, Z.H., Huang, Q.Y.: The coupling relationship between construction land expansion and high-temperature area expansion in China's three major urban agglomerations. Int. J. Remote Sens. **40**(17), 6680–6699 (2019)

4. Chen, Y., Chen, Z., Xu, G., Tian, Z.: Built-up land efficiency in urban China: insights from the general land use plan (2006–2020). Habitat Int. **51**, 31–38 (2016)

5. Zitti, M., Ferrara, C., Perini, L., Carlucci, M., Salvati, L.: Long-term urban growth and land use efficiency in Southern Europe: implications for sustainable land management. Sustainability **7**(3), 3359–3385 (2015)

6. Cui, X.G., Fang, C.L., Wang, Z.B., Bao, C.: Spatial relationship of high-speed transportation construction and land-use efficiency and its mechanism: case study of Shandong Peninsula urban agglomeration. J. Geograph. Sci. **29**(4), 549–562 (2019)

7. Liu, Y.S., Zhang, Z.W., Zhou, Y.: Efficiency of construction land allocation in China: an econometric analysis of panel data. Land Use Policy **74**, 261–272 (2018)

8. Ye, Y.Y., Li, S.F., Zhang, H.O., Su, Y.X., Wu, Q.T., Wang, C.J.: Spatial-temporal dynamics of the economic efficiency of construction land in the pearl river delta megalopolis from 1998 to 2012. Sustainability **10**(1), 63 (2018)

9. Zhang, Y., Jin, H.: Research on efficiency of urban construction land and the drive mechanism in the Mid-Yangtze river. Resour. Sci. **37**(7), 1384–1393 (2015)

10. Jing, S., Zhang, J.: Study on the convergence of Inter-provincial energy and environment efficiency in China. J. Shanxi Univ. Finan. Econ. **40**(1), 1–11 (2018)

11. Wu, Z., Tang, Q., Wang, Y.: Environmental efficiency evaluation of urban construction land in Jiangsu-Zhejiang-Shanghai area: based on three stage DEA and Bootstrap-DEA methods. Mod. Urban Res. **1**, 90–99 (2017)

12. Andersen, P., Petersen, N.C.: A procedure for ranking efficient units in data envelopment analysis. Manag. sci. **39**(10), 1261–1264 (1993)

13. Malmquist, S.: Index numbers and indifference surfaces. Trabajos de Estadistica **4**(2), 209–242 (1953)

14. Feng, M., Li, X.Y.: Evaluating the efficiency of industrial environmental regulation in China: a three-stage data envelopment analysis approach. J. Clean. Prod. **242**, 118535 (2020)

15. Xie, J.H., Zhou, S.C., Chen, Y.: Integrated data envelopment analysis methods for measuring technical, environmental, and eco-efficiencies. J. Clean. Prod. **238**, 117939 (2019)

16. Zhou, Z.X., Wu, H.Q., Song, P.F.: Measuring the resource and environmental efficiency of industrial water consumption in China: a non-radial directional distance function. J. Clean. Prod. **240**, 118169 (2019)

17. Zhong, Y., Lin, A.W., Zhou, Z.G.: Evolution of the pattern of spatial expansion of urban land use in the poyang lake ecological economic zone. Int. J. Environ. Res. Public Health **16**(1), 117 (2019)

Analysis of Potential Factors Influencing Ground-Level Ozone Concentrations in Chinese Cities

Pengfei Liu[1,3], Hongliang Li[2,3,4], Ziyun Jing[1],
and Hongquan Song[2,3,4(✉)]

[1] Key Research Institute of Yellow River Civilization and Sustainable
Development, Ministry of Education, Henan University, Kaifeng 475004, China
lpf@henu.edu.cn
[2] Laboratory of Geospatial Technology for the Middle and Lower Yellow River
Regions, Ministry of Education, Henan University, Kaifeng 475004, China
[3] Institute of Urban Big Data, College of Environment and Planning,
Henan University, Kaifeng 475004, China
[4] Henan Key Laboratory of Integrated Air Pollution Control and Ecological
Security, Henan University, Kaifeng 475004, China

Abstract. Based on Geodetector model, this study adopted the 2016 O_3 concentration data, meteorological factor data and anthropogenic emissions data to explore the driving factors of urban O_3 concentrations in China. The results show that: in general, meteorological factors dominate the urban ozone concentration in China, but there are some differences in the driving factors of ozone concentration at different scales. From the national scale, sunshine duration and temperature are the most significant driving factors, among which temperature is the most critical, while relative humidity and temperature have prominent interactions. From the seasonal perspective, the dominant factors for O_3 pollution in spring and winter are the duration of sunshine, the relative humidity in summer, and the temperature in autumn. In addition, the main interaction in spring, autumn and winter is the interaction between sunshine duration and temperature, while in summer it is air pressure and relative humidity. From the perspective of the zoning, the O_3 pollution in South China is dominated by the duration of sunshine, and the other regions are dominated by temperature. In terms of factor interaction, the strongest interactions in Central China, Northeast China, and Southwest China are combined with relative humidity and temperature. East China, North China, and Northwest China are the duration and temperature of sunshine, while South China is the duration of sunshine and wind speed. From the seasonal influence of different zones, the influencing factors of O_3 concentration changes in each region in different seasons are very different. The influence of evaluation factors on this scale is generally not obvious, and the driving factors and interactions are complicated.

Keywords: O_3 concentration · Geodetector model · Meteorological condition · Anthropogenic emissions precursors · Chinese cities

© Springer Nature Singapore Pte Ltd. 2020
Y. Xie et al. (Eds.): GSES 2019/GeoAI 2019, CCIS 1228, pp. 422–441, 2020.
https://doi.org/10.1007/978-981-15-6106-1_31

1 Introduction

O_3 (ozone) in the atmosphere mainly exists in the ozone layer in the lower part of the stratosphere. It absorbs ultraviolet light of a certain wavelength from outer space and protects the earth, including human beings. However, the tropospheric O_3, which is mainly derived from photochemical reactions, is a harmful secondary pollutant. Under continuous accumulation, it not only has a great impact on the human respiratory system, but also endangers human health and can cause the material to fade and age. It affects the growth of plants by affecting the soil, which ultimately affects the ecological environment [1, 2]. The main sources of tropospheric O_3 include natural sources and anthropogenic sources. Natural sources mainly have stratospheric input and photochemical reactions, while the anthropogenic sources are mainly industrial production, electricity supply, residential emissions, transportation and other sources. O_3 precursors such as NOx (nitrogen oxides), VOC (volatile organic compounds), and CO (carbon monoxide) emitted during these production and life processes will undergo photochemical reactions under certain environmental conditions to form O_3 [3].

At present, O_3 pollution is becoming more and more serious, and O_3 has already occupied the second place in China's urban air pollutants, second only to particulate matters [4]. The country and the people urgently need and expect pollution to be effectively governed and controlled. Therefore, in the process of green transformation under the guidance of the sustainable development concept in all walks of life, the research on O_3 driving factors in different cities and regions will provide crucial theoretical reference and basis. In recent years, scholars have adopted meteorological factors [5, 6], anthropogenic emissions [1, 3], regional transmission effects [7, 8], and O_3 generation sensitivity based on different scales of cities, regions, and countries [9, 10] and so on, through the construction of atmospheric photochemical box model, or using correlation analysis, principal component analysis, trajectory cluster analysis, potential source contribution factor analysis and other methods, to analyze and study the driving factors of O_3 concentration, and continue to contribute to urban air pollution control.

At present, the research on the influencing factors of O_3 concentration has provided sufficient theoretical basis and test methods for subsequent exploration, but the areas targeted by these studies are mainly the relatively serious air pollution situation in a certain urban area, city or nationwide, such as Beijing-Tianjin-Hebei, Pearl River Delta, and Yangtze River Delta regions. It is relatively lacking in the comprehensive regional and national scale to consider the driving force of O_3 concentration. Moreover, there have been many studies using correlation analysis methods. Although it is possible to detect the influence direction of factors, it cannot accurately explain the influence of factors on the change of O_3 concentration, which is what geography detectors are good at. This paper selects the O_3 concentration monitoring day data of 364 cities in 2016, combines several relevant meteorological factors data and anthropogenic emissions data of these cities, and uses Geodetector to detect the spatial differentiation characteristics and driving factors of O_3 concentration changes in Chinese cities. This paper provides scientific and effective theoretical guidance and decision-making basis for the development of national O_3 pollution awareness and prevention measures.

2 Materials and Methods

2.1 Data Source

O_3 precursors produced by humans in production and discharged into the atmosphere are the main premise for most of the near-surface O_3 formation, and in this process, natural conditions play a crucial role, and because of its equally important influence on the accumulation and diffusion of O_3, meteorological factors and O_3 concentration changes have a strong correlation [3, 4]. In this study, six of these meteorological factors, such as accumulated precipitation (PRE), surface pressure (PRS), 2-m relative humidity (RHU), sunshine duration (SSD), 2-m air temperature (TEM) and 10-m wind speed (WIN), which have important driving effects on O_3 concentration changes were selected, together with three O_3 precursors, CO (carbon monoxide), NOx (nitrogen oxides), VOC (volatile organic compounds), a total of nine evaluation factors to detect the driving factors of Chinese urban O_3 concentration.

The O_3 data is based on the maximum 8 h (sliding) average concentration data obtained from the national urban air quality real-time release platform (http://106.37. 208.233:20035/), and its time scale is from January 01, 2016 to December 31, 2016. A total of 367 national monitoring cities included in the 2016 O_3 concentration data (excluding Hong Kong, Macao and Taiwan and Sansha). Due to the serious lack of O_3 concentration data in Zhuji City, Haimen City and Haidong District also lacked more data, while the data of other cities were relatively complete, therefore, the O_3 concentration data of 364 cities nationwide were selected for research.

The meteorological data includes meteorological daily data such as precipitation, air pressure, relative humidity, sunshine duration, temperature, and wind speed. The time scale is also the year from January 01, 2016 to December 31, 2016. These data were obtained from the China Meteorological Data Network (http://data.cma.cn/) ground data.

The anthropogenic emissions data comes from the MEIC model of Tsinghua University (http://www.meicmodel.org/). The gridd emissions inventory provided by the model website includes pollutant discharge data is stored in sectors such as agriculture, industry, residents, electricity, transportation, etc., stored in raster data. This study selects the 2016 emissions inventory, which includes annual data and monthly data on pollutant emissions such as CO, NOx, and VOC.

2.2 Data Preprocessing

The data used in the study included: O_3 concentration data, meteorological data, anthropogenic emissions data, and other ancillary data. After the preliminary preparation of the data, the minimum time scale of the O_3 precursor data is monthly, and the monitoring sites of the O_3 concentration data and meteorological data are mainly distributed in the urban area. For the unified calculation of the caliber, the CO, NOx, and VOC emission data are extracted to the city scale. According to the Ambient Air Quality Standard (GB 3095-2012), calculate the arithmetic mean of the maximum 8 h (sliding) average concentration data on the O_3 day of each calendar month, and obtain the average concentration of O_3 month. The monthly average data of precipitation, air pressure, relative humidity, sunshine duration, temperature, and wind speed are also obtained by calculating the arithmetic mean of the respective daily value data in one calendar month. In addition, this paper divides the country into seven major geographical divisions, including East China, Central China, North China, South China, Northeast China, Northwest China and Southwest China, and organize the data by partition. Finally, the type of data, in the Geodetector, the dependent variable can be either a numerical quantity or a type quantity, but the independent variable only supports the type quantity, and each evaluation factor as an independent variable belongs to a numerical quantity. Therefore, the evaluation factors are discretized, and the values of each factor are divided into 10 levels using the quantile method [11, 12]. The distribution map of China's air quality monitoring site is shown in Fig. 1.

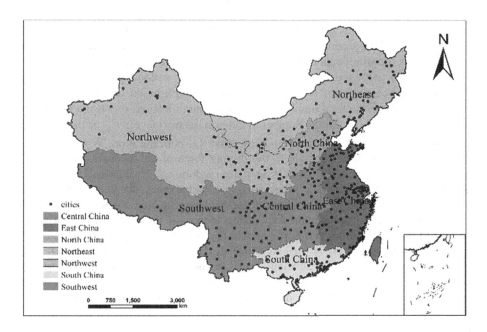

Fig. 1. Distribution map of China's air quality monitoring site

2.3   Geodetector Model

The Geodetector is an analysis method developed by Wang Jinfeng [11] to detect spatial differentiation characteristics and its driving mechanism, and has been compiled into an easy-to-use software based on Excel (http://www.geodetector.org/). The principle is that the influence factors of the typed processing are set as independent variables, and the geographical phenomenon as the research object is regarded as the dependent variable, assuming that the two variables have significant consistency in spatial distribution, and whether the two are related to each other. In recent years, Geodetector have been widely used in food crop production [13], land development and utilization [14], tourism development [15], ecological protection [16], etc. It also has certain applications in environmental pollution [17], providing a high-quality reference for the development of related research. Geography detectors include interaction detectors, ecological detectors, differentiation and factor detectors, and risk detectors. This paper mainly uses the functions of differentiation and factor detector and interaction detector to analyze the driving factors of urban O_3 concentration spatial distribution.

The differentiation and factor detector is mainly used to detect the explanatory power of each influence factor on the O_3 concentration. The expression is as follows:

$$P_{D,U} = 1 - \frac{1}{n\sigma_U^2} \sum_{i=1}^{m} n_{D,i}\sigma_{U_{D,i}}^2 \tag{1}$$

In the formula: $P_{D,U}$ is the driving force or explanatory force of the driving factor D to the change of O_3 concentration (equivalent to the q value used in the Geodetector software); U is the stratification of the driving factor D; n is the number of cities in the country that effectively monitor; σ_U^2 is the variance of the national effective monitoring of urban O_3 concentration; i is the sub-level stratification based on the difference of driving factors; m is the total number of secondary stratification; $n_{D,i}$ is the number of cities that effectively monitor the number of cities in the next level; $\sigma_{U_{D,i}}^2$ is the sub-level layer to effectively monitor the variance of urban O_3 concentration. When the value range of $P_{D,U}$ is [0, 1] and $P_{D,U} = 0$, the driving force of the evaluation factor is 0, indicating that the O_3 concentration is spatially randomized. Evaluation factor has no correlation with O_3 generation. When $P_{D,U} = 1$, it indicates that the change in O_3 concentration is completely dependent on the evaluation factor. Therefore, the larger the value of $P_{D,U}$, the stronger the explanatory power of the driving factor D to the change of O_3 concentration, that is, the greater the influence of O_3 generation [11, 12].

The interaction detector is used to detect the interpretative force of the O_3 concentration between the two evaluation factors in the interaction, that is, whether the driving force of the O_3 concentration is enhanced or weakened by the interaction of two factors. Or the impact of the two on the research object O_3 is independent and noninterference. The types of interaction between the two evaluation factors can be divided into five categories as shown in Table 1.

Table 1. The interactive categories of two factors and the interactive relationship

Description	Interaction
P(X1 ∩ X2) < min(P(X1), P(X2))	Weaken; univariate
min(P(X1), P(X2)) < P(X1 ∩ X2) < max(P(X1), P(X2))	Weaken; univariate
P(X1 ∩ X2) > max(P(X1), P(X2))	Enhanced; bivariate
P(X1 ∩ X2) = P(X1) + P(X2)	Independent
P(X1 ∩ X2) > P(X1) + P(X2)	Nonlinearly enhance

3   Driving Factors of O_3 Concentration in Chinese Cities

3.1   Analysis of Driving Factors of O_3 Concentration in Chinese Cities

According to the results of Geodetector differentiation and factor detection, the driving force of the nine evaluation factors passed the significance test on the time scale of 2016 (Table 2; Note:*, **, ***, indicates that the evaluation factors are significant at the levels of 0.1, 0.05, and 0.01, respectively) That is, meteorological factors and anthropogenic emissions factors have a significant impact on O_3 concentration changes. Among them, meteorological factors, especially temperature (0.335***) and sunshine duration (0.216***) have the greatest impact on O_3 concentration, and its explanatory power is far greater than the explanatory power of O_3 precursors. Among the other four meteorological factors, the driving force of precipitation (0.103***) and air pressure (0.123***) is also high. The relative humidity (0.035***) and wind speed (0.048***) are at the end of the meteorological factor. In contrast, anthropogenic emissions factors play a much less important role in the O_3 spatial difference generation process than these natural factors. However, it can also be found that, at the national scale, the explanatory power of VOC (0.009***) in the precursor is greater than the explanatory power of NOx (0.005***). This result is consistent with existing research on the sensitivity of urban O_3 generation. In addition, the influence of CO (0.015***) is greater than that of NOx and VOC, and it can be judged that O_3 generation in this large area of the country is CO-controlled. This is different from the view that NOx and VOC are generally considered to be the most important precursors of O_3 formation. In general, although the precursor is indispensable for the formation of O_3, its influence on the change of O_3 concentration is almost negligible compared with the influence of meteorological factors. Therefore, it can be broadly stated that meteorological factors dominate the influence of the spatial pattern of O_3 concentration. Specifically, temperature and duration of sunshine are the main driving factors of O_3 pollution in Chinese cities, with temperature being the most critical.

Table 2. The annual and seasonal p values of driving factors for O_3 concentration at the national scale.

Time	PRE	PRS	RHU	SSD	TEM	WIN	CO	NOx	VOC
Year	0.103***	0.123***	0.035***	0.216***	0.335***	0.048***	0.015***	0.005***	0.009***
Spring	0.089***	0.050***	0.140***	0.181***	0.123***	0.103***	0.010	0.013	0.013
Summer	0.087***	0.080***	0.159***	0.102***	0.102***	0.098***	0.020**	0.027***	0.026***
Autumn	0.091***	0.129***	0.107***	0.204***	0.439***	0.030***	0.012	0.007	0.011
Winter	0.063***	0.049***	0.086***	0.139***	0.110***	0.026***	0.013	0.008	0.007

After the interaction detector calculation, the interaction driving force between the evaluation factors of 2016 and the national urban O_3 concentration was obtained (Fig. 2). The interaction between relative humidity and temperature (0.581), and the interaction between sunshine duration and temperature (0.570) all reached 0.5 or higher, which is the highest among all factors. This is mainly because in the process of photochemical reaction of the troposphere to generate O_3, the light, temperature and moisture in the air are closely related to the reaction, and the water vapor also has a dilution effect on O_3. In addition, the two factors of temperature and sunshine duration are the largest in explaining the change of O_3 concentration, and the relative humidity is less explanatory, but the synergistic effect of relative humidity and temperature is the strongest. Therefore, the interaction between relative humidity and temperature is far stronger than the interaction between temperature and sunshine duration. In terms of the type of interaction, in addition to the synergy between precipitation and temperature, the synergy between air pressure and temperature, the synergistic effect of sunshine duration and wind speed are two-factor enhancement, the interaction between all other factors is nonlinear enhancement. Among them, the interaction between precipitation and relative humidity, the interaction between CO and VOC, the interaction between CO and NOx, and the interaction between NOx and VOC are the most significant nonlinear enhancements. The P value of the interaction between them is much larger than the result of adding the two P values. The interaction between CO and NOx is even close to 4 times the sum of their driving forces, and the interaction effect is extremely amazing. Although their synergistic driving force is not very high, but one of the factors change will have a greater impact on the O_3 concentration. Therefore, at the national scale, the interaction between CO and NOx on the spatial layout of Chinese urban O_3 is the most obvious, and the interaction between relative humidity and temperature is the most influential. The changes in these two sets of factors can be reflected to the O_3 concentration fastest and most violently.

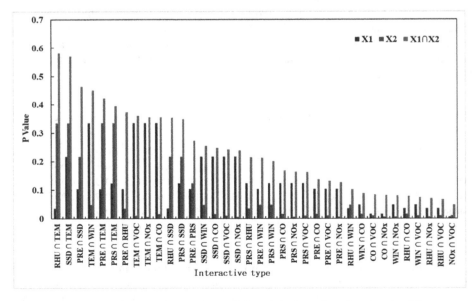

Fig. 2. The annual interactive p value and the original p value of each pair of factors.

3.2   Seasonal Difference Analysis of Driving Factors of O_3 Concentration in Chinese Cities

According to Table 2, although the influence of natural conditions and human conditions, the driving force of the artificially emitted O_3 precursors in the four seasons is all lower than that of the six meteorological factors. Still the influence of natural conditions is stronger. However, according to each evaluation factor, the driving force of each factor is very different between the whole year and the seasonal scale, and there are great differences in different seasons. As a result, the dominant factors of the spatial layout of Chinese cities in different seasons are different.

In terms of natural factors, the dominant factor in spring is the duration of sunshine (0.181***), while the relative humidity (0.140***) is second, and the wind speed (0.103***) is the strongest in the four seasons. The dominant factor in summer is relative humidity (0.159***), while the explanatory power of sunshine duration (0.102***) and temperature (0.102***) is at the end of the four seasons, indicating that the increase in the duration of sunshine and the rise in temperature are inversely proportional to their respective explanatory powers. The dominant factor in the fall is the temperature (0.439***), which is the highest in the four seasons with the influence of sunshine duration (0.204***), precipitation (0.091***), and air pressure (0.129***). The dominant factors in winter are sunshine duration (0.139***) and temperature (0.110***), their P value is greater than summer, and the remaining P values of the meteorological factors are the lowest in the four seasons.

The explanatory power of CO, NOx and VOC in spring, autumn and winter failed to pass the significance test, indicating that they have certain influence on the change of O_3 concentration in these three seasons, but from the national point of view, this effect is not obvious. In the summer, the explanatory power of CO (0.020**), NOx (0.027***), and VOC (0.026***) reached the highest level in the year, Especially NOx, which is more than 5 times the annual average. This may be related to higher temperatures and stronger, longer-lasting illumination in summer, indicating that to some extent meteorological factors influence the driving force of O_3 precursors on O_3 concentration distribution, and this effect is positive.

From the interaction between the two factors, due to the high P value of temperature (0.439***) and sunshine duration (0.204***) in autumn, the highest interactive driving force in autumn is much larger than the other three seasons. The interaction driving force and distribution trend between spring and summer are more consistent. In winter, the P value of the impact factor is generally low, resulting in the overall level of interaction driving force at the lowest position in the four seasons. Among them, the synergistic effect of sunshine duration and temperature has the highest explanatory power in spring, autumn and winter. The highest in summer is the interaction between air pressure and relative humidity, and the synergistic effect of relative humidity and temperature is second in all four seasons, which is consistent with the results of detection at the annual scale. Interactive driving force of O_3 concentration on national scale in different seasons is shown in Fig. 3.

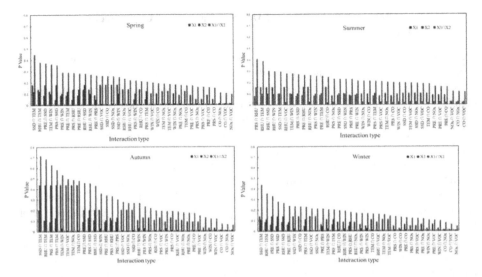

Fig. 3. Interactive driving force of O_3 concentration on national scale in different seasons

3.3   Regional Difference Analysis of Urban O_3 Concentration Driving Factors

According to China's seven geographical divisions, the paper divides the country into seven regions: Central China, East China, North China, Northeast China, Northwest China, South China and Southwest China. Based on this, the performance of the evaluation factors in different regions is detected (Table 3, Fig. 4).), to analyze the driving factors of spatial differences in urban O_3 in various regions.

Table 3. Annual p values of driving factors in 7 regions of China.

Area	PRE	PRS	RHU	SSD	TEM	WIN	CO	NOx	VOC
Central China	0.104***	0.216***	0.145***	0.542***	0.585***	0.055*	0.023	0.024	0.068***
East China	0.064***	0.346***	0.055***	0.325***	0.391***	0.035***	0.034***	0.009	0.030***
North China	0.441***	0.296***	0.033	0.342***	0.615***	0.054*	0.081**	0.130***	0.028
Northeast	0.364***	0.060**	0.093**	0.366***	0.472***	0.095***	0.084***	0.027	0.060***
Northwest	0.293***	0.041	0.090***	0.295***	0.485***	0.126***	0.039**	0.022	0.031*
South China	0.084	0.061	0.044	0.357***	0.210***	0.065	0.055	0.024	0.033
Southwest	0.070***	0.076***	0.114***	0.158***	0.183***	0.120***	0.014	0.051**	0.027

(1) Among the evaluation factors in Central China, precipitation (0.104***), air pressure (0.216***), relative humidity (0.145***), sunshine duration (0.542***), temperature (0.585***), wind speed (0.055*) and VOC (0.068***) all passed the

significance test. Among them, the driving force of sunshine duration and temperature is far greater than the driving force of other factors, indicating that sunshine duration and temperature are the dominant factors of O_3 spatial difference in Central China. In the O_3 precursor, VOC has the highest driving force, and CO and NOx fail to pass the significance test. In addition, the interaction detection results in the region showed that the interaction between relative humidity and temperature (0.829), and the interaction between sunshine duration and temperature (0.822) reached P value 0.8 or more, which had the strongest impact on Central China. Therefore, for O_3 pollution in Central China, based on the relative humidity, sunshine and temperature conditions for a period of time, reasonable control of VOC emissions will achieve relatively obvious effects in a short period of time.

(2) The evaluation factors in East China passed the significant test except for NOx. The air pressure (0.346***), the duration of sunshine (0.325***), and the temperature (0.391***) were the dominant factors in the region. The precipitation in East China (0.064***), South China (0.084), and Southwest (0.070***) has less impact on O_3 concentration changes than in other areas due to the high precipitation. In terms of anthropogenic pollution emissions, CO (0.034***) and VOC (0.030***) have a certain driving force for the formation of O_3. From the perspective of interaction, the top three influences in East China are the interaction between sunshine duration and temperature (0.711), the interaction between relative humidity and temperature (0.680), and the interaction between precipitation and sunshine duration (0.649). It can be found that although the explanatory power of precipitation is small, it can have a greater impact on the O_3 concentration after combining other factors. The pollution control in East China mainly considers the control of CO and VOC emissions under the influence of these three types of interaction.

(3) Relative humidity and VOC in North China failed to pass the significance test, but the explanatory factors of other factors were generally stronger. Precipitation (0.441***) and temperature (0.615***) are the dominant factors in the region, and their explanatory power is the highest among the 7 major divisions. The explanatory power of CO (0.081**) and NOx (0.130***) has reached an astonishing height. This is mainly due to the industrial development of the Beijing-Tianjin-Hebei region, and large-scale heavy pollution industries in Shanxi and central Inner Mongolia. The pollution situation is severe. The O_3 precursors have large emissions, and the local precipitation is less. The erosion ability of O_3 is weaker and the temperature is lower. The changes of these factors will be strongly and quickly feedback to the change of O_3 concentration. In addition, the interaction between the duration of sunshine and temperature (0.825), the interaction of temperature and wind speed (0.801) also has a crucial impact. The main focus of North China should be on industrial regulation, environmental protection, and economic development coordination, and pay special attention to CO and NOx emissions.

(4) The northeastern region includes the three northeastern provinces, which are the old industrial strong provinces and the eastern Inner Mongolia region. CO (0.084***) and VOC (0.060***), which passed the significant test in their

precursors, have strong driving force. The main driving factors in the region are precipitation (0.364***), sunshine duration (0.366***) and temperature (0.472***), which are basically consistent with North China, and the two regions have certain similarities. The strongest interaction in the Northeast is the interaction of relative humidity and temperature (0.720) and the interaction of relative humidity and precipitation (0.713). For the northeast region, the overall temperature is lower, the duration of sunshine is shorter, and the photochemical reaction is weaker. In the case of accumulation of precursors, a short temperature rise and a longer sunshine time will generate a large amount of O_3 pollution. The flushing of the atmosphere by precipitation and relative humidity is particularly important.

(5) The dominant factors in the northwestern region are precipitation (0.293***), duration of sunshine (0.295***), and temperature (0.485***). Air pressure and NOx did not pass the significance test, while the explanatory power of wind speed (0.126***) is the strongest in several regions. This is because the northwestern region is vast, the population and industrial distribution are far less dense than other regions, and atmospheric transport has a greater impact on O_3 concentration. The interaction between sunshine duration and temperature (0.626), the interaction between precipitation and sunshine duration (0.614) have the strongest interaction influence. Pollution control in the northwestern region needs to combine regional characteristics, consider precipitation, temperature, sunshine and wind speed, and delineate key areas to control CO and VOC emissions.

(6) In South China, only the sunshine duration (0.357***) and the temperature (0.210***) passed the significant test. Other factors have certain influence in the region, but the impact is not obvious enough. The type with the highest interaction driving force is the interaction between sunshine duration and wind speed (0.531) and the interaction between precipitation and sunshine duration (0.506). The visible part is located in the tropical South China region. Multiple typhoons, high temperatures, heavy rainfall and long sunshine provide extremely complex and powerful natural driving in the region, especially in the highly industrialized, urbanized and O_3 polluted areas of the Pearl River Delta.

(7) The sunshine duration (0.158***), temperature (0.183***), relative humidity (0.114***) and wind speed (0.120***) in the southwestern region have a greater impact on O_3 concentration. In terms of pollutant emissions, only NOx (0.051**) passed the significance test, but its explanatory power was only lower than North China in the 7 geographical divisions. In terms of interaction, the interaction between relative humidity and temperature in the region (0.444) is the strongest, but it is the weakest in the 7 regions. On the whole, the impact of O_3 concentration in the southwest on the national scale is also the smallest. The Sichuan Basin and the Yunnan-Guizhou Plateau in the southwest have many clouds, the water system is developed, and the atmosphere is in a state of high humidity all the year round. At the same time, the wind speed in some areas has a significant impact on the accumulation of precursors. These factors and the sunshine and temperature which affect photochemical reactions are the dominant factors in the spatial distribution of O_3 in Southwest China.

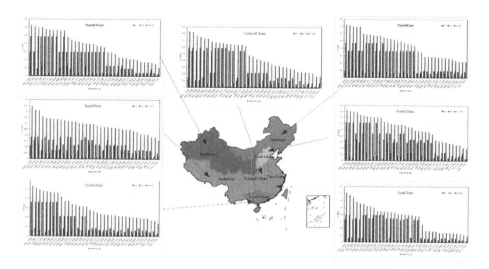

Fig. 4. Annual interaction p values between influencing factors over the 7 regions in China.

3.4 Seasonal Difference Analysis of Driving Factors of O_3 Concentration in Different Areas

Based on the overall analysis of the seven geographical divisions of Central China, East China, North China, Northeast China, Northwest China, South China and Southwest China, the time scale is further subdivided to detect the driving factors and differences of each region in different seasons in spring, summer, autumn and winter. The interactive detection results are shown in Fig. 5, Fig. 6, Fig. 7, and Fig. 8.

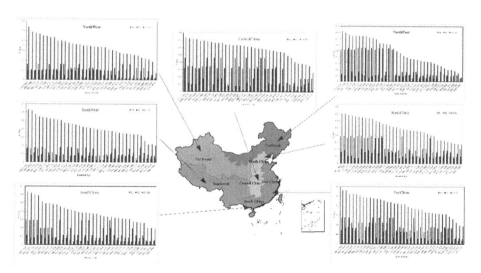

Fig. 5. The interaction p values between influencing factors in spring over the 7 regions in China.

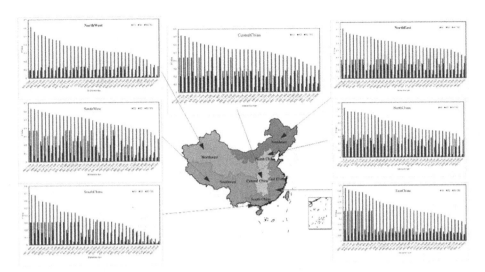

Fig. 6. The interaction p values between influencing factors in summer over the 7 regions in China.

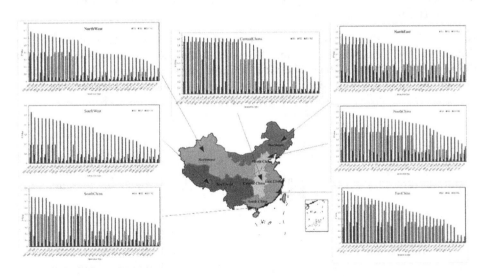

Fig. 7. The interaction p values between influencing factors in autumn summer over the 7 regions in China.

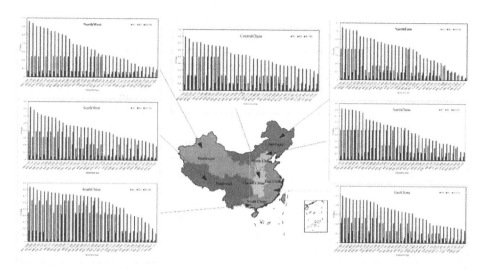

Fig. 8. The interaction p values between influencing factors in winter over the 7 regions in China.

(1) O_3 precursors discharged from Central China (Table 4) did not pass significant tests in all four seasons, and meteorological conditions remained the main driving factor in the region in different seasons. The dominant factor in spring is precipitation, relative humidity and sunshine duration. The strongest interaction type is the interaction between precipitation and sunshine duration. Wind speed is the only factor that passes the test in summer, which dominates the O_3 drive of the season. The combination of relative humidity and NOx has the strongest interaction. In the autumn, precipitation, air pressure, relative humidity, sunshine duration and temperature passed the significance test. The latter two had the greatest impact on O_3 pollution in Central China, and the synergistic interpretation of pressure and temperature was close to 1, and the interaction was extremely obvious. The dominant factor in winter is the duration of sunshine. Among the remaining factors, precipitation, relative humidity and temperature have also passed the significance test. The strongest type of interaction in this season is sunshine duration and NOx. It can be seen that although the influence of NOx itself on the Central China region is not obvious enough, it can greatly drive the change of O_3 concentration in the region under the synergy of other factors.

Table 4. Seasonal p values of driving factors in Central China

Season	PRE	PRS	RHU	SSD	TEM	WIN	CO	NOx	VOC
Spring	0.317^{**}	0.097	0.345^{***}	0.460^{***}	0.178	0.314	0.026	0.079	0.123
Summer	0.267	0.061	0.438	0.260	0.101	0.206^{*}	0.182	0.205	0.193
Autumn	0.111^{*}	0.203^{***}	0.559^{***}	0.846^{***}	0.857^{***}	0.072	0.042	0.043	0.063
Winter	0.225^{***}	0.039	0.223^{***}	0.498^{***}	0.208^{***}	0.110	0.028	0.087	0.116

(2)   In East China (Table 5), all factors except CO and NOx passed the significant test in spring, in which the relative humidity and sunshine duration have the highest driving force. The factors that pass the test in summer are relative humidity, wind speed and VOC. In spring and summer, we can start from controlling the emission of VOC and get better pollution control effect in a short period of time. Precipitation, air pressure, relative humidity, duration of sunshine, and temperature passed the significant test in the fall, with the latter two being the dominant factor. In winter, only the precipitation, relative humidity and duration of sunshine have passed the test, in which the duration of sunshine is the main drive. The interaction between precipitation and sunshine duration dominates in spring, autumn and winter, and the interaction between relative humidity and wind speed is the strongest in summer.

Table 5. Seasonal p values of driving factors in East China

Season	PRE	PRS	RHU	SSD	TEM	WIN	CO	NOx	VOC
Spring	0.142^{***}	0.176^{**}	0.221^{***}	0.274^{***}	0.166^{***}	0.128^{***}	0.038	0.052	0.064^{**}
Summer	0.081	0.111	0.271^{***}	0.084	0.070	0.102^{**}	0.054	0.057	0.071^{**}
Autumn	0.260^{***}	0.279^{***}	0.342^{***}	0.626^{***}	0.529^{***}	0.025	0.019	0.012	0.021
Winter	0.212^{***}	0.024	0.127^{*}	0.265^{***}	0.104	0.031	0.056	0.026	0.043

(3)   The performance of most evaluation factors in different seasons in North China (Table 6) is not obvious enough, but the explanatory power of the factors that passed the significance test has reached a high level. The dominant factors in spring are precipitation and temperature. The dominant type of interaction is precipitation and NOx. The scouring effect of precipitation on NOx is obvious. The dominant factor in summer is relative humidity. The dominant interaction type is the interaction between sunshine duration and temperature. The dominant factors in autumn are precipitation, sunshine duration and temperature. The dominant interaction type is the interaction between air pressure and temperature. The dominant factors in winter are relative humidity, sunshine duration and wind speed. The driving intensity of wind speed is the highest in the four seasons, indicating that the accumulation of O_3 and its precursors in winter is serious in North China, and the trans-regional transport of the atmosphere is particularly important. In addition, the combination of wind speed and CO is the dominant type of interaction for the season.

Table 6. Seasonal *p* values of driving factors in North China

Season	PRE	PRS	RHU	SSD	TEM	WIN	CO	NOx	VOC
Spring	0.381^{*}	0.211	0.102	0.147	0.359^{***}	0.187	0.105	0.161	0.074
Summer	0.018	0.159	0.330^{**}	0.282	0.037	0.191	0.257	0.143	0.240
Autumn	0.472^{***}	0.218	0.039	0.536^{***}	0.620^{***}	0.169	0.137	0.204	0.071
Winter	0.193	0.126	0.235^{***}	0.478^{***}	0.007	0.480^{***}	0.065	0.049	0.116

(4) Northeast region (Table 7) precipitation has the strongest influence in spring, and the temperature has the highest explanatory power in autumn, and the winter sunshine duration can produce the strongest drive. The relative humidity influences in summer and autumn are similar, with slightly higher in autumn. However, only the P value of summer CO emissions in the O_3 precursor passed the significance test, which makes the northeast region no longer completely dominated by meteorological conditions in summer, but is dominated by relative humidity, temperature and CO. From the aspect of the interaction of evaluation factors, the strongest driving types from spring to winter are precipitation and VOC interaction, relative humidity and VOC interaction, temperature and wind speed interaction, and sunshine duration and NOx interaction. It can be seen that the old industrial strong provinces have strong precursors.

Table 7. Seasonal p values of driving factors in Northeast China

Season	PRE	PRS	RHU	SSD	TEM	WIN	CO	NOx	VOC
Spring	0.519^{***}	0.101	0.108	0.056	0.548^{***}	0.005	0.160	0.084	0.117
Summer	0.051	0.275	0.266^{*}	0.114	0.207^{***}	0.120	0.194^{**}	0.110	0.193
Autumn	0.287	0.080	0.287^{*}	0.311^{***}	0.683^{***}	0.210	0.176	0.050	0.118
Winter	0.062	0.096	0.175	0.381^{***}	0.004	0.120	0.096	0.024	0.052

(5) In the northwestern region (Table 8), there are very few evaluation factors passed the significance test. In the spring and summer, there is no evaluation factor that passed the significance test. In autumn, only precipitation, sunshine duration and temperature passed the test. In winter, the relative humidity and the duration of sunshine passed the significance test. This may be because the northwest is inland, far from the coast, the economy is relatively backward, and the meteorological environment is complex and changeable. The distribution of O_3 concentration in different cities and even areas is quite different, and the driving factors are different, which fails to form a unified impact on the whole. This situation is particularly evident in spring and summer. In addition, the combinations of factor interactions that have the greatest impact from spring to winter are temperature and NOx, followed by wind speed and VOC, sunshine duration and temperature, wind speed and NOx.

Table 8. Seasonal p values of driving factors in Northwest China

Season	PRE	PRS	RHU	SSD	TEM	WIN	CO	NOx	VOC
Spring	0.123	0.016	0.023	0.091	0.162	0.051	0.104	0.104	0.096
Summer	0.027	0.009	0.016	0.133	0.118	0.091	0.131	0.094	0.098
Autumn	0.030	0.121^{**}	0.080	0.345^{***}	0.401^{***}	0.025	0.044	0.059	0.056
Winter	0.062	0.072	0.243^{***}	0.246^{***}	0.008	0.077	0.039	0.067	0.036

(6) In south China (Table 9), there is no evaluation factor passed the test in spring, and the interaction between the duration of sunshine and VOC has the greatest impact on the season. The dominant factors in summer are the duration of sunshine and wind speed. The dominant interaction is also the interaction of these two factors. The P value of wind speed is the highest in the four seasons, which is consistent with the impact of summer typhoon on the transport of O_3 pollution area in this area. The dominant factors in autumn are the duration of sunshine and temperature. The dominant interaction is the interaction between relative humidity and sunshine duration. The dominant factors in winter are precipitation, relative humidity, duration of sunshine and temperature. The dominant interaction is the interaction between relative humidity and wind speed. The emission of O_3 precursors in this area has not been significantly affected in the four seasons. The reason may be that the urbanization and industrialization levels of the Pearl River Delta region and Guangxi and Hainan are quite different, and the driving factors for O_3 concentration are different. On the whole, it failed to form a strong leading role.

Table 9. Seasonal p values of driving factors in South China

Season	PRE	PRS	RHU	SSD	TEM	WIN	CO	NOx	VOC
Spring	0.045	0.054	0.041	0.293	0.200	0.079	0.113	0.083	0.116
Summer	0.038	0.061	0.010	0.309***	0.021	0.305***	0.201	0.160	0.102
Autumn	0.167	0.100	0.235	0.501***	0.469***	0.108	0.081	0.057	0.020
Winter	0.568***	0.025	0.446***	0.634***	0.423***	0.170	0.031	0.037	0.144

(7) Precipitation, wind speed, CO and VOC in the Southwest (Table 10) have no significant effect on the overall O_3 concentration change in the region. The dominant factor in spring is the duration of sunshine, the dominant interaction is the duration of sunshine and VOC; the summer is dominated by air pressure, duration of sunshine, temperature, and NOx, among which the interaction between pressure and NOx is the strongest. In autumn, there is no factor passed the test; the winter dominant factors are air pressure, relative humidity and sunshine duration, and the interaction between sunshine duration and CO is the strongest, and the relative humidity driving force of this season is the highst in the four seasons.

Table 10. Seasonal p values of driving factors in Southwest China

Season	PRE	PRS	RHU	SSD	TEM	WIN	CO	NOx	VOC
Spring	0.081	0.051	0.096	0.164*	0.073	0.115	0.023	0.084	0.176
Summer	0.041	0.372***	0.104	0.245***	0.305***	0.176	0.048	0.195*	0.074
Autumn	0.064	0.029	0.024	0.063	0.195	0.038	0.034	0.091	0.048
Winter	0.029	0.212***	0.312***	0.388***	0.045	0.158	0.058	0.071	0.040

4 Conclusion

In general, meteorological factors dominate the change of O_3 concentration in Chinese cities, but there are some differences in the driving forces of different evaluation factors at different scales, different regions and different seasons. Therefore, in some cases, the driving force of the O_3 precursor is no less than that of some meteorological factors [18–21].

From the national scale, the driving factors of sunshine duration and temperature are the most significant, which is consistent with some previous studies [22]. Among them, temperature is the most critical, while the relative humidity and temperature groups have prominent interactions. TEM is one of the most important driving factors affecting ozone concentration. High temperature is a favorable meteorological condition for accelerating photochemical reaction and ozone formation. It has been verified in previous studies [23]. This study found that compared with other influence factors, the influence of TEM on ozone concentration is dominant in most of the region and seasonal scales, and it appears to be relatively stable.

By analyzing the national pollution from a seasonal perspective, it can be found that the dominant factors of O_3 pollution in spring and winter are the duration of sunshine, the relative humidity in summer and the temperature in autumn. RHU is one of the key factors affecting ozone concentration in China, which is consistent with some previous studies [21]. On the one hand, water vapor in the atmosphere can affect solar radiation, photochemical reactions are weakened by the extinction mechanism; on the other hand, higher humidity favors O_3 wet removal. In addition, the main interactions in spring, autumn and winter are sunshine duration and temperature. In summer, it is air pressure and relative humidity.

From the perspective of the seven geographical divisions, the temperature dominates the variation and distribution of O_3 concentrations in the six regions of Central China, East China, North China, Northeast China, Northwest China and Southwest China. The O_3 pollution in South China is dominated by the duration of sunshine. In terms of factor interaction, the strongest interactions in Central, Northeast, and Southwest regions are combined with relative humidity and temperature. The interaction between sunshine duration and temperature dominates in East China, North China, and Northwest China, while South China has the strongest interaction between sunshine duration and wind speed. It can be found that there are large differences between South China and the other six regions.

From the seasonal influences of different sub-areas, the influencing factors of O_3 concentration changes in each region in different seasons are very different. The influence of evaluation factors on this scale is generally not obvious, and the driving factors and interactions are complicated. The dominant factor for spring and winter in Central China is the duration of sunshine. The driving of wind speed is the main influence in summer. The dominant factor in autumn is temperature, and the winter is the length of sunshine. The strongest types of interactions from spring to winter are precipitation and sunshine duration, relative humidity and NOx, air pressure and temperature, and sunshine duration and NOx. The dominant factor in the spring, autumn and winter of East China is the duration of sunshine. The dominant interaction is the interaction between precipitation and sunshine duration. The dominant factor in summer is relative humidity. The dominant interaction combination is relative humidity and wind speed. The dominant factors from spring to winter in North China are precipitation, relative humidity, temperature and wind speed, while the dominant interactions are combined with precipitation and NOx, relative humidity and CO, pressure and temperature, wind speed and CO. The dominant factors in spring and autumn in Northeast China are temperature, summer is dominated by relative humidity, and winter is sunshine duration. The dominant interactions of the four seasons are precipitation and VOC, relative humidity and VOC, temperature and wind speed, sunshine duration and NOx. In the northwestern region, all the factors in the spring and summer failed to pass the significance test, so it is impossible to judge the dominant factors of the two seasons, and the factors that mainly affect autumn and winter are the temperature and the duration of sunshine. Temperature and NOx, wind speed and VOC, sunshine duration and temperature, wind speed and NOx are the main interactions of the four seasons. The spring evaluation factors in South China have not passed the test. The dominant factors in summer, autumn and winter are the sunshine duration. The dominant interactions of the four seasons are sunshine duration and VOC, sunshine duration and wind speed, relative humidity and sunshine duration, relative humidity and wind speed. The evaluation factor of Southwest China did not pass the significant test in autumn, the factors that dominate spring and winter are the duration of sunshine, and the summer is the pressure. The dominant combinations of interactions in the four seasons are sunshine duration and VOC, pressure and NOx, sunshine duration and temperature, sunshine duration and CO.

Some limitations of the study should be clarified. Different types of anthropogenic precursor produced by different emission sectors have different types and quantities, and the effects on ozone concentration are also different. These potential effects are not fully considered, and the results of the research may be uncertain. Therefore, in future research, it is necessary to refine the emission inventory data to sub-sectors to obtain emission data with higher spatial resolution, so as to help us improve the accuracy and efficiency of comprehensive assessment of the effects of anthropogenic precursor on ozone concentration. It can also provide a basis for the government to develop more precise air pollution control policies.

Acknowledgements. This study was financially supported by the National Science Foundation of China (41401107) and Henan Basic Frontier and Technology Research Project (162300410132) and 2017 Key Research Project of Henan Province Higher Education (17B170003).

References

1. Fishman, J., Crutzen, P.J.: The origin of ozone in the troposphere. Nature **274**(5674), 855–858 (1978)
2. Lou, S., Zhu, B., Liao, H.: Impacts of O_3 precursoronsurfaceO_3 concentration over China. Trans. Atmos. Sci. **33**(4), 451–459 (2010)
3. Dong, S., Li, B.: Sources and characteristics of ozone pollution and the significance of ozone monitoring. Environ. Dev. **31**(02), 189+191 (2019)

4. Pan, B., Cheng, L., Wang, J., et al.: Characteristics and source attribution of ozone pollution in Beijing-Tianjin-Hebei region. Environ. Monit. China **32**(05), 17–23 (2016)
5. Cheng, L., Wang, S., Gong, Z., et al.: Spatial and seasonal variation and regionalization of ozone concentrations in China. China Environ. Sci. **37**(11), 4003–4012 (2017)
6. He, W., He, T.: Relationship between Atinospheric ozone concentration and meteorological conditions in Wujing district. Anhui Agric. Sci. Bull. **24**(24), 91–93 (2018)
7. Zhou, X., Liao, Z., Wang, M., et al.: Characteristics of ozone concentration and its relationship with meteorological factors in Zhuhai during 2013—2016. Acta Sci. Circumst. **39**(01), 143–153 (2019)
8. Chen, Z., Zhuang, Y., Xie, X., et al.: Understanding long-term variations of meteorological influences on ground ozone concentrations in Beijing during 2006–2016. Environ. Pollut. **245**, 29–37 (2019)
9. Liang, Y., Liu, Y., Wang, H., et al.: Regional characteristics of ground-level ozone in Shanghai based on PCA analysis. Acta Sci. Circum. **38**(10), 3807–3815 (2018)
10. Wu, K., Kang, P., Yu, L., et al.: Pollution status and spatio-temporal variations of ozone in China during 2015–2016. Acta Sci. Circumst. **38**(06), 2179–2190 (2018)
11. Fu, Z., Dai, C., Wang, Z., et al.: Sensitivity analysis of atmospheric ozone formation to its precursors in summer of Changsha. Environ. Chem. **38**(3), 531–538 (2019)
12. Wang, J., Xu, C.: Geodetector: principle and prospective. Acta Geogr. Sin. **72**(01), 116–134 (2017)
13. Huang, X., Zhao, J., Cao, J., et al.: Spatial-temporal variation of ozone concentration and its driving factors in China. Environ. Sci. **40**(03), 1120–1131 (2019)
14. Ye, Y., Qi, Q., Jiang, L., et al.: Impact factors of grain output from farms in Heilongjiang reclamation area based on geographical detector. Geogr. Res. **37**(01), 171–182 (2018)
15. Zhao, X., Li, Y., Zhao, Y., et al.: Spatiotemporal differences and driving factors of land development degree in China based on geographical detector. Resour. Environ. Yangtze Basin **27**(11), 2425–2433 (2018)
16. Liang, Q., Huang, J., Xie, X., et al.: Study on tourism spatial differentiation characteristics and influencing factors in northern slope of Tianshan mountains based on geodetector. J. Nortwest Norm. Univ. (Nat. Sci.) **54**(06), 82–88 (2018)
17. Wang, Y., Hu, B.: Spatial and temporal differentiation of ecological vulnerability of Xijiang river in Guangxi and its driving mechanism based on GIS. J. Geo-inf. Sci. **20**(7), 947–956 (2018)
18. Zhou, L., Zhou, C., Yang, F., et al.: Spatio-temporal evolution the influencing factors of PM2.5 in China between 2000 and 2015. J. Geogr. Sci. **29**(02), 253–270 (2019)
19. Wang, W., Cheng, T., Gu, X., et al.: Assessing spatial and temporal patterns of observed ground-level ozone in China. Sci. Rep. **7**, 3651 (2017)
20. Wang, Y., Du, H., Xu, Y., et al.: Temporal and spatial variation relationship and influence factors on surface urban heat island and ozone pollution in the Yangtze river delta, China. Sci. Total Environ. **631**, 921–933 (2018)
21. Chen, C., Yan, R., Ye, H., et al.: Researh on the characteristics of ozone pollution in Hangzhou. Environ. Pollut. Control **41**(03), 339–342 (2019)
22. Maji, K., Ye, W., Arora, M., Nagendra, S.S., et al.: Ozone pollution in Chinese cities: assessment of seasonal variation, health effects and economic burden. Environ. Pollut. **247**, 792–801 (2019)
23. Camalier, L., Cox, W., Dolwick, P.: The effects of meteorology on ozone in urban areas and their use in assessing ozone trends. Atmos. Environ. **41**, 7127–7137 (2007)

Spectral Characteristics and Model Inversion of Common Greening Plants in Guangzhou Under Dust Pollution

Chen Zhang[1,2], Xia Zhou[1,2(✉)], Yong Li[1,2], Ji Yang[1,2],
Chuanxun Yang[1,2], Kaixiang Wen[1,2], Yuchan Chen[1,2],
and Libo Yang[3]

[1] Key Lab of Guangdong for Utilization of Remote Sensing
and Geographical Information System, Guangdong Open Laboratory
of Geospatial Information Technology and Application,
Guangzhou Institute of Geography, Guangzhou 510070, China
zhouxia@gdas.ac.cn
[2] Southern Marine Science and Engineering Guangdong Laboratory
(Guangzhou), Guangzhou 511458, China
[3] State Grid Hunan Maintenance Company, Hengyang 421000, China

Abstract. Hyperspectral remote sensing provides a highly efficient, convenient, and non-destructive technical means for the quantitative study of the dust retention content (DRC) of plants. Based on the common greening plants in Guangzhou, this study used hyperspectral technology to explore the spectral characteristics of different plant leaves under dust pollution, and used the Pearson correlation method to construct the DRC estimation models of plants. The results show that the spectral characteristics of the dust-retention effect of plants are mainly concentrated in the visible and near-infrared wavelengths (350–1360 nm), which are important band intervals for the study of spectral characteristics and the inversion band selection. The first derivative of spectral reflectance is more suitable for constructing the DRC estimation model than the spectral reflectance, and CF has a good fitting effect, and the deviation between the Model estimated DRC and the measured DRC is small, and its R^2 is 0.55 and the RMSE is 0.300 g/m^2. Through this research, the application of hyperspectral remote sensing technology in the dust retention effect research of plants can be effectively promoted, which can help to quickly and non-destructively evaluate the DRC of plants, and provide a scientific basis for rational selection of green plant types and urban air quality monitoring, and lay a foundation for large-scale DRC estimation.

Keywords: Greening plants in Guangzhou · Hyperspectral remote sensing · Dust-retention effect · Spectral curve characteristics · DRC estimation model

1 Introduction

With the rapid development of industrialization and urbanization, urban air pollution is becoming increasingly serious [1]. Urban greening plants can effectively retain dust in the air and improve the quality of urban ecological environment [2–5]. At present, the dust-retention effect of plants has become an important indicator for selecting urban

© Springer Nature Singapore Pte Ltd. 2020
Y. Xie et al. (Eds.): GSES 2019/GeoAI 2019, CCIS 1228, pp. 442–450, 2020.
https://doi.org/10.1007/978-981-15-6106-1_32

green plants [6]. How to assess the dust-retention effect of plants has become the focus of current research.

Hyperspectral remote sensing provides a highly efficient, convenient, and non-destructive technical means for the quantitative study of the dust retention content (DRC) of plants [7]. Domestic and foreign scholars have carried out a lot of research on the dust-retention effect of plants, but the application of hyperspectral remote sensing technology in the dust-retention effect research of plants needs to be further improved. As a first-tier city in China, Guangzhou has become increasingly serious in environmental pollution problems [8–17]. There are few studies on the dust-retention effect of plants of Guangzhou, and the related quantitative research results are not perfect. Therefore, this study selected the common and representative green plants in Guangzhou as the research object, including *Ficusmicrocarpa L. f. cv Golden leaves* (GL) and *Cordyline fruticosa (L.) A. Cheval* (CF). By using the hyperspectral remote sensing technology, the spectral characteristics of different plant leaves under dust pollution were studied. Through qualitative analysis of the dust-retention effect of plants, the range of bands suitable for dust-retention research and inversion is selected, and the DRC estimation models of plants are constructed to provide a scientific basis for rational selection of green plant types and urban air quality monitoring, and lay a foundation for large-scale DRC estimation.

2   Study Area

Guangzhou is an important central city in China, located in the core area of the Pearl River Delta, and has obvious characteristics of composite atmospheric pollution. The annual average temperature is between 21.5 °C and 22.2 °C , and there is abundant rain. The rainy season is from March to September, the weather is hot from July to September with many typhoons, and the dry season from October to February is less precipitation. The climatic conditions in Guangzhou provide a good ecological environment for the growth of plants, and there are many biological species. Investigations show that the common plants in Guangzhou are *Ficusmicrocarpa L. f. cv Golden leaves, Loropetalum chinensis (R.Br) Oliv. Varrubrum Yieh, and Cordyline fruticosa (L.) A. Cheval.*

3   Data and Methods

3.1   Data Acquisition

In this study, the DRC of unit leaf area was used to indicate the DRC of the plant, which was determined by the DRC and leaf area of plant leaves. The mass difference method was used to measure the data of DRC, and the LI-3100C leaf area instrument was used to measure the data of leaf area. The spectral reflectance of the leaves was measured using an ASD Field-Spec 3 spectrometer from American ASD Corporation. The spectral range of this spectrometer was 350–2500 nm and with a sampling interval of 1 nm. The front sides of five leaves were measured per bag and one leaf is repeated five times. Part of the measured spectral reflectance is shown in Fig. 1.

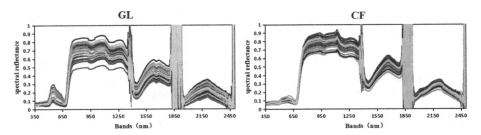

Fig. 1. Spectral reflectance measured by ASD field-spec 3 spectrometer (partial)

3.2   Methods

Spectral Data Process. Before spectral feature analysis, the collected spectral data needs to be processed to reduce the effects of noise and non-sensitive bands. The specific steps are as follows:

(1) First, to ensure the accuracy of the spectral measured data, the mean value was calculated of five-times repeated spectral data of each point after removing the evident deviated curves. And in order to make the spectral reflectivity better describe the concentration trend of the spectral data without changing the original characteristics of the curve, the median of the five spectral reflectance data measured per bag is taken to obtain the spectral reflectance corresponding to the dust retention.

(2) Second, to ensure the comparability of the measured data from different times and conditions and eliminate errors caused by the experimental environment, we divided it by the white board reflectance because the spectral reflectance should follow the principle of proximity.

(3) Then, we removed the bands that were assimilated by water vapor because water vapor assimilation has a great effect on the spectral curve, and this wavelength band range is meaningless in botany spectral research. To simplify the following data process, the wave bands were removed directly during this research.

(4) Finally, to compress the influence of background noise on the target information [18, 19], the spectral reflectance is spectrally transformed, that is, the first derivative (dR) of the spectral reflectance is calculated. The calculation equation is as follows:

$$dR(\lambda_i) = \frac{R(\lambda_{i+1}) - R(\lambda_{i-1})}{\lambda_{i+1} - \lambda_{i-1}} \tag{1}$$

λ_{i+1}, λ_i, and λ_{i-1} are the adjacent wavelengths; $dR(\lambda_i)$ is the first derivative of wavelength λ_i; $R(\lambda_{i+1})$, $R(\lambda_i)$, and $R(\lambda_{i-1})$ are the original reflectance of the wave lengths λ_{i+1}, λ_i, and λ_{i-1}, respectively.

Inversion Model Construction. This study used the most frequently used methods, such as Pearson Correlation Coefficient (Pearson), the coefficient of determination (R^2), the root-mean-square error (RMSE), the mean absolute error (MAE) and Bias to construct and verify the DRC estimation model. Among them, R^2 represents the degree of linear correlation between the two data, MAE and RMSE are used to estimate the overall level of error, and Bias reflects the deviation degree between the estimated data and the measured data.

$$\text{Pearson}(r, x) = \frac{\sum_{i=1}^{i=N}(r_i - \bar{r})(x_i - \bar{x})}{\sqrt{\sum_{i=1}^{i=N}(r_i - \bar{r})^2 \sum_{i=1}^{i=N}(x_i - \bar{x})^2}} \tag{2}$$

$$R^2 = \frac{\left\{ \sum_{i=1}^{N}\left[(x_i - \bar{x})\left(x_i' - \overline{x'}\right)\right] \right\}}{\sqrt{\left[\sum_{i=1}^{N}(x_i - \bar{x})^2\right]}\sqrt{\left[\sum_{i=1}^{N}\left(x_i' - \overline{x'}\right)^2\right]}} \tag{3}$$

$$\text{RMSE} = \sqrt{\sum_{i=1}^{i=N}(x_i - x_i')^2 / n} \tag{4}$$

$$\text{MAE} = \sum_{i=1}^{N}\left|\left(x_i - x_i'\right)\right| / n \tag{5}$$

$$\text{Bias} = \frac{\sum_{i=1}^{N} x_i'}{\sum_{i=1}^{N} x_i} - 1 \tag{6}$$

r_i is the spectral reflectance, $\bar{r}$ is the average of the spectral reflectance, x_i is the DRC of the plant, x_i' is the DRC of the plant predicted by model, $\overline{x'}$ is the average of DRC of the plant, x' is the average of DRC of the plant predicted by model, and N is the number of samples.

4   Results and Discussion

4.1   Spectral Characteristics of Plants Under Dust Pollution

Spectral Reflectance of Different DRC. By analyzing the spectral reflectance curves of different DRC (Fig. 2), it is found that the spectral reflectance corresponding to different DRC is different, and different laws are exhibited in different bands. In the visible wavelengths (350–760 nm), the spectral reflectance of the two plants increased with the increase of DRC, and the two were positive correlated. In the near-infrared wavelengths (760–1360 nm), the spectral reflectance of the two plants decreased with the increase of DRC, and the two were negatively correlated.

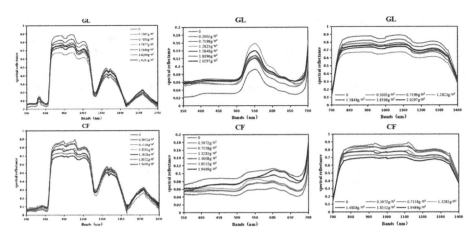

Fig. 2. The spectral reflectance curves of different DRC

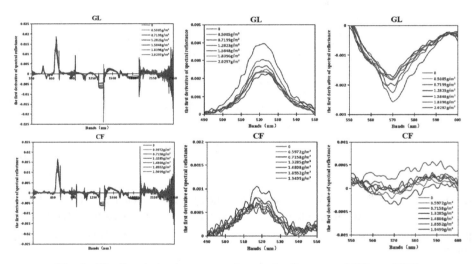

Fig. 3. The first derivative curves of spectral reflectance of different DRC

The First Derivative of Spectral Reflectance of Different DRC. The spectral characteristics of plants cannot be fully displayed by the spectral reflectance. In this case, the spectral reflectance needs to be transformed. One of the more common methods is the first derivative of spectral reflectance [18]. By analyzing the first derivative curves of spectral reflectance of different DRC (Fig. 3), it is found that the trend of the first derivative of spectral reflectance before and after dust retention is basically the same. The first derivative of spectral reflectance has two peaks and two valleys in the visible and near-infrared wavelengths, wherein the peaks are located between 490–550 nm and 670–760 nm; the troughs are located between 550–600 nm and 1050–1250 nm. The first derivative values of the peaks and troughs of different plants are different, and in the 490–550 nm, the first derivative of the two plants is GL > CF.

The First Derivative of Spectral Reflectance of Different DRC. The spectral characteristics of plants cannot be fully displayed by the spectral reflectance. In this case, the spectral reflectance needs to be transformed. One of the more common methods is the first derivative of spectral reflectance [18]. By analyzing the first derivative curves of spectral reflectance of different DRC (Fig. 3), it is found that the trend of the first derivative of spectral reflectance before and after dust retention is basically the same. The first derivative of spectral reflectance has two peaks and two valleys in the visible and near-infrared wavelengths, wherein the peaks are located between 490–550 nm and 670–760 nm; the troughs are located between 550–600 nm and 1050–1250 nm. The first derivative values of the peaks and troughs of different plants are different, and in the 490–550 nm, the first derivative of the two plants is GL > CF.

Similar to the spectral reflectance, the first derivative of spectral reflectance corresponding to different DRC is also different, and different laws are exhibited in different bands. In the range of 490–550 nm, the first derivative of spectral reflectance is expressed as dust leaves > clean leaves, and is proportional to the DRC. In the range of 550–600 nm, the first derivative of spectral reflectance is expressed as dust leaves < clean leaves, and is inversely proportional to the DRC. It is the most significant in GL.

According to the above conclusions, the spectral characteristics of the dust-retention effect of plants are mainly concentrated in the visible and near-infrared wavelengths (350–1360 nm). Therefore, the visible and near-infrared wavelengths are important band intervals for the study of spectral characteristics and the inversion band selection.

4.2   Construction of DRC Estimation Model

The correlation analysis method is the most commonly method used for selecting the inversion band of the DRC estimation model [20]. Therefore, this study uses the Pearson correlation coefficient to select the most relevant band to determine the inversion band to construct the DRC estimation model with the highest fitness. In the visible and near-infrared wavelengths, the Pearson correlation coefficients of spectral reflectance and spectral reflectance's first derivative and DRC were calculated respectively, and the maximum correlation value and the maximum correlation band were selected. The results are shown in Table 1.

Table 1. The maximum correlation coefficient between spectral and DRC

Plant	The maximum correlation value (spectral reflectance)	The maximum correlation band (spectral reflectance)	The maximum correlation value (the first derivative of spectral reflectance)	The maximum correlation band (the first derivative of spectral reflectance)
GL	−0.32	1065 nm	−0.66	761 nm
CF	0.64	675 nm	0.77	628 nm

The results show that the maximum correlation value and the maximum correlation band of different spectral data are different, and the correlation between the first derivative of spectral reflectance and DRC is stronger than the spectral reflectance. It can be seen that the first derivative of spectral reflectance is more suitable for constructing the DRC estimation model. In this study, the first derivative data of spectral reflectance and the maximum correlation band is selected to construct the optimal model for estimating DRC, with the 70% part used to train the model, and the 30% part set aside for testing. Figure 4 shows the results of model inversion and verification.

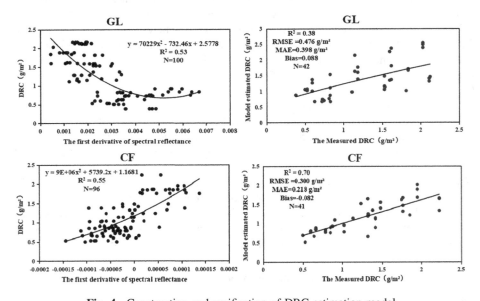

Fig. 4. Construction and verification of DRC estimation model

The results show that the quadratic curve is the best fitting curve. The fitting effect of CF in the two plants is better, and the deviation between the Model estimated DRC and the measured DRC is small. The model has good inversion effect and high verification accuracy, and its R^2 is 0.55 and the RMSE is 0.300 g/m^2.

5   Results and Discussion

Hyperspectral remote sensing provides a highly efficient, convenient, and non-destructive technical means for the quantitative study of the DRC of plants. Based on the common greening plants in Guangzhou, this study used hyperspectral technology to explore the spectral characteristics of different plant leaves under dust pollution. The range of bands suitable for dust-retention research and inversion is selected, and the

DRC estimation models of plants are constructed by Pearson correlation method. The main findings are as follows:

(1) The spectral characteristics of the dust-retention effect of plants are mainly concentrated in the visible and near-infrared wavelengths (350–1360 nm). In the visible wavelengths, the spectral reflectance of the plants increased with the increase of DRC, it is reversed in the near-infrared wavelengths. The first derivative of spectral reflectance has two peaks and two valleys in the visible and near-infrared wavelengths. With the increase of DRC, the peaks and troughs remain basically unchanged. And the first derivative of spectral reflectance exhibits different increases and decreases in the two regions of 490–550 nm and 550–600 nm.

(2) The correlation between the first derivative of spectral reflectance and DRC is stronger than the spectral reflectance, and the first derivative of spectral reflectance is more suitable for constructing the DRC estimation model.

(3) Visible and near-infrared wavelengths are important band intervals for the study of spectral characteristics and the inversion band selection. Based on the characteristic bands of visible and near-infrared wavelengths and the first derivative data of spectral reflectance, it is possible to establish a high-precision DRC estimation model.

Through this research, the application of hyperspectral remote sensing technology in the dust retention effect research of plants can be effectively promoted, which can help to quickly and non-destructively evaluate the DRC of plants, and provide a scientific basis for rational selection of green plant types and urban air quality monitoring, and lay a foundation for large-scale DRC estimation.

Acknowledgments. This study was jointly supported by the National Natural Science Foundation of China (41401430); Guangdong Provincial Science and Technology Program (2017B010117008); Guangzhou Science and Technology Program (201806010106, 201902010033); the National Natural Science Foundation of China (41976189,41976190); the Guangdong Innovative and Entrepreneurial Research Team Program (2016ZT06D336); the Southern Marine Science and Engineering Guangdong Laboratory (Guangzhou) (GML2019ZD0301); the GDAS's Project of Science and Technology Development (2016GDASRC-0211, 2018GDASCX-0403, 2019GDASYL-0301001, 2017GDASCX-0101, 2018GDASCX-0101).

References

1. Jiang, S.L., Jin, H.X., Xu, X.L.: Overview of research on dust-retention function of garden plants. Forest. Sci. Technol. Dev. **25**(6), 5–9 (2011)
2. Wang, H., Guo, J.P., Wang, Z.M., et al.: Simulation test of dust retention potential of main tree species in highway greening. Shanxi Forest. Sci. Technol. **40**(2), 13–16 (2011)
3. Liu, M.M., Yang, L.X., Zhang, J., et al.: Investigation and analysis of the dust-retention effects of main greening plants on university campuses—taking the greening plants of the three universities in Shenyang as an example. J. Shenyang Agric. Univ.: Soc. Sci. Ed. **14**(1), 115–118 (2012)

4. Liu, Y., Li, C.Z., Xing, W.Y., et al.: Study on the dust-retention effect of greening plants in urban traffic roads. North. Hortic. (3) 77–81 (2015)
5. Shi, Y.Y., Zhang, Y.M., Lu, Y.X., et al.: Study on leaf dust retention ability of several common greening tree species in Baoding City. Hebei Forest. Res. **30**(3), 289–294 (2015)
6. Chai, Y.X., Zhu, N., Han, H.J.: The dust-retention effect of urban greening tree species—taking Harbin city as an example. Chin. J. Appl. Ecol. **13**(9), 1121–1126 (2002)
7. Tong, Q.Y., Zhang, B., Zheng, L.F.: Principles, Techniques and Applications of Hyperspectral Remote Sensing. Higher Education Press, Beijing (2006)
8. Wang, F.Z., Li, N., Hu, K.W.: Study on the dust-retention effect of landscape plants. Mod. Gard. 33–37 (2006)
9. Jiang, S.L., Jin, H.X., Zhou, J.F., et al.: Study on the dust-retention effect of common street trees in Hangzhou. Chin. Agric. Sci. Bull. **28**(10), 282–288 (2012)
10. Kretinin, V.M., Selyanina, Z.M.: Dust retention by tree and shrub leaves and its accumulation in light chestnut soils under forest shelter belts. Eurasian Soil Sci. **39**(3), 334–338 (2006)
11. Gao, J.H., Wang, D.M., Zhao, L., et al.: Study on the law of dust-retention in plant leaves—taking Beijing as an example. J. Beijing Forest. Univ. **29**(2), 94–99 (2007)
12. Li, H.M., Wang, W.: Study on dust-retention ability of five greening plants in Chengyang district of Qingdao city. Shandong Forest. Sci. Technol. (3), 34–36 (2009)
13. Liao, L.T., Su, X., Li, X.L., et al.: Overview of research on dust-retention efficiency and dust influencing factors of urban greening plants. Forest Eng. **30**(2), 21–24 (2014)
14. Prajapati, S.K., Tripathi, B.D.: Seasonal variation of leaf dust accumulation and pigment content in plant species exposed to urban particulates pollution. J. Environ. Qual. **37**(3), 865–870 (2008)
15. Wu, C.Y., Wang, X.F.: Study on the spectral variation characteristics of Ginkgo biloba leaves based on leaf dust. J. Southwest Forest. Univ. **36**(1), 91–99 (2016)
16. Saaroni, H., Chudnovsky, A., Ben-Dor, E.: Reflectance spectroscopy is an effective tool for monitoring soot pollution in an urban suburb. Sci. Total Environ. **408**(5), 1102–1110 (2010)
17. Li, W.T., Wu, J., Chen, T.S., et al.: Estimation model of leaf dust-retention content based on hyperspectral data. Trans. Chin. Soc. Agric. Eng. **32**(2), 180–185 (2016)
18. Yu, X., Zhang, F.S., Liu, Q., et al.: Spectral reflection characteristics analysis of typical mangrove. Spectrosc. Spectr. Anal. **33**(2), 454–458 (2013)
19. Qu, Y., Liu, S.H., Li, X.W.: Reflectance spectrum analysis of time-varying law of corn leaf based on function fitting. Spectro. Spectr. Anal. **33**(1), 131–135 (2013)
20. Wang, Y.F., Zhao, W.J., Chen, F.T., et al.: Inversion of dust-retention content in different vegetation leaves in Beijing area. Chin. Sci. Technol. Pap. **11**(3), 329–335 (2016)

Author Index

Printed in the United States
By Bookmasters